Communications in Computer and Information Science 1335

More information about this series at http://www.springer.com/series/7899

Jianhua Qian · Honghai Liu ·
Jiangtao Cao · Dalin Zhou (Eds.)

Robotics and Rehabilitation Intelligence

First International Conference, ICRRI 2020
Fushun, China, September 9–11, 2020
Proceedings, Part I

Springer

Editors
Jianhua Qian
Liaoning Shihua University
Fushun, China

Honghai Liu ⓘ
University of Portsmouth
Portsmouth, UK

Jiangtao Cao ⓘ
Liaoning Shihua University
Fushun, China

Dalin Zhou ⓘ
University of Portsmouth
Portsmouth, UK

ISSN 1865-0929   ISSN 1865-0937   (electronic)
Communications in Computer and Information Science
ISBN 978-981-33-4928-5   ISBN 978-981-33-4929-2   (eBook)
https://doi.org/10.1007/978-981-33-4929-2

This Springer imprint is published by the registered company Springer Nature Singapore Pte Ltd.
The registered company address is: 152 Beach Road, #21-01/04 Gateway East, Singapore 189721, Singapore

Preface

The First International Conference on Robotics and Rehabilitation Intelligence (ICRRI 2020) was held at Liaoning Shihua University, China. ICRRI is a newly-established international conference sponsored by Springer, IEEE SMCS Japan Chapter, and IEEE SMCS Portsmouth Chapter, focusing on the advanced development of Intelligence in Robotics and Rehabilitation Engineering. ICRRI 2020 covered both theory and applications in robotics, rehabilitation, and computational intelligence systems, and was successful in attracting a total of 188 submissions addressing state-of-the-art development and research covering topics related to Human-robot Interaction, Robotic Vision, Multi-agent Systems and Control, Robot Intelligence and Learning, Robot Design and Control, Robot Motion Analysis and Planning, Medical Robot, Robot Locomotion, Mobile Robot and Navigation, Biomedical Signal Processing, Artificial Intelligence in Rehabilitation, Computational Intelligence in Rehabilitation, Systems Modeling and Simulation in Rehabilitation, Neural and Rehabilitation Engineering, Wearable Rehabilitation Systems, Rehabilitation Education, Biomedical and Health Informatics, Policy on Healthcare Innovation and Commercialization, Advanced Control Theory and Applications, Artificial Intelligence, Big Data, and Optimization Methods. Following the rigorous reviews of the submissions, a total of 60 papers (31.9% acceptance rate) were selected to be presented in the conference during September 9–11, 2020. We hope that the published papers of ICRRI 2020 will prove to be technically constructive and helpful to the research community. We would like to express our sincere acknowledgment to the attending authors and the distinguished plenary speakers. Acknowledgment is also given to the ICRRI 2020 Program Committee members for their efforts in the rigorous review process. Special thanks are extended to Jane Li in appreciation of her contribution in the technical support throughout ICRRI 2020. Last but not least, the help from Jane Li and Celine Chang of Springer is appreciated for the publishing.

We hope that the readers will find this volume of great value for reference.

September 2020

Jianhua Qian
Honghai Liu
Dalin Zhou
Jiangtao Cao

Organization

General Chairs

Jianhua Qian	Liaoning Shihua University, China
Naoyuki Kubota	Tokyo Metropolitan University, Japan
Honghai Liu	University of Portsmouth, UK

General Co-chairs

Jiangtao Cao	Liaoning Shihua University, China
Junyou Yang	Shenyang University of Technology, China
Dalin Zhou	University of Portsmouth, UK

Program Chairs

Yonghui Yang	University of Science and Technology Liaoning, China
Qiang Zhao	Liaoning Shihua University, China

Special Session Chairs

Xiaofei Ji	Shenyang Aerospace University, China
Gongfa Li	Wuhan University of Science and Technology, China
Hongyi Li	Bohai University, China
Hongwei Gao	Shenyang Ligong University, China

Publication Chairs

Yinfeng Fang	Hangzhou Dianzi University, China
Kairu Li	Shenyang University of Technology, China
Chengli Su	Liaoning Shihua University, China
Guoliang Wang	Liaoning Shihua University, China

Publicity Chairs

Kaspar Althoefer	Queen Mary University of London, UK
Jangmyung Lee	National Pusan University, South Korea
Qiang Liu	Liaoning Shihua University, China
Nick Savage	University of Portsmouth, UK
Dalai Tang	Inner Mongolia University of Finance and Economics, China

Award Chairs

Nanshu Lu	The University of Texas at Austin, USA
Gaoxiang Ouyang	Beijing Normal University, China
Yuichiro Toda	Okayama University, Japan

Secretaries

Yue Wang	Liaoning Shihua University, China
Taoyan Zhao	Liaoning Shihua University, China

Contents – Part I

Intelligent Control and Perception

Smart Remanufacturing and Industrial Intelligence

Intelligent Control of Integrated Energy System

Contents – Part II

Robot Design and Control

Robotic Vision and Machine Intelligence

Optimization Method in Monitoring

Rehabilitation Robotics and Safety

A Novel Approach to Abnormal Gait Recognition Based on Generative Adversarial Networks

Zixuan Song[1(✉)], Shuoyu Wang[2], Junyou Yang[1], and Dianchun Bai[1]

[1] School of Electrical Engineering, Shenyang University of Technology,
Shenyang 110870, China
songzxuan@qq.com
[2] Department of Intelligent Mechanical Systems Engineering,
Kochi University of Technology, Kami, Kochi 7828502, Japan

Abstract. It is crucial to accurately recognize abnormal gaits of users with weak motion capability during the process of interacting with a robot. However, the method of recognizing abnormal gaits through wearable sensors has limitations when the elderly and disabled are assisted. Meanwhile, the non-contact recognition method of abnormal gaits based on deep learning requires a large number of labeled samples to solve underfitting and overfitting problems. In this paper, we propose a novel approach based on the combination of generative adversarial networks and deep convolutional neural networks to recognize abnormal gaits. Firstly, to obtain rich and varying abnormal gait images, we propose AGR-GAN, which classifies generated gait images in a supervised way and learns interpretable representations between latent variables and abnormal gait images in an unsupervised way. Secondly, we screen out the latent variables related to high-quality generated gait images. We input the latent variables and category labels into the trained AGR-GAN to obtain gait images of specific abnormal gait types with diverse postures and varying perspectives. Finally, we use the transfer-learning AGR-GAN discriminator as a gait recognition network to recognize multiple abnormal gait images. Through verification and comparison of real gaits, we obtain an accuracy rate of 88.9%. Therefore, our proposed abnormal gait recognition method based on generative adversarial networks increases the gait types and scale of the dataset, improving the accuracy of multiple gait recognition.

Keywords: Abnormal gait · Generative adversarial network · Transfer learning

1 Introduction

Abnormal gait refers to patients' reel, fall, and drag while using rehabilitation equipment. It is essential to understand the laws of human movement and the coordination of the excavated limbs through the identification and analysis of abnormal gaits. Abnormal gait is also of important value in clinical diagnosis and rehabilitation. Quick and accurate recognition of abnormal gaits can effectively prevent secondary injuries and help patients improve the rehabilitation process. However, abnormal gait recognition has some trouble. Significant changes in patients' clothing, the perspective, posture and obscured patient's body hinder gait recognition.

© Springer Nature Singapore Pte Ltd. 2020
J. Qian et al. (Eds.): ICRRI 2020, CCIS 1335, pp. 3–15, 2020.
https://doi.org/10.1007/978-981-33-4929-2_1

Glowinski S [1] uses the wavelet transform of signals obtained from six inertial ProMove mini sensors. He estimates the translational acceleration of angular velocity data measured by the gyro sensors to analyze gaits and use one of the wavelets transforms to indicate a characteristic feature. Gao Y [2] presents a real-time algorithm based on a wireless inertial sensor placed on the shank for gait-even detection which combines the cycle-extremum and the updating threshold method to detect data. Caldas R [3] proposes fuzzy c-means and self-organizing maps to simplify the interpretation of gait kinematic and kinetic data provided by inertial sensors. The disadvantage of the wearable sensor detection method is that the patient has to wear the sensor devices for a long time, hindering the patient's actions. Edu I [4] proposes a complex tele-monitoring system for human fall detection based on a miniaturized inertial measurement unit. He processes and analyzes signals received from the navigation unit and provides an estimation of the patients' indoor activities through a wavelet filtering mechanism. Uddin M [5] utilizes Local Directional Patterns (LDP) for the local feature extraction of depth silhouettes and DBN to recognize gait posture. Recognition of abnormal gait based on room facilities is simple during data collection and analysis, and the obtained data is single, convenient for processing and analysis. However, the recognition is limited to a fixed room and the accuracy is low. Tuan N [6] detects and classifies abnormal gaits by using depth images and skeleton joints of the human subjects detected from the images. Nieto-Hidalgo M [7] proposes a vision-based gait analysis method to work with frontal view sequences to classify normal and abnormal gaits. The video-based gait recognition method requires powerful real-time computing capabilities. The disadvantages are the high cost and that the factors such as human wear, lighting, and layout have a bad impact on recognition accuracy [8].

Recognition methods based on deep convolutional neural networks face the problems of underfitting and overfitting. The best way to solve these problems is to increase the type and number of gait samples. In the object recognition, data augmentation methods such as cropping, flipping and adding noise can only expand samples with a single pattern such as size and location, which is not suitable for expanding gait samples. The gaits recognition classifies the different postures and needs gait-posture images. Generative Adversarial Network [9] generates high-quality images that are different from the real images, significantly expanding the dataset. However, the GAN is prone to mode collapse. Radford et al. propose Deep Convolutional GAN [10], which combines a deep convolutional network with GAN. DCGAN improves the generalization ability of GANs and mode collapse, including a series of tricks to enhance the stability of model training. Odena et al. propose ACGAN [11], which can generate samples and classify them simultaneously. InfoGAN [12] proposed by Xi et al. can control some features of the generated samples in an unsupervised manner. Zhang M uses the improved GAN model DeepRoad [13] to generate autonomous driving scenes in severe weather, and uses the severe weather pictures captured on the video to migrate the severe weather to the real normal weather conditions. Li S [14] proposes Cycle-consistent Attentive Generative Adversarial Networks to generate gait images consistent with the view and realistic images for cross-view gait recognition.

In response to the shortcomings of traditional methods, we propose a novel abnormal gait recognition method based on generative adversarial networks with weak limitations for the scene and high accuracy. This method collects gait images through a fixed camera in the room by opencv, recognizes abnormal gait images through deep convolutional neural networks and uses generative adversarial networks for data augmentation to improve recognition accuracy and generalization ability. Through experimental verification, AGR-GAN generates realistic abnormal gait images based on abnormal gait sample types and generates abnormal gait features such as posture and perspective by the control of latent variables. Our approach achieves high accuracy for non-contact recognition of abnormal gaits.

The rest of the paper is organized as follows. Section 2 describes the proposed method. Experiments and evaluation are presented in Sect. 3. The last section, Sect. 4, gives the conclusions.

2 Abnormal Gait Recognition Based on Generative Adversarial Networks

In order to increase the number of abnormal gait images, such as posture and perspective, GAN is used to generate abnormal gait images. Abnormal gait images are divided into three categories: rehabilitation equipment drag, reel and fall (A fall occurs when the body of patient is tilted at an angle greater than 45°). We propose an abnormal gait recognition generative adversarial network (AGR-GAN) that generates realistic abnormal gait images different from the original abnormal gait images.

AGR-GAN based on DCGAN combines the network structures of ACGAN and InfoGAN. ACGAN generates specific categories of images. InfoGAN matches interpretable representation with latent variables in an unsupervised manner and controls disentangled representation of the generated images through latent variables. During training, the abnormal gait pictures and tags that are divided into drag, reel and falls are used as input for AGR-GAN. AGR-GAN generates and classifies abnormal gait images. The classified gait images are used together with the original images to train for gait recognition. This is a semi-supervised learning process. AGR-GAN uses the model of InfoGAN to match the latent variables with the representation of the generated images such as perspective and posture in an unsupervised manner. Unsupervised learning of disentangled representation is applicable to both continuous and discrete latent variables. We propose a recognition network based on a deep convolutional neural network as a gait recognition network to get high accuracy. We will describe the network frameworks in detail in the following subsections.

2.1 Abnormal Gait Images

The abnormal gait images are collected by a camera with a 60° viewing angle and 1 million pixels. The original image size is 480 * 480. The RGB images become binary images after image processing by mog2, denoising, corrosion, and dilation. The binary images reduce the influence of the patients' body and appearance on the experiment. The real abnormal gait images are provided in Fig. 1.

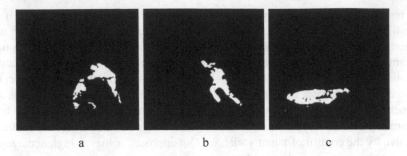

Fig. 1. The abnormal gait images collected in the laboratory. **a** Drag, **b** Reel, **c** Fall.

2.2 InfoGAN: Interpretable Representation Learning by Information Maximizing Generative Adversarial Nets

In general, learned representation is entangled and encoded in a data space in a complicated manner. When a representation is disentangled, it would be more interpretable and easier to apply to tasks. We need to match known semantic information with the original sample and does supervised training to achieve this goal. However, labeling a large number of samples will greatly increase the experiment time. In InfoGAN proposed by Xi et al. [12], the semantic information of the original sample exists and is unknown, coded by c. InfoGAN [12] provides a prior distribution for c and infers the posterior distribution $p(c|x, z)$ from the real samples, where z represents the basic entangled noise and z in GAN can be used in a highly entangled way. The InfoGAN structure is illustrated in Fig. 2.

In order to make some dimensions of z represent some salient feature of the training data, InfoGAN maximizes the mutual information $I(C|X)$ between the latent code C and the generated data X. c is interpretable to the generated data $G(z, c)$. Then c and $G(z, c)$ should have a high correlation, that is, the mutual information between them is relatively large. Mutual information is a measure of dependence between two random variables. The larger the mutual information, the lower the loss of latent code c when the generator generates data. The loss function of GAN is defined as

$$\min_G \max_D V(D, G) = E_{x \sim p_{data}(x)}[log(D(x))] + E_{z \sim p_z(z)}[log(1 - D(G(z)))] \qquad (1)$$

We hope to maximize mutual information $I(c; G(z, c))$ between c and $G(z, c)$. Therefore, the objective function of the model becomes:

$$\min_G \max_D V_1(D, G) = V(D, G) - \lambda I(c : G(z, c)) \qquad (2)$$

However, during the calculation of the mutual information between c and $G(z, c)$, the real $P(c|x)$ is difficult to obtain. Therefore, in the specific optimization process, the idea of variational inference is put forward and variational distribution $Q(c|x)$ is introduced to approximate $P(c|x)$. It is an iterative solution to the lower bound of optimal mutual information. The loss of InfoGAN is defined as

$$\min_{G} \max_{D} V_{InfoGAN}(D, G, Q) = V(D, G) - \lambda L_1(G, Q) \qquad (3)$$

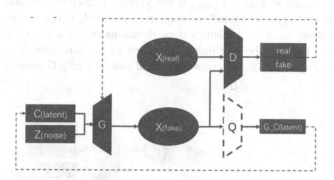

Fig. 2. The structure of InfoGAN

2.3 ACGAN: Auxiliary Classifier Generative Adversarial Nets

It is very troublesome to label a large number of abnormal gait samples. ACGAN [11] classifies samples while generating gait images. ACGAN [11] uses an auxiliary classifier to generate specific types of abnormal gait images. The inputs of ACGAN's generator and discriminator are images and labels. The output of the discriminator adds three abnormal gait categories. The loss function of ACGAN is divided into discriminator loss and classification loss. The discriminator loss function is defined as

$$L_S = E[\log P(S = real|X_{real})] + E[\log P(S = fake|X_{fake})] \qquad (4)$$

The classification loss function is defined as

$$L_C = E[\log P(C = c|X_{real})] + E[\log P(C = c|X_{fake})] \qquad (5)$$

2.4 AGR-GAN: Abnormal Gait Recognition Generative Adversarial Network

GAN [9] is composed of two neural networks, generator (G) generating samples and discriminator (D) discriminating generated samples and real samples. The purpose of the generator is to make the generated sample output 1 by the discriminator. Deep Convolutional GAN (DCGAN) [10] uses a convolutional neural network to replace the multilayer perceptron in GAN, improving the generative ability of the generator. The AGR-GAN proposed in this paper is based on the improvement of DCGAN. AGR-GAN adopts the sum of the loss function of InfoGAN [12] and ACGAN [11]. The loss function of AGR-GAN is defined as

$$\min_{G} \max_{D} V_{AGR-GAN}(D,G,Q,C) = V(D,G) - \lambda L_1(G,Q) + L_S + L_C \qquad (6)$$

The structure of the Abnormal Gait Recognition Generative Adversarial Network structure is illustrated in Fig. 3. $C_{(class)}$ is the vector of categories of abnormal gaits. There are three categories, named as drag, reel and fall. $C_{(latent)}$ represents latent variables that need to know semantics through unsupervised learning. $Z_{(noise)}$ is the entangled noise. $X_{(fake)}$ is the fake image generated by the generator. $X_{(real)}$ is the real abnormal gait images. D is the discriminator. Real and fake help G training in the form of 0 and 1.

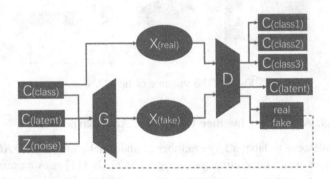

Fig. 3. Abnormal gait recognition generative adversarial network structure.

Three types of abnormal gaits are used in this article, so $C_{(class)}$ is one-dimensional. $C_{(class)}$ is converted into a 3×50 feature dimension when it inputs to a generator. The dimension of the noise $Z_{(noise)}$ is 60. The dimension of the latent variable $C_{(latent)}$ is 40. $X_{(real)}$ is binary abnormal gait images with a size of 480 * 480. The generator uses a deconvolution structure as illustrated in Fig. 4. Each layer of Transposed Convolution is followed by BatchNormalization, Relu activation function. The output is activated by tanh.

The discriminator uses a convolution structure. Compared to the GAN and CNN, the pooling layers and the fully connected layers are removed, and the global pooling layer is used instead. All layers are activated with LeakyReLU. The discriminator framework of AGR-GAN is illustrated in Fig. 5. The inputs are real abnormal gait images and generated images. The output is not activated.

2.5 AGR-GAN Recognition Network

We propose an AGR-GAN recognition network to recognize gaits. There are four types of recognition network outputs: normal gaits, drag, fall and reel. In this paper, the transfer learning method is used. The feature extraction layers of the trained discriminator are used as the feature extraction layers of the recognition network, and the

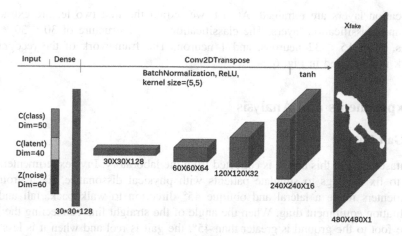

Fig. 4. The framework of the generator of AGR-GAN.

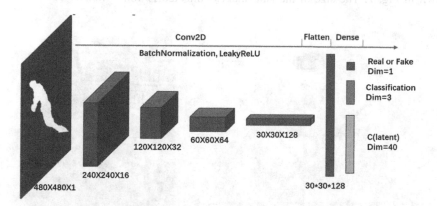

Fig. 5. The discriminator framework of AGR-GAN.

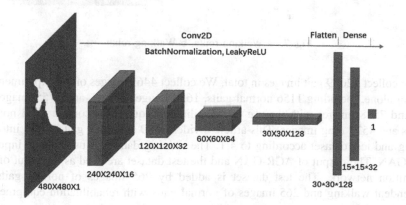

Fig. 6. The framework of AGR-GAN recognition network

classification layers are retrained. At last, we retrain the first two feature extraction layers and classification layers. The classification uses a structure of 30 * 30 * 128 neurons, 15 * 15 * 32 neurons, and 1 neuron. The framework of the recognition network is illustrated in Fig. 6.

3   Experiments and Analysis

3.1   Gait Dataset

The dataset used in this article is collected from the laboratory. Five experimenters use straps to fix the legs to simulate patients with physical dissonance. Each group of experimenters made a lateral and oblique 45° direction to walk back, fall and the rehabilitation equipment drag. When the angle of the straight line connecting the head and the foot to the ground is greater than 45°, the gait is reel and when it is less than 45°, the gait is fall. The data on the walking route and walking robot collected are shown in Fig. 7. The size of the gait images collected is 480 * 480.

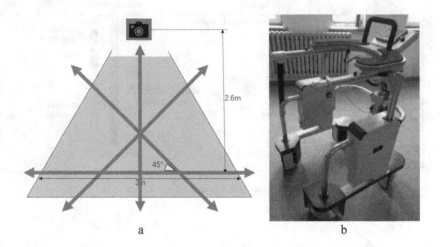

a   b

Fig. 7. a Walking route, **b** Walking robot

We collect 12009 gait images in total. We collect 4461 images of the experimenters walking alone, including 1156 normal gaits, 1650 images of reel, and 1755 images of fall. And 7548 images with the use of rehabilitation equipment contain 2973 normal images and 4575 drag images. This article divides 7980 abnormal gait images into the training and test dataset according to 4:1. The training dataset is used as the input of AGR-GAN. The output of AGR-GAN and the test dataset are used as the input of the recognition network. The test dataset is added by 265 images of normal gaits of independent walking and 265 images of normal gaits with rehabilitation equipment.

3.2   Experimental Design

In this paper, the images generated by AGR-GAN during the training process are discarded. We use images output by the trained AGR-GAN network to ensure the quality of abnormal gait images.

However, we do not know the difference between the abnormal gaits generated by AGR-GAN. It is important to know whether the generated abnormal gait images can be used as training data. First, we train the recognition network with real abnormal gait images and test the accuracy, and use the generated abnormal gait images to test the accuracy of the recognition network. Second, we train the recognition network with the generated abnormal gait images and test the accuracy with real abnormal gait images.

AGR-GAN generates 5000 images of reel, 5000 images of fall and 2000 images of drag. These images together with the original gait images are used as new training datasets to train the recognition network and test the accuracy. The test accuracy is compared with the recognition network trained by the real dataset.

The AGR-GAN generator is a network that gradually expands the image. Here, each convolution in this paper expands the length and width by 2 times, using a lot of tricks provided by DCGAN [10]. The parameters of AGR-GAN training are shown in Table 1. The optimizers of generator and discriminator all adopt Adam and the learning rate adopts $1e-5$. Because too high and too low learning rate will make the network run poorly and train slowly. We set the batch size to 32 in order to make full use of the limited computing power. The bias makes GAN training unstable. To determine the size of the generated images, we set Padding equal to Same. The prefetching method is applied to input images for speeding up the training process, which are automatically adjusted by the computer according to the demand. The classification of the identification network uses the softmax function.

Table 1. Parameters of the AGR-GAN training

Configuration	Parameters
Optimizers	Adam
Learning rate	0.00001
Batch size	32
Strides	(2, 2)
Use bias	False
Padding	Same
Prefetch	AUTOTUNE
Epoch	4000

3.3   Experiments

AGR-GAN has a powerful ability to generate abnormal gait images. The abnormal gait images generated by AGR-GAN are shown in Fig. 8. By setting a uniform distribution for $C_{(latent)}$, we gradually control disentangled representation of gait images, such as posture and legs. The recognition network's recognition capabilities are compared with

12   Z. Song et al.

Vgg16 and Xception as shown in Table 2. Firstly, we use the same training and test dataset to verify the recognition ability of the network, which can judge underfitting. The recognition network has good gait recognition ability. Secondly, we use cross-validation of true and fake (generated) gait images to prove that the recognition networks trained by true and fake gait images can recognize each other. The network trained by real gait images or fake gait images obtains an accuracy of 30% when test input is fake or real. Generated abnormal gait images can be used as a training dataset to recognize real gait images. The accuracy is higher when the input is fake gait during training, because the number of fake gait images is much larger than that of the real gait images. Finally, we use the real gait images different from the training dataset as the test dataset to measure the accuracy of the recognition with rich dataset. The accuracy of the recognition network reaches 88.9%, which is 36.1% higher than the network trained with real gait images.

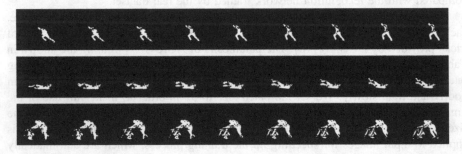

Fig. 8. Generated abnormal gait images including reel, fall and drag

Table 2. The accuracy of different training set and test set under AGR-GAN recognition network, Vgg16 and Xception network for 400 epochs.

Network	Train input	Test input	Accuracy
Recognition network	Real gait	Real gait	0.953
	Fake gait	Fake gait	0.957
	Real gait	Fake gait	0.275
	Fake gait	Real gait	0.352
	Real gait	Test real gait	0.528
	Fake gait + Real gait	Test real gait	0.889
Vgg16	Real gait	Real gait	0.966
	Fake gait	Fake gait	0.979
	Real gait	Fake gait	0.276
	Fake gait	Real gait	0.379
	Real gait	Test real gait	0.624
	Fake gait + Real gait	Test real gait	0.823

(*continued*)

Table 2. (*continued*)

Network	Train input	Test input	Accuracy
Xception	Real gait	Real gait	0.960
	Fake gait	Fake gait	0.987
	Real gait	Fake gait	0.285
	Fake gait	Real gait	0.404
	Real gait	Test real gait	0.645
	Fake gait + Real gait	Test real gait	0.837

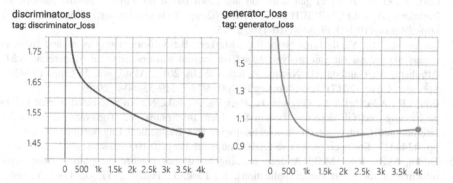

Fig. 9. The loss of discriminator and generator.

However, the loss function of AGR-GAN is not optimal. The discriminator loss has room for the decline, and the generator loss is overfitted. The discriminator loss and generator loss are shown in Fig. 9.

4 Conclusions

In this paper, we propose a novel approach to abnormal gait recognition based on generative adversarial networks. We build an abnormal gait recognition generative adversarial network (AGR-GAN), which combines ACGAN and InfoGAN on the basis of DCGAN. Compared with GANs, AGR-GAN adds an auxiliary classification layer and a disentangled representation layer. We establish the gait recognition network by transfer learning of the discriminator to classify abnormal gaits. Finally, through the test of accuracy of the real gait images and the generated abnormal gait images, the generated images are qualified to be training data for gait recognition. The experimental results show that our approach to abnormal gait recognition based on generative adversarial networks riches the gait recognition dataset and achieves a high accuracy of multiple gait recognition in a short time.

However, a number of improvements are needed in the future. At present, the gait collection scene is relatively single. In the future the image localization will be added to make the application scene more extensive. The parameters of the recognition network need improvement to obtain higher accuracy. The posture of the generated gait images is not rich enough, and we will try more latent variables in the future. We believe that GANs will improve abnormal gait recognition in various scenarios.

References

1. Glowinski, S., Blazejewski, A., Krzyzynski, T.: Inertial sensors and wavelets analysis as a tool for pathological gait identification. In: Gzik, M., Tkacz, E., Paszenda, Z., Piętka, E. (eds.) Innovations in Biomedical Engineering. AISC, vol. 526, pp. 106–114. Springer, Cham (2017). https://doi.org/10.1007/978-3-319-47154-9_13
2. Gao, Y., et al.: A novel gait detection algorithm based on wireless inertial sensors. In: Badnjevic, A. (ed.) CMBEBIH 2017. IP, vol. 62, pp. 300–304. Springer, Singapore (2017). https://doi.org/10.1007/978-981-10-4166-2_45
3. Caldas, R., Hu, Y., de Lima Neto, F.B., Markert, B.: Self-organizing maps and fuzzy c-means algorithms on gait analysis based on inertial sensors data. In: Madureira, A.M., Abraham, A., Gamboa, D., Novais, P. (eds.) ISDA 2016. AISC, vol. 557, pp. 197–205. Springer, Cham (2017). https://doi.org/10.1007/978-3-319-53480-0_20
4. Edu, I.R., Adochiei, F.C., Grigorie, L., Pasarica, A., Jula, N.: An automated inertial indoor positioning and fall detection system for elder. In: Sontea, V., Tiginyanu, I. (eds.) 3rd International Conference on Nanotechnologies and Biomedical Engineering. IP, vol. 55, pp. 424–427. Springer, Singapore (2016). https://doi.org/10.1007/978-981-287-736-9_100
5. Uddin, M.Z., Kim, M.R.: A deep learning-based gait posture recognition from depth information for smart home applications. In: Park, J., Pan, Y., Yi, G., Loia, V. (eds.) CSA/CUTE/UCAWSN - 2016. LNEE, vol. 421, pp. 407–413. Springer, Singapore (2017). https://doi.org/10.1007/978-981-10-3023-9_64
6. Tuan, N.V.A., Vo Van, T., Hau, N.V.D., Thang, N.D.: Abnormal gait detection and classification using depth camera. In: Vo Van, T., Nguyen Le, T., Nguyen Duc, T. (eds.) BME 2017. IP, vol. 63, pp. 749–754. Springer, Singapore (2018). https://doi.org/10.1007/978-981-10-4361-1_128
7. Nieto-Hidalgo, M., Ferrández-Pastor, F.J., Valdivieso-Sarabia, R.J., Mora-Pascual, J., García-Chamizo, J.M.: Vision based gait analysis for frontal view gait sequences using RGB camera. In: García, C.R., Caballero-Gil, P., Burmester, M., Quesada-Arencibia, A. (eds.) UCAmI 2016. LNCS, vol. 10069, pp. 26–37. Springer, Cham (2016). https://doi.org/10.1007/978-3-319-48746-5_3
8. Wu, Z., Huang, Y., Wang, L., et al.: A comprehensive study on cross-view gait based human identification with deep CNNs. IEEE Trans. Pattern Anal. Mach. Intell. 39(2), 209–226 (2017)
9. Goodfellow, I., Pouget-abadie, J., Mirza, M., et al.: Generative adversarial nets. In: Ghahramani, Z., Welling, M., Cortes, C., Lawrence, D., Weinberger, Q. (eds.) 2014 International Conference on Neural Information Processing, NIPS, vol. 2, pp. 2672–2680. MIT Press, Montreal (2014)
10. Radford, A., Metz, L., Chintala, S., et al.: Unsupervised representation learning with deep convolutional generative adversarial networks. In: Bengio, Y., LeCun, Y. (eds.) 2016 International Conference on Learning Representations 2016, pp. 1–12 (2016). arXiv CS arXiv:1511.06434

11. Yao, Z., Dong, H., Liu, F., Guo, Y.: Conditional image synthesis using stacked auxiliary classifier generative adversarial networks. In: Arai, K., Kapoor, S., Bhatia, R. (eds.) FICC 2018. AISC, vol. 887, pp. 423–433. Springer, Cham (2019). https://doi.org/10.1007/978-3-030-03405-4_29

12. Chen, X., Duan, Y., Houthooft, R., et al.: InfoGAN: interpretable representation learning by information maximizing generative adversarial nets. In: Daniel, D., Masashi, S. (eds.) 2016 International Conference on Neural Information Processing, NIPS, vol. 1, pp. 2180–2188. MIT Press, Montreal (2016)

13. Zhang, M., Zhang, Y., Zhang, L., et al.: DeepRoad: GAN-based metamorphic testing and input validation framework for autonomous driving systems. In: Marianne, H., Christian, K., Gordon, F. (eds.) 2018 ACM/IEEE International Conference on Automated Software Engineering, ASE, vol. 1, pp. 132–142. ACM, New York (2018)

14. Li, S., Liu, W., Ma, H., et al.: Beyond view transformation: cycle-consistent global and partial perception GAN for view-invariant gait recognition. In: Jay, K., Nguyen, T., Zeng, W. (eds.) 2018 IEEE International Conference on Multimedia and Expo, ICME, vol. 1, pp. 987–1006. IEEE, New York (2018)

A Novel Non-contact Recognition Approach of Walking Intention Based on Long Short-Term Memory Network

Lili Lv[1]([✉]), Junyou Yang[1], Donghui Zhao[1], and Shuoyu Wang[2]

[1] School of Electrical Engineering, Shenyang University of Technology,
Shenyang 110870, China
yizhiniu@189.cn
[2] Department of Intelligent Mechanical Systems Engineering,
Kochi University of Engineering, Kami, Kochi 7828502, Japan

Abstract. To daily assisted walking and walking rehabilitation training scenarios, it is especially important that the robot accurately recognizes user's walking intentions. To accurately identify walking intention in the process of operating walking rehabilitation training robot, a novel non-contact walking intention recognition method based on LSTM (Long and Short-Term Memory network) is proposed under this paper. Firstly, this paper introduces the mechanical structure of walking rehabilitation training robot and establishes the kinematics model of the robot. Secondly, the distance information of the left and right legs is detected by a multi-channel proximity sensor, and speed information of the left and right legs is obtained using a distance-speed conversion algorithm. The distance information of the left and right legs and speed information of the left and right legs is used as the input of the LSTM algorithm. The multivariable LSTM is utilized to predict the user's walking intention to obtain the desired movement speed of the robot. Finally, the algorithm is tested for walking intention recognition experiment. The experiment shows that the algorithm is suitable for diverse users. Users walk at constant speed and variable speed walking, walking rehabilitation training robot has a speedy recognition and reaction ability. The multi-channel proximity sensor used in this paper can detect information in a non-contact manner, which provides users with a comfortable and unconstrained walking experience.

Keywords: Walking rehabilitation training robot · Multi-channel proximity sensor · Multivariable LSTM · Walking intention recognition

1 Introduction

Nowadays, the extension of human life expectancy has led many countries to gradually enter an aging society. Cognitive skills and walking ability of the senior will gradually descend with age. Falling is a usual occurrence. Therefore, to satisfy their routine life, walking rehabilitation training robots has become a heated issue in the field of robot research [1]. How to improve the user's comfort during using the walking rehabilitation training robot, walking intention recognition has become a more prominent issue in

J. Qian et al. (Eds.): ICRRI 2020, CCIS 1335, pp. 16–32, 2020.
https://doi.org/10.1007/978-981-33-4929-2_2

domestic and foreign research. There are many methods to recognize walking intention at home and abroad. For example, Pennsylvania State University designs the RPB robot [2], which uses magnetometers and gyroscopes to obtain the posture deviation between humans and robots, and reduces the deviation by continuously adjusting the robot, the results show the robot has good recognition ability, but this method requires a lot of calculation and operation is too complex. Kyung Hee University in South Korea designs an intelligent walking robot with active control of human-computer interaction [3], which uses lidar sensors and inertial measurement unit (IMUs) sensors to obtain the direction of movement and angular velocity, thereby actively controlling its speed and direction of movement. However, the error of gait intention obtained is too large, reaching up to 30 cm. Guangdong university of technology designs MAMAW walker [4], which uses six-axis force/torque sensor to obtain the interaction force between human and robot, and uses fuzzy control algorithm to control the robot's speed, this method can achieve good control performance, but six axis force/torque sensor is too expensive, causing unnecessary waste. Huazhong university of science and technology designs a flexible and safe walking robot [5], which uses force sensors and laser sensors to integrate the motion intention of the upper and lower limbs to obtain the user's walking intention, so as to control the speed and direction of the robot, the results show that this method can achieve more accurate and more compliant the user's intention speed, but this method is not applicable to different walking habits of different users.

To solve the above problems, the distance information of the left and right legs is detected by a multi-channel proximity sensor, and speed information of the left and right legs is obtained using a distance-speed conversion algorithm. The distance information of the left and right legs and speed information of the left and right legs is used as the input of the LSTM algorithm, and the multi-variable LSTM is used to predict the user's walking intention. The results show that the algorithm can precisely recognize the user's walking intentions.

The structure of this paper is as follows. The second part introduces walking rehabilitation training robot; the third part describes the method used to identify walking intention; in the fourth part, LSTM method is used to simulate the walking intention under the condition of constant speed and variable speed, and the feasibility of the method is proved by experiments.

2 Walking Rehabilitation Training Robot

2.1 Structure Introduction of Walking Rehabilitation Training Robot

Walking rehabilitation training robot is shown in Fig. 1, the main components are four omnidirectional wheels, operation panel, support board, and multi-channel proximity sensor. The omnidirectional wheels have the ability of in-situ rotation and lateral movement, which can provide users with comfortable walking support. The operation panel has two control modes, one is the manual mode, the other is the PC mode. This paper uses the PC mode, which uses multi-channel proximity sensors to obtain distance information, and uses the trained LSTM algorithm to identify the user's walking intention speed to control the movement of the walking rehabilitation training robot. The support plate provides users with additional stabilizing equipment, thereby reducing the burden on users while helping users maintain upper body stability. The multi-channel proximity sensor is shown in Fig. 2, which is composed of eight proximity sensors on the front side and eight proximity sensors on the left side. This paper uses eight proximity sensors on the front side to detect the distance information of the left and right legs.

Fig. 1. Walking rehabilitation training robot.

Fig. 2. Multi-channel proximity sensor.

2.2 Kinematics Model

The kinematics model of walking rehabilitation training robot is essential for accurately identifying the user's walking intention speed. The moving speed of walking rehabilitation training robot is determined by the speed of the four omnidirectional wheels. The kinematic diagram of walking rehabilitation training robot is shown in Fig. 3, XOY is the static coordinate system of the robot. The angle between the robot moving direction and the positive x-axis direction is γ. a is the length from the geometric center of the robot to the center of one of the omnidirectional wheels. From the physical structure of the robot, $\beta = \beta_4$ can be taken.

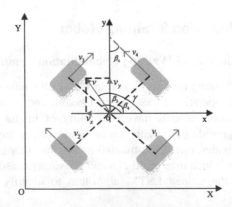

Fig. 3. Kinematic diagram of walking rehabilitation training robot.

The kinematics equation of the robot is as follows:

$$\begin{bmatrix} v_1 \\ v_2 \\ v_3 \\ v_4 \end{bmatrix} = \begin{bmatrix} cos\beta & sin\beta & a \\ -sin\beta & cos\beta & -a \\ cos\beta & sin\beta & -a \\ -sin\beta & cos\beta & a \end{bmatrix} \begin{bmatrix} v_x \\ v_y \\ \dot{\beta} \end{bmatrix} \tag{1}$$

where the transfer matrix is $Q^T = \begin{bmatrix} cos\beta & sin\beta & a \\ -sin\beta & cos\beta & -a \\ cos\beta & sin\beta & -a \\ -sin\beta & cos\beta & a \end{bmatrix}$.

Equation (1) can obtain the kinematic constraint relationship between omnidirectional wheels:

$$v_3 + v_4 = v_1 + v_2 \tag{2}$$

The pose of the robot is $p = (X, Y, \beta)^T$, and $\dot{p} = (\dot{X}, \dot{Y}, \dot{\beta})^T = (v_x, v_y, \dot{\beta})^T$. The speed of the four omnidirectional wheels are $v^* = (v_1, v_2, v_3, v_4)^T$. The robot's moving speed is v. Let $K = (QQ^T)^{-1}Q$, the kinematics model of walking rehabilitation training robot is:

$$\dot{p} = Kv \tag{3}$$

3   Walking Intention Recognition Method

The flow chart of identifying walking intention based on LSTM is illustrated in Fig. 4.

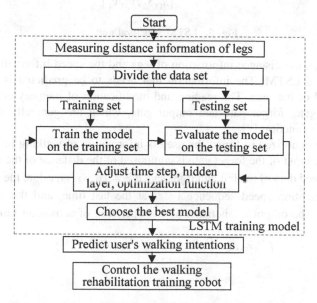

Fig. 4. Flow chart of user's walking intention recognition based on LSTM algorithm.

The figure shows that the measured leg distance information is trained by the LSTM model to predict the user's walking intention to control the robot.

3.1 LSTM Neural Network

LSTM neural network is a time-recurrent neural network with strong memory and learning ability, and has outstanding effectiveness and adaptability in the analysis of multivariable input prediction problems [6]. Considering that the measured distance data and the speed data have continuity in time series, in this paper, the LSTM model with circular memory and reasoning ability to establish a deep learning model based on the time series to predict the user's walking intention.

Compared with ordinary RNN, LSTM has added a storage unit and three control gates [7], including the input gate, output gate and forget gate [8]. LSTM structural unit is shown in Fig. 5.

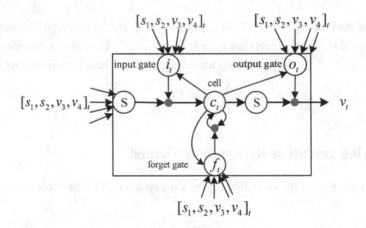

Fig. 5. LSTM structural unit.

In this paper, the distance information of legs and the speed information of legs is used as input of LSTM. The information that needs to be predicted is the walking intention speed of the user. The reading and modification of memory cells in LSTM rely on input gate, forget gate and output gate, which are generally described by sigmoid or tanh function [9].

Forget gate is utilized to forget the useless information of walking intention speed in the past. The input of the forget gate is composed of the distance of the left and right legs and the speed data of the left and right legs $g^{(t)} = [s_1, s_2, v_3, v_4]_t$ at the current time, the walking intention speed sequence $v^{(t-1)}$ at the last time, and the old cell state variable $c^{(t-1)}$. The output $f^{(t)}$ is taken by the operation of activation function μ.

$$f^{(t)} = \mu(A^{(f)}g^{(t)} + B^{(f)}v^{(t-1)} + C^{(f)}c^{(t-1)}) \tag{4}$$

The range of the output of the forget gate is 0 to 1, representing the probability of forgetting to hide the state of the cell at the last moment.

Input gate is used to remember some current information of walking intention speed. The input gate combines the past memory with the present memory and uses the activation function to output.

$$i^{(t)} = \mu(A^{(i)}g^{(t)} + B^{(i)}v^{(t-1)} + C^{(i)}c^{(t-1)}) \tag{5}$$

$$\tilde{r}^{(t)} = tanh(A^{(c)}g^{(t)} + B^{(c)}v^{(t-1)}) \tag{6}$$

The cell state $c^{(t-1)}$ at the last moment is updated to the new cell state $c^{(t)}$. The new cell state $c^{(t)}$ at the current time consists of two parts. The first part is the product of output $f^{(t)}$ of the forget gate and the old cell state $c^{(t-1)}$ at the last moment. The second part is the product of $i^{(t)}$ and $\tilde{r}^{(t)}$.

$$c^{(t)} = f^{(t)} \circ c^{(t-1)} + i^{(t)} \circ \tilde{r}^{(t)} \tag{7}$$

Output gate is used to establish the final output of walking intention speed. The update of the hidden state consists of two parts [10]. The first part, the output of the output gate $o^{(t)}$ is obtained from the distance of the left and right legs and the speed data of the left and right legs $g^{(t)}$ at the current time, the walking intention speed sequence $v^{(t-1)}$ at the last time, and the cell state variable $c^{(t-1)}$ through the sigmoid activation function. In the second part, the updated cell state $c^{(t)}$ is multiplied by $o^{(t)}$ through the tanh function.

$$o^{(t)} = \mu(A^{(o)}g^{(t)} + B^{(o)}v^{(t-1)} + C^{(o)}c^{(t)}) \tag{8}$$

$$v^{(t)} = o^{(t)} \circ tanh(c^{(t)}) \tag{9}$$

where $A^{(f)}, B^{(f)}, C^{(f)}, A^{(i)}, B^{(i)}, C^{(i)}, A^{(c)}, B^{(c)}, A^{(o)}, B^{(o)}, C^{(o)}$ are the weights of LSTM method.

The output of the sequence index predicted walking intention speed $\hat{v}^{(t)}$ at the current time is:

$$\hat{v}^{(t)} = \mu(C \cdot v^{(t)} + c) \tag{10}$$

where C is the coefficient of linear function, and c is the linear constant.

3.2   Training Method of LSTM Neural Network

This paper uses this time-expanded backward error propagation algorithm (BPTT) to train the LSTM network. BPTT algorithm has the advantages of simple operation and remarkable computational efficiency [11] The basic idea of the BPTT algorithm is: the LSTM network is expanded into a deep network according to the time sequence, and then the classical back propagation (BP) is utilized to train the expanded network to reduce errors [12]. Like the standard BP algorithm, BPTT also requires repeated application of the chain rule [13]. For LSTM network, the loss function (LF) is not simply related to the output layer, but also related to the hidden layer before and after the time point [14].

The loss function is as (11):

$$l^{(t)} = f\left(\hat{v}^{(t)}, v^{*(t)}\right) = \left\|\hat{v}^{(t)} - v^{*(t)}\right\|^2 \tag{11}$$

where $\hat{v}^{(t)}$ represents the user's speed output of walking intention recognition at the current moment, and $v^{*(t)}$ represents the desired speed of walking intention recognition at the current moment [15].

The loss function of globalization is as (12):

$$L = \sum_{t=1}^{T} l(t) \tag{12}$$

The gradient algorithm is utilized to decrease the value of the loss function [16]. Propagation training of LSTM mainly relies on $v^{(t)}$ and $c^{(t)}$ to reduce errors.

$$\delta_v^{(t)} = \frac{\partial L}{\partial v^{(t)}} \tag{13}$$

$$\delta_c^{(t)} = \frac{\partial L}{\partial c^{(t)}} \tag{14}$$

When back propagation training, $\delta_c^{(t)}$ is propagating in the reverse direction, and $\delta_v^{(t)}$ is calculated in the current layer [17]. The back propagation training model is shown in Fig. 6.

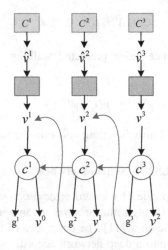

Fig. 6. The back propagation training model.

Gradient is used to update each weight [18], which is expressed by formula (15):

$$\Delta \psi^m = \lambda \Delta \psi^{m-1} - \delta \frac{\partial L}{\partial \psi^m} \tag{15}$$

The above is the whole cyclic process of LSTM structure's cyclic neural network.

3.3 Model Design Based on Walking Intention Recognition

The distance data of the legs measured by the multi-channel proximity sensor and the speed of the legs are used as input feature quantity. The LSTM algorithm is used to establish a dynamic event model to predict the user's walking intention speed based on multivariate time series. The structure diagram of walking intention recognition based on LSTM method is shown in Fig. 7. The structure is made up of an input layer, three hidden layers F1, F2, F3 and an output layer. T1 is the LSTM layer, which is used to process sequence information. T3 is the fully connected layer, which is used to merge the output of each LSTM unit. For the convenience of argument, the parameters T1, T2 and T3 are described as the number of hidden nodes of each unit in F1, the number of hidden nodes of each unit in F2 and the number of hidden nodes of each unit in F3.

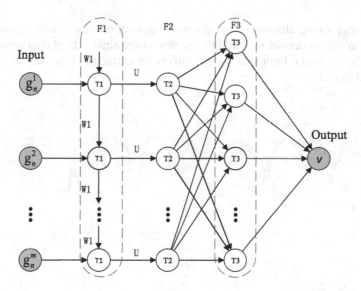

Fig. 7. The structure diagram of walking intention recognition based on LSTM method.

The multi-channel proximity sensor gives real-time feedback to the user's movement information and predicts the user's walking intention based on LSTM, so as to satisfy the user's comfort in the process of human-computer interaction.

The basic steps of the LSTM model are:

1) Extraction of dataset information

The multi-channel proximity sensor measures eight sets of data distances from the anterior eight sensors to the left and right legs. The four sensors on the left and the four sensors on the right automatically select the distance data of the left leg and the right leg according to the shortest distance of the leg from the sensor when walking, this distance information can more accurately reflect the user's real movement. The plan view of the legs and sensors is shown in Fig. 8.

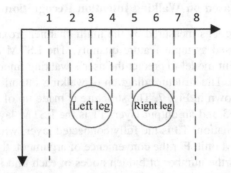

Fig. 8. The plan view of the legs and sensors.

As the legs swing alternately, the distance information has certain regularity and periodicity in the process of walking. Figure 9(a) is the eight sets of data measured by the eight sensors on the front side. Figure 9(b) is the extracted distance information of the left and right legs.

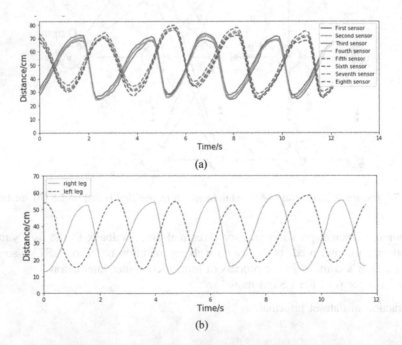

Fig. 9. The distance information of the left and right legs.

The distance information of the left and right legs obtains the speed information of the left and right legs v_L according to formula (16). The speed information of the left and right legs is shown in Fig. 10.

$$v_L = \frac{s_{i+1} - s_i}{T} (i = 1, 2, 3, \ldots)$$

(16)

where s_i is the distance measured at time i, and $T = 100$ ms is the sampling time.

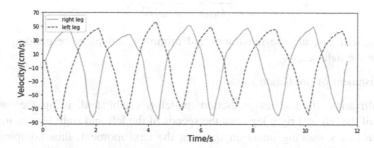

Fig. 10. The speed information of the left and right legs.

The user's walking intention speed v_R is recorded by a proximity sensor, and the conversion relationship from the distance to speed is obtained. Compare the distance between the left leg distance and the right leg distance of each sampling point, and take the small value S_j. The walking intention speed v_R is obtained from formula (17). The user's walking intention speed v_R is shown in Fig. 11.

$$v_R = 40(1 - \frac{S_j}{60}) \quad (j = 1, 2, 3, \ldots)$$

(17)

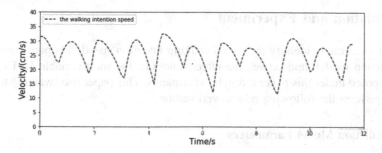

Fig. 11. The user's walking intention speed.

The distance information of the left and right legs, speed information of the left and right legs, and speed information of the walking intention is used as the data set.

2) Data preprocessing

It can solve the problem of noise and inconsistency in the original data set. Therefore, normalize the measured distance of the legs, the speed of the human legs, and the speed of user's walking intention, and reduce its value to the interval $[-1, 1]$:

$$\tilde{g} = \frac{g - (g_{max} + g_{min})/2}{(g_{max} - g_{min})/2} \tag{18}$$

where g_{max} and g_{min} are the maximum and minimum values corresponding to each variable in the data set.

3) Establishment of prediction model

A multivariate time series prediction model is established, using the measured distances of the left and right legs and the speeds of the left and right legs as inputs, to predict the user's walking intention speed at the next moment, thus completing the precise control of the walking training robot.

4) Model evaluation

This paper uses root mean square error (RMSE) to assess the results of predicting the user's walking intention speed. The smaller the value of RMSE, the higher the accuracy of the result of predicting the user's walking intention speed.

$$\text{RMSE} = \sqrt{\frac{1}{k} \sum_{i=1}^{k} (\hat{v}_i - v^*)^2} \tag{19}$$

where $\hat{v}_i$ is the predicted walking intention speed data in the test set; v^* is the real walking intention speed data in the test set; k is the test set size.

4 Simulation and Experiment

In order to accurately identify the user's walking intention speed, this paper optimizes the prediction model, and proves the effectiveness of the multivariable LSTM algorithm proposed under this paper through experiments. This paper uses walking training robot to perform the following related verification.

4.1 Prediction Model Parameters

This paper considers that the user walks at a constant speed and variable speed in a straight state. In the constant speed state, 771 data sets are used as the training data of the LSTM prediction model, and 436 data sets are used as the testing data of the LSTM prediction model. In the variable speed state, 671 sets of data are used as the training data of the LSTM prediction model, and 332 sets of data are used as the testing data of the LSTM prediction model.

The LSTM network that predicts the user's walking intention speed mainly needs to determine five hyper-parameters: the dimension of the input layer, the dimension of the output layer, the number of hidden layers, the dimension of each hidden layer, the time step.

The most easily determined parameters are the dimension of the input layer and the output layer. Since the input in this paper is 4 variables, the dimension of the input layer of the model is 4. The output is a function of historical information to predict the user's walking intention speed, so the dimension of the output layer is 1.

As the number of LSTM layers increases, the performance of the model also increases, the training difficulty of the model also increases [19]. For the prediction of walking intention speed, the hidden layer is gradually increased from 1 in the experiment. It is found that the performance did not improve much, but the training time became much slower. Therefore, this paper sets the hidden layer to 1. After many trials, the dimension of each hidden layer is set to 20 to obtain better prediction results.

The time step is the length of the time series used to predict the user's walking intention speed. This parameter must consider both the completeness of initial knowledge and the validity of model training. Too short a historical sequence length will reduce the accuracy of prediction; too long a historical sequence length will increase the training difficulty of the model [20]. The mean square error of time steps from 1 to 20 steps is illustrated in Fig. 12. This paper sets the time step to 12 steps.

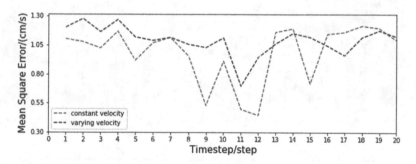

Fig. 12. The mean square error of time steps from 1 to 20 steps.

4.2   Result Analysis

Figure 13 is a diagram of the recognition of walking intention speed based on LSTM when walking at a constant speed in a straight state. Figure 13(a) is the distance information of the left and right legs at a constant speed. Figure 13(a) shows regular changes because the walking speed is uniform. Figure 13(b) shows the speed of the left and right legs and the walking intention speed when a person walks. The broken line in Fig. 13(c) is the walking intention speed, and the solid line is the predicted walking intention speed. Figure 13(c) shows the predicted walking intention speed and the walking intention speed are in excellent agreement, and the user's walking intention speed can be accurately recognized.

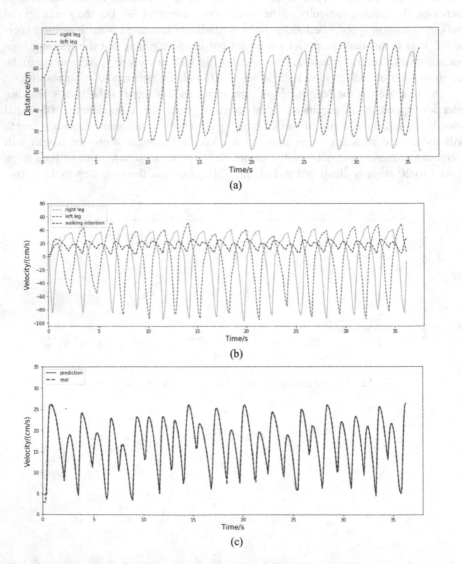

(a)

(b)

(c)

Fig. 13. Recognition of walking intention based on LSTM when walking at constant speed.

Figure 14 is a diagram of the recognition of walking intention speed based on LSTM when walking in variable speed in the straight state. Figure 14(a) is the distance information of the left and right legs when changing speed. It can be seen in the figure that the period of human leg exchange becomes shorter when the person walks accelerating, while the period of human leg exchange becomes longer when the person walks decelerating. Figure 14(b) shows the speed of the left and right legs and the walking intention speed when a person walks. The broken line in Fig. 14(c) is walking intention speed, and the solid line is the predicted walking intention speed. Figure 14 shows the predicted walking intention speed and the walking intention speed are in

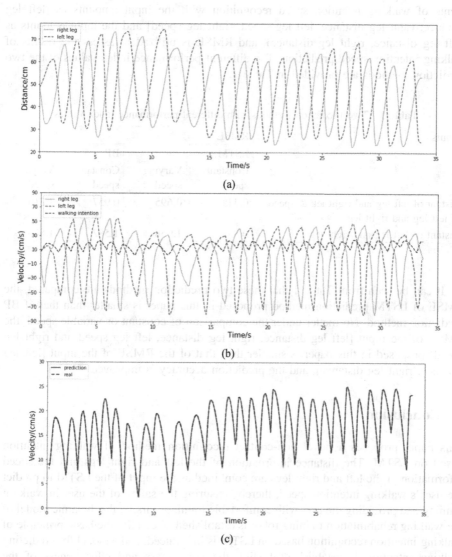

(a)

(b)

(c)

Fig. 14. Recognition of walking intention based on LSTM when walking at variable speed.

great agreement, and the walking training robot can accurately recognize the walking intention speed of the user. It can be seen that regardless of whether the user is at a constant speed or variable speed. The proposed walking intention recognition based on LSTM can accurately identify the user's walking intention speed.

4.3   Performance Evaluation of Prediction Models

To verify the advantages of the recognition of walking intention speed based on multi-variable LSTM proposed in this paper, it is compared with the BP neural network prediction algorithm. These two prediction methods are used to perform the experiments of walking intention speed recognition with the input amounts as [left leg distance, right leg distance, left leg speed, right leg speed] and the input amounts as [left leg distance, right leg distance], and RMSE is utilized to assess the results of walking intention speed prediction. The different performance index values of the two prediction methods are provided in Table 1.

Table 1. Two prediction methods with different performance index values.

Inputs	RMSE			
	LSTM		BP	
	Constant speed	Varying speed	Constant speed	Varying speed
Distant of left leg and right leg & speed of left leg and right leg	0.445	0.695	0.957	1.216
Distant of left leg and right leg	0.838	1.058	1.582	1.865

It can be seen from Table 1 that these two prediction methods are compared, the RMSE of LSTM prediction method proposed in this paper is smaller than that of BP prediction method. No matter under the comparison of constant or variable speed, the RMSE of the input [left leg distance, right leg distance, left leg speed and right leg speed] proposed in this paper is smaller than that of the RMSE of the input [left leg distance, right leg distance], and the prediction accuracy is improved.

5   Conclusion

This paper proposes a novel non-contact recognition method of walking intention based on LSTM. The distance information of the left and right legs and the speed information of the left and right legs are combined as the input of the LSTM to predict the user's walking intention speed, thereby ensuring the safety of the user in walking training and providing the user with comfortable training. Firstly, the dynamic model of the walking rehabilitation training robot is established; secondly, the basic principle of walking intention recognition based on LSTM is introduced, and a model for predicting walking intention is established; finally, the correctness and effectiveness of the walking intention recognition method based on LSTM proposed in this paper are verified by experiments. The proposed method can accurately identify the user's walking intention under constant speed and variable speed, and improve the user's safety and comfort in the process of human-computer interaction. In the following work, the user rotation angle recognition of the walking rehabilitation training robot will be researched and combined with the user's intention speed recognition in this paper to realize omnidirectional translation.

References

1. Hu, J., Hou, Z., Chen, Y., Zhang, F., Wang, W.: Lower limb rehabilitation robots and interactive control methods. Acta Autom. Sin. **40**(11), 2377–2390 (2014)
2. Leutenegger, S., Lynen, S., Bosse, M.: Keyframe-based visual-inertial odometry using nonlinear optimization. Int. J. Robot. Res. **34**(3), 314–334 (2015)
3. Ihn-Sik, W.: Intelligent robotic walker with actively controlled human interaction. ETRI J. **40**(4), 522–530 (2018)
4. Chen, A., Wang, H.: Real-time control of intelligent walker based on physical human-computer interaction. Appl. Res. Comput. **34**(5), 1362–1366 (2017)
5. Xu, W., Huang, J., Yan, Q.: Research on walking-aid robot motion control with both compliance and safety. Acta Autom. Sin. **42**(12), 1859–1873 (2016)
6. Song, E., Soong, F.K., Kang, H.G.: Effective spectral and excitation modeling techniques for LSTM-RNN-based speech synthesis systems. IEEE/ACM Trans. Audio Speech Lang. Process. **25**(11), 2152–2161 (2017)
7. Chen, Y., Liu, S., He, S., Liu, K., Zhao, J.: Event extraction via bidirectional long short-term memory tensor neural networks. In: Sun, M., Huang, X., Lin, H., Liu, Z., Liu, Y. (eds.) CCL/NLP-NABD -2016. LNCS (LNAI), vol. 10035, pp. 190–203. Springer, Cham (2016). https://doi.org/10.1007/978-3-319-47674-2_17
8. Gao, L., Guo, Z., Zhang, H.: Video captioning with attention-based LSTM and semantic consistency. IEEE Trans. Multimedia **19**(9), 2045–2055 (2017)
9. Schak, M., Gepperth, A.: Robustness of deep LSTM networks in freehand gesture recognition. In: Tetko, I.V., Kůrková, V., Karpov, P., Theis, F. (eds.) ICANN 2019. LNCS, vol. 11729, pp. 330–343. Springer, Cham (2019). https://doi.org/10.1007/978-3-030-30508-6_27
10. Soutner, D., Müller, L.: Application of LSTM neural networks in language modelling. In: Habernal, I., Matoušek, V. (eds.) TSD 2013. LNCS (LNAI), vol. 8082, pp. 105–112. Springer, Heidelberg (2013). https://doi.org/10.1007/978-3-642-40585-3_14
11. Chen, K., Huo, Q.: Training deep bidirectional LSTM acoustic model for LVCSR by a context-sensitive-chunk BPTT approach. IEEE/ACM Trans. Audio Speech Lang. Process. **24**(7), 1185–1193 (2016)
12. Zhu, Q., Li, H., Wang, Z.: Ultra-short-term prediction of wind farm power generation based on long short-term memory network. Power Grid Technol. **41**(12), 3798–3802 (2017)
13. Zhang, X., Zou, Y.Y., Li, S.Y., Xu, S.H.: Product yields forecasting for FCCU via deep bi-directional LSTM network. In: Proceedings of the 37th Chinese Control Conference, pp. 8013–8018. IEEE, Wuhan (2018)
14. Zhou, P., Shi, W., Tian, J., Qi, Z.Y.: Attention-based bidirectional long short-term memory networks for relation classification. In: Proceedings of the 54th Annual Meeting of the Association for Computational Linguistics, pp. 207–212. ASSOC, Berlin (2016)

15. Ertam, F.: An effective gender recognition approach using voice data via deeper LSTM networks. Appl. Acoust. (156), 351–358 (2019)
16. Huo, Y.J., Wong, D.F., Ni, L.M.: Knowledge modeling via contextualized representations for LSTM-based personalized exercise recommendation. Inf. Sci. (523), 266–278 (2020)
17. Cortez, B., Carrera, B., Kim, Y.J.: An architecture for emergency event prediction using LSTM recurrent neural networks. Expert Syst. Appl. **97**, 315–324 (2018)
18. Xu, Z., Zou, Y.: A weighted auto regressive LSTM based approach for chemical processes modeling. Neurocomputing (367), 64–74 (2019)
19. Amin, U., Khan, M.: Activity recognition using temporal optical flow convolutional features and multi-layer LSTM. IEEE Trans. Ind. Electron. **66**(12), 9692–9702 (2019)
20. Wu, Y., Yuan, M., Dong, S.: Remaining useful life estimation of engineered systems using vanilla LSTM neural networks. Neurocomputing **275**, 167–179 (2018)

A Novel Optimization Approach Oriented to Auxiliary Transfer Based on Improved Reinforcement Learning

Han Bao[1]([✉]), Junyou Yang[1], Donghui Zhao[1], and Shuoyu Wang[2]

[1] School of Electrical Engineering, Shenyang University of Technology,
Shenyang 110870, China
1757880104@qq.com
[2] Department of Intelligent Mechanical Systems Engineering,
Kochi University of Technology, Kami, Kochi 7828502, Japan

Abstract. Aiming to provide an adaptive transfer method for people with weak motion capability, a novel optimization approach oriented to auxiliary transfer based on improved reinforcement learning is proposed. Firstly, the function of excretory support robot and the limitation of daily autonomous behavior are discussed. To obtain the transfer point of the user groups with different mobility, a two-link model of human body during sitting down is proposed and the mapping relationship of human body position and sitting point is established. According to the difference of human sitting posture, a fuzzy-reinforcement learning algorithm is proposed to calculate the appropriate transfer point. Integrating fuzzy reasoning method into the reward mechanism of reinforcement learning, the mapping relationship is continuously optimized and the appropriate sitting point is obtained in multiple transfer process. Finally, the transfer experiments and simulation are carried out. Experiments show that the transfer optimization method based on improved fuzzy-reinforcement learning proposed in this paper effectively obtains the optimal transfer point according to different motion capability and different transfer postures of users during the transfer process. The approach has strong popularization characteristics, which has the potential to be applied to nursing homes, hospitals and other rehabilitation sites.

Keywords: Excretory support robot · Transfer · Reinforcement learning · Fuzzy reasoning

1 Introduction

With the development of robot technology, robots are expected to play a greater role in caring for the elderly and the disabled [1]. For the elderly and the disabled, it is an important problem to realize daily life behavior assistance, such as getting up, standing and going to the toilet by robot transfer. At present, researchers have developed many nursing robots, such as intelligent wheelchair [2], and intelligent bed [3, 4]. Different sensor modules are installed on it to achieve different functions, including voice recognition, visual processing, physiological index detection and other technologies, which greatly reduces the burden of nursing staff. Although intelligent nursing bed with

© Springer Nature Singapore Pte Ltd. 2020
J. Qian et al. (Eds.): ICRRI 2020, CCIS 1335, pp. 33–49, 2020.
https://doi.org/10.1007/978-981-33-4929-2_3

excretory function has been developed [5], toilet is installed on the bed. If not cleaned up for a long time, it is easy to cause hygiene problems. However, the transfer method that specifically solve the toilet problems of the elderly and the disabled has not been developed. In recent years, as one of the hotspots in the fields of artificial intelligence and machine learning, reinforcement learning has attracted extensive attention. Many domestic and foreign scholars widely apply reinforcement learning to robot path optimization [6], and real-time obstacle avoidance [7]. Luy [8] applies reinforcement learning to implement intelligent tracking control of wheeled mobile robot. Jaradat [9] and others adopt Q-learning algorithm to study the navigation problem of robot in dynamic environment. Compared with the previous artificial potential field method, this method is more reasonable in path selection, and the time required to reach the target point is significantly reduced.

In this paper, a fuzzy-reinforcement learning algorithm is proposed and applied to auxiliary excretion transfer. Combined with the two-link model of sitting down, an appropriate human-robot collaboration strategy for various user groups with different mobility and physical conditions has been established. According to the fuzzy reasoning algorithm, the objective and subjective evaluation are combined to judge the suitability of sitting point. The excretory support robot learn the sitting point and adjust the action strategy continuously through the improvement of reward signal in reinforcement learning. Finally, the robot moves to the appropriate position and achieve auxiliary transfer. The method proposed in this paper has the following innovations:

(1) A novel optimization approach for auxiliary transfer based on the combination of fuzzy reasoning and reinforcement learning is proposed. It effectively implements transfer task planning according to the characteristics of user groups with different motion capability.
(2) The quantitative evaluation of pressure sensors and subjective evaluation of people are integrated through fuzzy reasoning algorithm. The reward mechanism of reinforcement learning is improved, which effectively realizes the accurate evaluation of the transfer process.
(3) Integrating fuzzy reasoning method into the reward mechanism of reinforcement learning, a self-learning evaluation strategy is established. The excretory support robot adjusts the action strategy in the continuous human-robot interaction, and finally obtain the optimal transfer point.

2 The Transfer Control System of Excretory Support Robot

The overall design scheme of excretory support control system in the transfer process is shown in Fig. 1. Firstly, the human leg link model is established, and a mapping relationship of leg position (d, α) and corresponding sitting point is calculated. Meanwhile, the kinematic model of the robot is established. Secondly, the position information of the human leg is detected by laser radar. The data is filtered and matched with the mapping relationship to obtain the moving target of the robot. According to the position and movement of the leg, a better path is planned, and the robot moves to the designated position. The fuzzy reasoning algorithm and reinforcement learning method are combined to establish the self-learning evaluation mechanism. After the user sits down, the subjective evaluation of the user and the objective evaluation reflected by the pressure sensor are adopted to evaluate the position of the robot, and then the evaluation is taken as the reward return value of reinforcement learning. The robot adjusts action strategy according to the output of evaluation system, and finally obtain a more appropriate sitting point from multiple human-robot cooperation tasks through the self-learning evaluation mechanism.

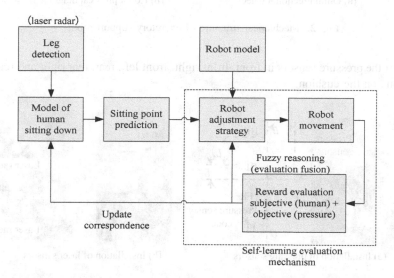

Fig. 1. Overall framework of auxiliary excretion transfer control system

2.1 Excretory Support Robot

In this paper, the excretory support robot developed by our laboratory is used as the experimental platform, and its mechanical structure is shown in Fig. 2. The robot motion mechanism adopts omnidirectional moving chassis, and its structure is shown in Fig. 2(a). The robot support is in a regular triangle structure, and the angle between each other of the three omnidirectional wheels is 120°. It is evenly distributed and installed on the chassis support of the excretory support robot [10], and the distance between the center of each wheel and the chassis center of the excretory support robot is equal, as shown in Fig. 2(b).

The design of excretory support robot is similar to the toilet with hollow in the middle. When people sit on the robot in an inappropriate position, it have a great impact on the comfort of the toilet. Therefore, six groups of pressure sensors are installed evenly in the cushion position of the robot, which are used to detect the pressure distribution after people sit down, and evaluate the motion position of the robot. The sensor installation distribution diagram and the label of sensor size are shown in Fig. 3(a), which are used to calculate the center of gravity position after people sit down in subsequent chapters. FF, FR, FL, BB, BR and BL respectively

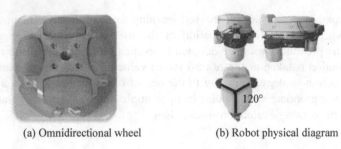

(a) Omnidirectional wheel   (b) Robot physical diagram

Fig. 2. Mechanical structure of excretory support robot

represent the pressure sensors in front, front right, front left, rear, rear right and rear left directions on the cushion.

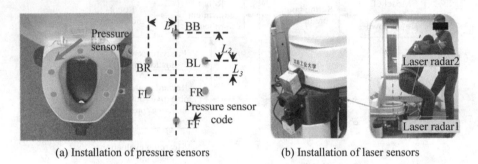

(a) Installation of pressure sensors   (b) Installation of laser sensors

Fig. 3. Distribution of sensors

In the transfer process, the primary task of the excretory support robot is to obtain the posture information of the human body. The current sensors for detecting human behavior and posture mainly include posture sensors and infrared depth cameras. The posture sensor is a contact measurement, which is not convenient for people with physical disabilities. The depth camera is able to detect the human posture when the body parts of each person must be in the vision of the camera without overlapping. With the cooperation of nurses, the body posture information detected by depth camera is inaccurate. Laser radar has the advantages of large detection range and high detection accuracy. To better detect the posture information of the human body, two laser sensors are installed on the robot and their distribution is shown in Fig. 3(b). In this paper, an enriched human posture data set is obtained through laser radar detection, which is shown in the source archive.

Aiming to study the motion characteristics of the robot and control the robot, the position and posture of the robot should be described firstly. The robot is in a plane motion, the global coordinate system is $\{o_a\}$, and the local coordinate system of the robot is $\{o_r\}$. The robot posture is defined as follows: the world coordinate system $\{o_a\}$ is the robot motion plane coordinate system, and its origin o_a needs to be determined according to the actual environment, and is usually taken as the starting point of the robot motion. When a mobile robot moves in the world coordinate system, it is regarded as a particle, and the position of the robot in the world coordinate system is represented by (x, y). The angle between the x_r axis of the robot coordinate system and the x_a axis of the world coordinate system is the robot motion direction angle θ, and the overall motion speed of the robot is v. The speeds of the three omnidirectional wheels are v_i, $i = 1, 2, 3$ respectively representing the tangential speeds of the ith wheel. The distance between the three wheels and the geometric center of the robot is l, and the three wheels are installed symmetrically at $120°$ to each other (see Fig. 4).

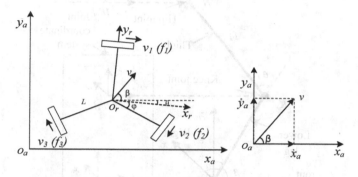

Fig. 4. Robot model

Through the analysis of the relationship between the overall speed v of the robot and the tangential speed v_i of the three wheels, the kinematic equation of the omni-directional mobile robot is expressed as follows:

$$
\begin{aligned}
v_1 &= v_x + \dot{\theta}l \\
v_2 &= -v_x\cos\tfrac{\pi}{3} - v_y\sin\tfrac{\pi}{3} + \dot{\theta}l \\
v_3 &= -v_x\cos\tfrac{\pi}{3} + v_y\sin\tfrac{\pi}{3} + \dot{\theta}l
\end{aligned}
\tag{1}
$$

Wherein:

$$
\begin{aligned}
v_x &= v\cos\theta \\
v_y &= v\sin\theta
\end{aligned}
\tag{2}
$$

The matrix form is obtained:

$$
\begin{bmatrix} v_1 \\ v_2 \\ v_3 \end{bmatrix}
=
\begin{bmatrix}
1 & 0 & l \\
-\frac{1}{2} & -\frac{\sqrt{3}}{2} & l \\
-\frac{1}{2} & \frac{\sqrt{3}}{2} & l
\end{bmatrix}
\begin{bmatrix} v_x \\ v_y \\ \dot{\theta} \end{bmatrix}
\tag{3}
$$

2.2   The Two-Link Model of Human Lower Limb

In this paper, the lower leg and thigh are regarded as a link on a two-dimensional plane. The two-link model and the ankle joint are regarded as a universal joint, it is, the lower leg is able to move forward and backward along the y-axis, or left and right along the x-axis (see Fig. 5).

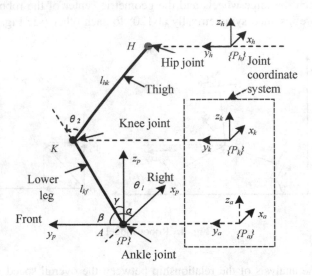

Fig. 5. Two-link model of human lower limb

Because the hip joint point is involved in three-dimensional space movement, the method of homogeneous coordinate transformation of three-dimensional space points by fourth-order square matrix is used to carry out kinematic analysis of human body model. The ankle joint, knee joint and hip joint are set as key points A, K and H respectively. And establish coordinate system $\{P_a\}$, $\{P_k\}$ and $\{P_h\}$ with A, K and H points respectively. Note that the translation vector of $\{P_k\}$ relative to $\{P_a\}$ is $q_k^a(x^a, y^a, z^a)$, note that the translation vector of $\{P_h\}$ relative to $\{P_k\}$ is $q_h^k(x^k, y^k, z^k)$. Then it is concluded that:

$$\begin{cases} x^a = l_{kf}\cos\beta \\ y^a = l_{kf}\cos\alpha \\ z^a = l_{kf}\cos\left(\frac{\pi}{2} - \alpha\right) \end{cases} \tag{4}$$

$$\begin{cases} x^k = 0 \\ y^k = l_{hk}\cos\theta_2 \\ z^k = l_{hk}\sin\theta_2 \end{cases} \tag{5}$$

Set the y-axis direction as the front of people and the x-axis direction as the right side of people. Where α, β, and γ are the angles between the lower leg l_{kf} and the x, y, and z axes of the coordinate system $\{P_a\}$, respectively. According to the knowledge of the range of motion of the lower limb joints, $\alpha \in (80° \sim 110°)$ represents the left and right movements of the leg, and $\beta \in (70° \sim 120°)$ represents the front and back movements of the leg $\gamma \in (0° \sim 20°)$. According to the actual experiments, it is found that according to the angle of α and β, the movement of the lower leg is fully characterized. γ angle has a corresponding coupling relationship with α, β angle. In this paper, the change of α, β angle is proposed to characterize the movement of the leg.

Since no rotation transformation exists between each coordinate system, the translation transformation of the coordinates is adopted to transform the coordinates of the joint points. The description transformation matrix of $\{P_k\}$ relative to $\{P_a\}$, $\{P_h\}$ relative to $\{P_k\}$ is respectively recorded as follows:

$$T_k^a = \begin{bmatrix} l_{3\times3} & q_k^a \\ 0 & 1 \end{bmatrix} \tag{6}$$

$$T_h^k = \begin{bmatrix} l_{3\times3} & q_h^k \\ 0 & 1 \end{bmatrix} \tag{7}$$

Therefore, the description of the coordinate system $\{P_h\}$ relative to $\{P_a\}$, it is, the transformation matrix of the ankle to hip coordinate is:

$$T_h^a = T_k^a T_h^k \tag{8}$$

As shown in Fig. 5, the coordinate of hip joint is the origin of coordinate system, and the coordinate vector $q_h^a = [x, y, z, 1]^T$ of point H under $\{P_a\}$:

$$q_h^a = T_h^a \tag{9}$$

A two-link model of human body during sitting down is established based on the above analysis. If the position coordinates of human feet are known, the spatial position of hip joint and the position after sitting are calculated according to the model.

3 An Optimization Transfer Method Based on Fuzzy-Reinforcement Learning

The sitting points on the excretory support robot are different because of the sitting habits of different users. It is difficult to accurately establish a mathematical model of sitting down in this situation. In this paper, the objective evaluation of the pressure sensor and the subjective evaluation of people are integrated through fuzzy reasoning theory, and a fuzzy evaluation system is established to evaluate the suitability of the excretory support sitting point.

According to the output results of the evaluation system, the reinforcement learning idea is adopted to control the robot to adjust the action strategy. Finally, the robot obtain the appropriate sitting point by self-learning and realize the optimal transfer. The specific learning strategies are shown in Fig. 6.

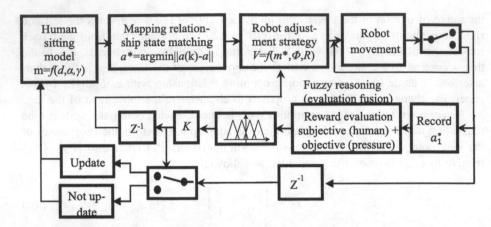

Fig. 6. Transfer point optimization method based on fuzzy-reinforcement learning

3.1 Reinforcement Learning

Reinforcement learning is different from other machine learning. It focuses on generating training data through interaction with environment. The learning system mainly determines the next action according to the reward signal and the current state of the environment, and finally realizes the learning goal. This kind of reward signal is often easy to obtain in practice, so reinforcement learning has a wide application prospect in solving complex decision-making and optimization problems that not accurately establish mathematical model in advance [11]. This section will combine the actual situation of this system to improve the reward signal of reinforcement learning.

In the main algorithm of reinforcement learning, iterative technique is adopted to update the estimated value of value function to obtain the optimal action strategy. The value function under the optimal strategy is obtained from Bellman formula, as shown in Eq. (10):

$$V^*(s) = \max_{a \in A(s)} E\{r_{t+1} + \gamma V^*(s_{t+1}) | a_t = a, s_{t+1} = s'\}$$
$$= \max_{a \in A(s)} \sum_{s'} P^a_{ss'} [R^a_{ss'} + \gamma V^*(s')] \tag{10}$$

Where $V^*(s)$ is approximated by Bellman iteration:

$$V(s) = (1 - \alpha)V(s) + \alpha\left(r^a_{ss'} + \gamma V(s')\right) \tag{11}$$

Where $\alpha \in (0, 1]$ is the learning rate.

The general idea of reinforcement learning algorithm is that every action of interaction between robot and environment will produce a new environment for robot. Then robot will give an evaluation according to the environment, and robot will make a new exploratory action again. According to the reward, a series of actions are selected as the optimal motion strategy of the robot. In this paper, it will be given a final evaluation after the robot moves to a position. Therefore, it is not necessary to consider the action process of the reinforcement learning robot at every moment and the given action evaluation. Therefore, according to the characteristics of the human-robot cooperation task in this paper, the idea of reinforcement learning is proposed to search the position that the robot should move to.

3.2 Fuzzy Evaluation System

The quantitative evaluation of sensors and the subjective evaluation of people are taken as a comprehensive evaluation because of the differences between different users, which accurately guide the robot to complete the human-robot cooperation task. Fuzzy reasoning is the main part of approximate reasoning. The intelligent control scheme of fuzzy control system is robust to some extent. It deal with the concepts of fuzzy and qualitative, and is accurately described by membership function [12, 13]. Behavior control based on fuzzy rules is able to directly map the environmental information to a more appropriate control output behavior, which has the characteristics of quick response and strong real-time performance.

Figure 7(a) shows the change waveforms of the six pressure sensors of the excretory support robot after normalization, and the normalization formula is:

$$P_i = \frac{p_i - p_{mini}}{p_{maxi} - p_{mini}} \tag{12}$$

Where p_i is the pressure value before normalization of pressure data, i = 1, 2, ...,6 respectively represent 6 groups of sensors, p_{mini}, p_{maxi} respectively represent the minimum and maximum values of each group of sensors, P_i is the value of each group of sensors after normalization.

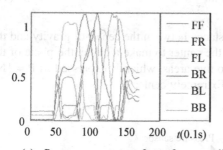 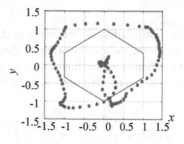

(a) Pressure sensor waveform after normalization (b) Trajectory of center of gravity

Fig. 7. Waveform of pressure sensor and change of center of gravity position

Assuming that the mass of a stationary object is M, the coordinate of its center of gravity in the plane is (x_m, y_m). If the support force of each point in a plane is P_i, and the coordinate is (x_i, y_i), according to the moment balance, the center of gravity of the object on the horizontal plane is expressed as:

$$x_m = \frac{1}{Mg} \sum_{i=1}^{6} P_i x_i \tag{13}$$

$$y_m = \frac{1}{Mg} \sum_{i=1}^{6} P_i y_i \tag{14}$$

When people sit on the robot, the body swings around in a circle. The change of the center of gravity after the calculation of the center of gravity formula is shown in Fig. 7(b). When the upper body moves in a circle, the position of its center of gravity changes, but the position of the buttocks is unchanged. It may appear that the position of the buttocks sitting on the robot is different from the position of the center of gravity measured in actual measurement. In the actual experiment, it is found that when people sit on the toilet, the body incline back and forth to a large extent, while the left and right inclinations are rarely found. The position of the buttocks on the robot will coincide with the position of the measured center of gravity when and only when the upper body of the human remains upright. Therefore, using pressure sensors alone to evaluate whether the robot has moved to a suitable position has certain errors, which needs to consider the subjective factors of human beings.

Through the calculation of the position of the gravity center of the human body after sitting down, the position of the gravity center of the human body in the cushion plane of the robot after sitting down is described quantitatively. The distance between the center of gravity (x_m, y_m) and the origin of the robot center is defined as:

$$d_m = \sqrt{x^2 + y^2} \tag{15}$$

The direction is the angle between the positive direction of the x-axis, set as:

$$\varphi = atan\frac{y_m}{x_m} \tag{16}$$

As shown in Fig. 8, according to the distance between the center of gravity and the origin o as the radius, the origin is used as the center to make a circle, the plane of the robot cushion is approximately divided into five areas, which is expressed as P = {No center, Litter center, Center, Very center, Extremely center}.

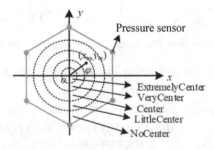

Fig. 8. Pressure zoning of robot

The membership function of the pressure sensor is defined according to Fig. 9(a). The horizontal axis coordinate represents the normalized d_m, $d_m \in [0, 1]$. $d_m = 1$ represents the center of gravity is located in the center of the robot pressure pad. $d_m = 0$ represents the center of gravity is located outside the robot pressure sensor, and the vertical axis represents the degree of the center of gravity close to the center, and its range is 0–1.

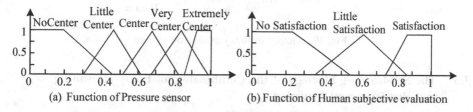

(a) Function of Pressure sensor   (b) Function of Human subjective evaluation

Fig. 9. Membership function of fuzzy evaluation system

For human subjective factors, the feedback evaluation of robot tasks is implemented by using the percentage system, which is divided into three levels, represented by the set S = {No Satisfaction, Little Satisfaction, Satisfaction}, and its membership function is shown in Fig. 9(b). The horizontal axis represents the normalized fraction range of 0–1. After the membership function is set, the next step is to set the rules. The fuzzy rules are shown in Table 1:

Table 1. Fuzzy rule table

Human\pressure sensor	No center	Little center	Center	Very center	Extremely center
No satisfaction	No center	No center	Little center	NULL	NULL
Little satisfaction	Little center	Little center	Center	Center	Very center
Satisfaction	NULL	NULL	Center	Very center	Extremely center

The system output as shown in Fig. 10 is obtained after the human subjective factors and the objective factors of the sensor are calculated according to the rules in Table 1. The human-robot cooperation task of the robot is better evaluated by combining the subjective and objective evaluation.

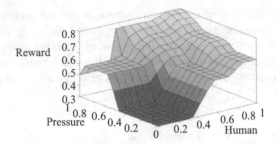

Fig. 10. Output of fuzzy evaluation system

3.3   The Self-learning Strategy of the Combination of Fuzzy Reasoning and Reinforcement Learning

Through the discussion of the previous two sections, combined with the quantitative description of sensors and human subjective factors, the quality of robotic human-robot cooperation tasks is evaluated. From the above brief description of the current reinforcement learning idea, it is found that the reward mechanism, the selection of action space and the selection of state space in reinforcement learning algorithm have a certain impact on the quality of robot learning results. At present, the exploration way of reinforcement learning is to match the state of the external environment with the state space $\{S\}$ defined by the system. After the matching is completed, actions are randomly selected in the action space $\{A\}$ to learn the trial and error mechanism. Every trial and error has corresponding rewards and penalties, and then the optimal action strategy π^* is determined by taking the extreme value function.

In this paper, we adopt the previously established model of human sitting down (when different people use it, the model is inaccurate, so the mapping relationship of the model is updated through learning), which has a mapping relationship of $m = f(d_1, \alpha, \gamma)$, where m is the final center of gravity coordinate $m(x_m, y_m)$, d_1 is the distance between the legs and the laser sensor below, α, γ are the angles of the lower legs. The specific process is to first calculate a mapping relationship by using the human sitting down model. Suppose that the input vector $a_z = \begin{bmatrix} d_{1z}, \alpha_z, \gamma_z \end{bmatrix}^T$ is a set of data that the robot measures when the human body is ready to sit down, a is the data in the mapping relationship, a^* is a set of vectors obtained by matching a and $a_z(k)$ through certain criteria. The specific matching criteria are as follows:

$$a^* = \operatorname{argmin} \left\| \sum_{i=1}^{3} \sqrt{\left(a_z(i)^2 - a(i)^2\right)} \right\| \tag{17}$$

After obtaining $a^* = \begin{bmatrix} d_1^* \alpha^* \gamma^* \end{bmatrix}^T$, the output $m(x_m, y_m)$ in the mapping relationship is correspondingly taken out, which is used to give the running target of the robot. According to the kinematic model of the robot, the control quantity of the robot is $v_x, v_y, \dot{\theta}$. The output of the evaluation system is $R \in [0, 1]$, and the reward R indicates the quality of the cooperation task. The larger the R is, the more appropriate the position the robot moves to in this task, the smaller the adjustment range of the robot next time. Similarly, the smaller R is, the greater the robot will adjust next time. The robot control quantity is given by (16):

$$v_x = (1 - R)d_m cos \Psi$$
$$v_y = (1 - R)d_m sin \Psi \quad\quad\quad (18)$$
$$\dot{\theta} = \alpha$$

Where α is the angle between the center vertical line of the feet of the user and the positive y-axis in the laser radar coordinate system. The y-axis direction of the robot is always taken as the front of the robot, and α is taken as the rotation component of the robot, which is used to control the head orientation of the robot, as shown in Fig. 11.

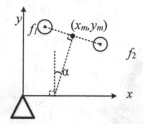

Fig. 11. Schematic diagram of robot heading angle control

According to the given control quantity of the robot, the robot runs to the designated position, and judges whether a person sits down by whether the pressure of the pressure sensor is zero. If not, the robot always adjusts its motion according to the position of the person. After people sit down, record the calculated a^*, and calculate the center of gravity position, and output a comprehensive evaluation (reward) of the robot cooperation task combined with the evaluation of 0–100 points given by the user. Finally, according to the size of the reward, determine whether to update the recorded state a^*, it is, update the data mapping relationship of $m = f(d_1, \alpha, \gamma)$. The reward and the center of gravity position are adopted to guide the robot adjustment strategy of the next task and determine the position to which the robot should move next time.

4 Simulation and Experimental Verification

The method proposed in this paper is a robot transfer optimization method based on fuzzy-reinforcement learning fusion. Simulation and experiment are carried out in MATLAB 2017 environment to verify the comfort and reliability of this method. Using C++ as programming language, vs2013 as software platform to control robot operation. According to the adjustment strategy in Fig. 6, the experimental results in Fig. 12 are obtained. In the figure, the x-axis represents the distance D from the human leg to the robot, where $D \in [0, 35]$; the y-axis represents the deviation angle α of the human leg in the left-right direction; $\alpha \in [-20, 20]$, it is, the lower leg may be 20° to the left and right respectively; the z-axis represents the Euclidean distance d between the sitting point and the geometric center of the robot calculated by the model when D and α take different values respectively. The initial value assumes that the geometric center of the robot is the most comfortable position for people to sit down. Before learning, as shown in Fig. 12(a), when $\alpha = 0$, $d = 7$ cm, people sit in the center of the robot after sitting down. The darker the color in the figure, the smaller the value, and the closer the sitting point is to the more comfortable point. The minimum value in the z-axis represents the value of D and α at the moment when a person is sitting in the geometric center of the robot. After a certain number of times of learning, according to the sitting position and habits of different people, the sitting points will also change.

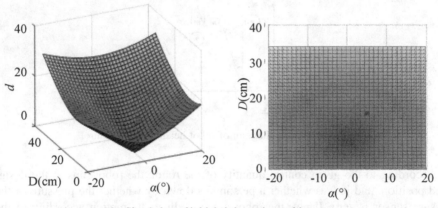

(a)   Simulation of initial sitting point position

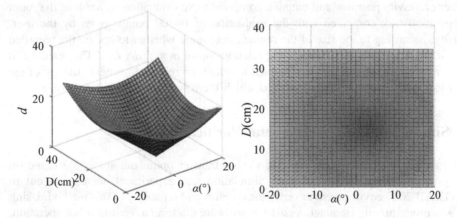

(b)   Simulation of optimal sitting point position

Fig. 12. Human-robot cooperative sitting point position simulation

Figure 13(a) and Fig. 13(c) show the simulation and experimental verification of the position of the robot when the human body sits down normally. Figure 13(b) and Fig. 13(d) show the adjustment direction that the robot should adjust according to the human sitting down model in case of left-right deviation, so that the human body sit on the more appropriate point on the robot. The experimental results in Fig. 13(d) also show that the simulation results are correct and reasonable.

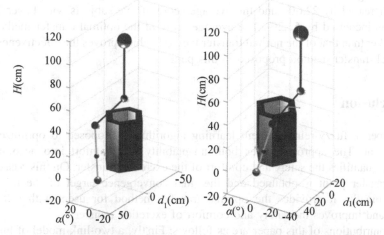

| (a) Simulation of robot normal position | (b) Simulation of robot adjusting position |

| (c) Experiment of robot normal position | (d) Experiment of robot adjusting position |

Fig. 13. Transfer experiments and simulation

To further verify the effectiveness and adaptability of the optimal transfer method, 10 test subjects are selected for the transfer experiment. Subjects 1–5 are 5 healthy subjects, and subjects 6–10 are installed fixation devices on their knees to imitate people with lower limb dysfunction. The test subjects scored the transfer method before and after the improvement on a ten-point scale based on the quality of the transfer experience. The results of the questionnaire are shown in Table 2.

Table 2. Score comparison before and after transfer optimization

Test subject	1	2	3	4	5	6	7	8	9	10
Score (normal transfer)	6	7	6	5	7	5	4	5	4	5
Score (optimal transfer)	7	8	8	7	8	7	7	8	6	7

The results show that the scores of the 10 test subjects for the optimal transfer method are higher than the normal transfer method. The average score of five healthy subjects increased by 22.6%, and the average score of five subjects with lower limb dysfunction increased by 52.2%. The average score of the optimal transfer method is 35.2% higher than that of the normal transfer method, which proves the effectiveness of the optimal transfer method proposed in this paper.

5   Conclusion

In this paper, a fuzzy-reinforcement learning algorithm is proposed to optimize the transfer point. The approach takes both adaptability and comfort into account. It effectively quantifies the safety and comfort of the excretion transfer. On this basis, the optimal transfer point is obtained and the final convergence target of the robot is adjusted. Finally, it provides the optimal transfer method for users with different mobility, and improves the safety and comfort of excretion transfer.

The contributions of this paper are as follows: Firstly, a two-link model of human sitting down is established to obtain the transfer point. Secondly, combining fuzzy reasoning algorithm with reinforcement learning method, a fuzzy-reinforcement learning algorithm is proposed. The mapping relation is continuously optimized in multiple transfer process, and the appropriate transfer point is obtained according to different sitting postures of users. Finally, the transfer method proposed in this paper has strong practicability and generalization. The approach is applied to realize excretion assistance in daily life, which has great popularization characteristics in pension centers and hospitals.

In the next step of the work, the safety and rapidity of people with different motion capability in the robot transfer process will be studied in depth.

References

1. Haibin, W., Jianmin, Y.: Research on the safety progress of robots in the process of human-computer interaction. Chin. J. Saf. Sci. **21**(11), 1003–3033 (2011)
2. Nakajima, S.: RT-Mover: a rough terrain mobile robot with a simple leg-wheel hybrid mechanism. Int. J. Robot. Res. **30**(30), 1609–1626 (2011)
3. Xiuzhi, L., Xingnan, L., Songmin, J.: Automatic docking of intelligent wheelchair bed based on visual measurement. J. Instrum. **40**(4), 189–197 (2019)
4. Minya, C., Liangliang, L.: Application of intelligent bed monitoring system in nursing quality control and management. J. Med. Inform. **37**(5), 35–37 (2016)

5. Chang, G., Hou, G., Guo, X.: Design and implementation of intelligent nursing bed system based on Ethernet. Electronic Devices **38**(4), 893–897 (2015)
6. Zhang, Q., Li, M., Wang, X.: Reinforcement learning in robot path optimization. J. Softw. **7**(3), 161–166 (2012)
7. Faust, A., Chiang, H.T., Rackley, N.: Avoiding moving obstacles with stochastic hybrid dynamics using pearl: preference appraisal reinforcement learning. In: Kragic, D. (ed.) 2016 IEEE International Conference on Robotics and Automation, ICRA, vol. 1, pp. 484–490. IEEE, New York (2016)
8. Luy, N.T., Thanh, N.T., Tri, H.M.: Reinforcement learning-based intelligent tracking control for wheeled mobile robot. Trans. Inst. Measur. Control **36**(7), 171–176 (2014)
9. Jaradat, M.A.K., Al-Rousan, M., Quadan, L.: Reinforcement based mobile robot navigation in dynamic environment. Robot. Comput. Integr. Manuf. **27**(1), 135–149 (2011)
10. Donghui, Z., Junyou, Y., Dianchun, B.: Transfer method of multiple welfare-robots based on minimal fuzzy system. Robot **41**(6), 813–822 (2019)
11. Cutler, M., Walsh, T.J., How, J.P.: Real-world reinforcement learning via multifidelity simulators. IEEE Trans. Robot. **31**(3), 655–671 (2017)
12. Singh, N.H., Thongam, K.: Mobile robot navigation using fuzzy logic in static environments. Procedia Comput. Sci. **125**(3), 11–17 (2018)
13. Dias, L.A., de Oliveira Silva, R.W.: Application of the fuzzy logic for the development of autonomous robot with obstacles deviation. Int. J. Control Autom. Syst. **16**, 823–833 (2018). https://doi.org/10.1007/s12555-017-0055-9

Analysis of Continuous Motion Angle for Lower Limb Exoskeleton Robot Based on sEMG Signal

Liuwen Jing[✉], Tie Liu, Haoming Shi, Yinming Shi, Shiyu Yao, Junyou Yang, Xia Yang, and Dianchun Bai

Shenyang University of Technology, No. 111, Shenyang Economic and Technological Development Zone, Shenyang, Liaoning, China
ytnnjingliuwen@163.com

Abstract. In this paper, the state of real-time motion angle corresponding to the surface electromyography signal (sEMG) in the process of exoskeleton robot assisted motion is analyzed, and a new control analysis method for exoskeleton robot is proposed. The method of wavelet decomposition is used to analyze the characteristics of sEMG signals, which can accurately extract the eigenvalues of sEMG signals from different angles. The extracted eigenvalues are used as the input of SVM classification and recognition, pattern recognition is carried out, and the relationship between the eigenvalues and the corresponding motion angle is established. At the same time, the surface EMG signals of the start and stop time in the process of continuous movement of the lower limbs are analyzed to obtain a more accurate relationship between the start and stop state and the eigenvalue. By comparing the analysis results with the actual movement angle state information, the correctness of the analysis results is verified, which provides a theoretical basis for the follow-up research of the lower extremity exoskeleton robot.

Keywords: sEMG signal · Continuous movement of lower limbs · Lower limb exoskeleton robot · Wavelet feature extraction · SVM support vector machine

1 Introduction

With the development of national economy, science and technology, and the improvement of people's living standards, promoting and improving the medical welfare of disabled people has become a very concerned issue of the government. As a representative human-computer interaction robot, wearable human exoskeleton robot has been highly valued in military, civil and rehabilitation treatment. Among them, the wearable exoskeleton robot applied to the lower limbs of human body can enhance the movement ability of the lower limbs under the control of the wearer. Exoskeleton mechanical system further calculates and controls the output in real time by estimating and predicting the motion state and intention of human body, so as to realize real-time and synchronous strength enhancement and assistance, enhance human body function, and enable human body to complete many tasks with the help of machinery [1].

© Springer Nature Singapore Pte Ltd. 2020
J. Qian et al. (Eds.): ICRRI 2020, CCIS 1335, pp. 50–63, 2020.
https://doi.org/10.1007/978-981-33-4929-2_4

Historically, the typical system is a series of exoskeletons built by m.vukobratovic and others. They have developed different types of exoskeletons, powered by hydraulic actuators, pneumatic actuators and DC servo motors, and verified the theoretical results. However, the control system is relatively backward, which can not achieve the precise and free control of the user exoskeleton robot [2]. At present, most mature Rehabilitation Exoskeleton robots are developed by foreign companies or institutions. For example, Hal exoskeleton robot developed by the University of Tsukuba in Japan can help the elderly and the disabled walk; eLeg exoskeleton system of belike bionics company in the United States can help paraplegic patients completely get rid of wheelchairs to walk independently, and can also help people who can't stand to train lower limb muscles and nerves. The research of domestic and foreign skeleton robots is mainly concentrated in universities and research institutes. Yang Canjun's team of Zhejiang University has developed a prototype of lower limb exoskeleton with pneumatic actuator, which uses the adaptive fuzzy artificial neural network algorithm to input the pressure sensor data of the wearer's sole, calculate the expected action of the exoskeleton robot, and drive the lower limb hip joint and knee joint Realize the control of exoskeleton robot [3]. Harbin University of technology designed the lower limb dynamic exoskeleton machine to control through the sensing data and the robot's motion position. Gait prediction is realized by the pressure sensor placed on the sole of the foot. However, the collection of foot pressure information often lags behind the actual action of the user, which leads to the action delay of the exoskeleton robot [4]. A wearable lower limb exoskeleton developed by the Chinese Academy of Sciences, which drives the hip and knee joints, is controlled by predefined trajectory, input the joint angle of normal people, and reproduce it on the exoskeleton. But the predefined orbit is often difficult to meet the user's movement diversity in actual use. Most of them mainly study the assisted exoskeleton robot, but few research the Rehabilitation Exoskeleton Robot. In addition, the research and development of the rehabilitation exoskeleton is still in the primary stage, most of which are treadmill type structures. The patients are passively trained after wearing the equipment, which is a certain distance from the practical application and active training [5]. Although the existing identification methods have achieved good results, there are still many problems. Most of the studies are less concerned with feature selection and usually adopt some combination of features directly. If these features are directly applied to gait recognition, the classification effect is poor. The present algorithm has poor performance in real-time control and long response time. In order to solve the problems existing in the above methods, a new algorithm for the continuous motion state of lower extremity exoskeleton robot based on surface emG signal is proposed in this paper, and continuous kick motion is taken as an example for analysis and verification. Type of rehabilitation of exoskeleton robot research and electromyographic signal analysis of lower limb movement condition, can get lower limb under different angles in the process of movement of electrical information, using the MATLAB data simulation verify its rationality at the same time, and get the reasonable motor driven data, for subsequent the electromyographic signal data analysis as well as the mechanical structure and drive system and design to provide the reference [6–10].

2   Experiment

2.1   Subjects

20 healthy volunteers were recruited, including 12 male volunteers and 8 female volunteers. There was no sports injury within 7 days before the experiment. There were no motor nerve diseases, no sprains, sports injuries, fractures and other injuries affecting the motor function in the lower limbs before the experiment; no strenuous exercise in the two days before the experiment; no muscle soreness and discomfort. The subjects were between 22 and 26 years old, with an average age of 24.7 ($\pm$1.08) years, height of 174.1 ($\pm$3.72) cm, and weight of 68.6 ($\pm$7.26) kg. In the definite experiment Start the experiment after the content [11].

2.2   Experiment Process

The equipment used in this experiment includes: electromyographic sensor, Tektronix oscilloscope, AgCl electrode, WSSS motion sensing system, etc. [12]. The target muscles detected in the experiment are the outer, rectus, inner and semitendinosus muscles of the lower limbs of normal human body. Before the experiment, the skin of the experimenter was wiped with a scrub and medical alcohol to reduce the influence of skin impedance on the experimental results.

The surface electrode was pasted on the abdomen of the outer thigh [13], rectus femoris, inner thigh and semitendinosus muscles, and the direction of the surface electrode was parallel to the longitudinal axis of the muscle fiber as far as possible; WSSS motion sensor was pasted on the skin surface of the midline of the outer thigh of the subject, and the height of the motion sensor was consistent with the height of the fingertip when the hands were released and lowered, and the y-axis direction of the sensor was kept at 0°. In order to reduce the artificial error of the experiment, the subjects are required to adopt the same standing posture, feet together, hands naturally drooping and relaxing. During the experiment, the subjects swung naturally with one leg according to the requirements, and set the position of standing with both legs together as the initial position. According to the requirements, the subjects began to swing forward from the initial position to the highest point of natural swing, and then changed to swing backward. After the initial position, they continued to swing backward to the highest point of natural swing, and changed to swing forward to the initial position [14]. The whole process is continuous and each group of actions is repeated five times. The action time of each group should not be less than 6 s, and the rest should be at least 30 s at the end of every five groups of actions, so as to prevent muscle fatigue from affecting the authenticity of experimental data. The specific experimental process is shown in Fig. 1.

Fig. 1. The experimental schematic.

3 Data Analysis and Experimental Results

3.1 Analysis of Starting and Stopping Points of Movement

The collected sEMG signal and movement angle data are input into the computer, and the simulation experiment is carried out by using MATLAB [15–17]. According to the corresponding relationship of the time information of sEMG signal in the movement angle data, the starting point of each group of actions is determined, and the sEMG signal located in the position interval between the starting point and the ending point of each group of actions is recorded, which is used for the four movement stages of each group of sEMG signals divide. According to the movement angle data and time, the collected EMG signals of each group of continuous movement are divided into four segments, which are: the initial position moves to the front highest point, the front highest point to the initial position, the initial position to the back highest point and the back highest point to the original position. In order to ensure the accuracy of the follow-up analysis, the synchronization of the angle data and the surface EMG signal data is realized to the greatest extent, and the accurate starting and stopping points of each action part are obtained. Figure 2 shows the complete angle data of five groups of actions in an experiment. According to the data initial value, maximum value and minimum value, the starting point and stopping point of a group of actions can be analyzed, and then the exact time corresponding to the starting point and stopping point can be obtained according to the corresponding relationship between the time measured by the Angle sensor and the Angle. Figure 3 shows the raw EMG data of a complete experiment. By comparing the time data of EMG data, the corresponding starting and stopping points of emg signals were found, so that the complete five consecutive EMG signals obtained in the experiment were decomposed into the motion of a single group of continuous emg signals. Figure 4 shows the continuous EMG signal after the extraction of the active segment [18].

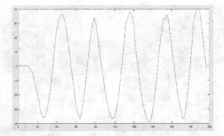

Fig. 2. Movement angle data of one group of subjects

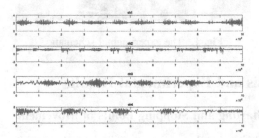

Fig. 3. Original sEMG signal.

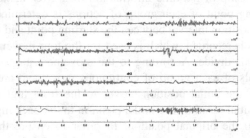

Fig. 4. Continuous sEMG signal after active segment extraction.

3.2 Analysis of Starting and Stopping Points of Movement

Wavelet transform (WT) is a time-frequency analysis method which has the same window size (area) but variable window shape, time window and frequency window. Wavelet transform has lower time resolution and higher frequency resolution in the application of low-frequency signals, but higher time resolution and lower frequency resolution in the analysis of high-frequency signals, which makes wavelet transform have good adaptability in the analysis of different sEMG signals [19].

The essence of wavelet analysis method is to decompose the signal s (t) into sub signals in different frequency bands of the basis function, and analyze the changes of surface EMG signal in the process of muscle dynamic contraction from the perspective of frequency domain. Wavelet has the characteristics of multi-resolution [20], which can gradually observe the signal from coarse to fine. Wavelet transform can be understood as a function of scale factor a and translation factor b. in the process of wavelet transform, changing the value of b only affects the position of the window on the time axis, while scale a not only affects the position of the window on the frequency axis, but also affects the shape of the window [21].

The continuous wavelet transform of the input signal of basic wavelet or parent wavelet is defined as the function family $\psi(a, b)$ generated by scale factor a and translation factor b:

$$\psi_{a,b}(t) = \frac{1}{\sqrt{a}}\psi\left(\frac{t-b}{a}\right) \, a, b \in, a \neq 0 \tag{1}$$

It is called analytical wavelet or continuous wavelet (CWT). The coefficients in the formula are used to realize the normalization of energy in the expansion process. The definition of continuous wavelet transform of s (t) is as follows:

$$W_s(a,b) = \int_{-\infty}^{+\infty} s(t)\psi_{a,b}^*(t)dt \tag{2}$$

Figure 5 is a comparison between the original surface EMG signal of four muscles in five groups of exercise and the signal after noise reduction by two methods.

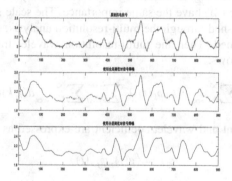

Fig. 5. Surface electromyography of a subject after noise reduction during continuous exercise

Through the method of unbiased likelihood estimation based on Stein, the noise reduction command is used to complete the signal noise reduction. There are two ways to select the threshold: global threshold and layered threshold. The global threshold applies the SURE principle, and the layered threshold applies and displays its advantages through data. From Fig. 5, it can be seen that the signal denoised by the global threshold and layered threshold method retains the high-frequency characteristics of the signal well. Between the two, the layered threshold loses the performance of the signal (compared with the original signal) Compared with the global threshold, the layered threshold reduces the details of the signal and loses some information.

3.3 Feature Extraction

The main purpose of the feature extraction module is to obtain the relationship between the muscle signals and time-frequency domain features during arm movement [22].

In order to ensure the real-time performance of the system, while ensuring the accuracy, the time-frequency domain feature with small amount of computation and rapid acquisition is used as the information measure.

In this paper, the method of wavelet analysis is used to decompose surface emg signals into low-frequency and high-frequency components. After several transformations, the data volume and frequency of each transformation are halved, and the multi-resolution decomposition of the original signal is finally realized.

In the process of wavelet decomposition, let the total frequency band occupied by the original emg signal $f(t)$ be the space V_0. After the first stage decomposition, V_0 is divided into the scale signal space V_1 of low frequency and the wavelet signal space W_1 of high frequency. Therefore, V_1 and W_1 are the subspaces of V_0 and can be expressed as $V_1 \oplus W_1 = V_0$. In order to decompose downward, multiple spatial decomposition can be obtained, that is, the multi-resolution analysis of wavelet.

When the low-frequency scale signal and the high-frequency wavelet signal are used to reconstruct the emg signal, it can be seen that the low-frequency signal component is relatively rough, while the high-frequency signal is relatively fine, so it can effectively represent the different components of the signal, which is conducive to the signal analysis and processing. In the orthogonal wavelet analysis, the scale function V_j and the wavelet function W_j have the same importance. The scale equation of the scale function $\varphi(t)$ is established through the multi-resolution analysis, and then the wavelet function $\psi(t)$ is obtained. Then the multi-resolution wavelet transform equation of function $g(t)$ is as follows:

$$g(t) = \sum_{k=-\infty}^{\infty} c(k)\varphi_k(t) + \sum_{j=0}^{\infty} \sum_{k=-\infty}^{\infty} d(j,k)\psi_{j,k}(t) \tag{3}$$

The scale coefficient and wavelet coefficient are expressed as

$$c_{j,k} = \langle g(t), \varphi_{j,k}(t) \rangle \tag{4}$$

$$d_{j,k} = \langle g(t), \psi_{j,k}(t) \rangle \tag{5}$$

Therefore, the surface emg signal is decomposed through the wavelet coefficients of different scales. The wavelet coefficients of each level are taken as the characteristics of the signal, and the signal can be accurately described by a few coefficients. Due to the non-stationarity of surface emg signals, it is important to choose a suitable wavelet base for the effective classification and recognition of signals. Symlets wavelet function system is an approximate symmetric wavelet function, which is an improvement on db function. The Symlets system of functions is usually represented as symN $(N = 2, 3, \cdots, 8)$. The support range of symN wavelet is $2N - 1$, the vanishing moment is N, and it also has good normality. Compared with dbN wavelet, this wavelet is consistent with dbN wavelet in terms of continuity, support length, filter length, etc., but symN wavelet can reduce phase distortion during signal analysis and reconstruction to some extent. In this paper, the well-classified orthogonal Sym3 wavelet basis function is selected to decompose the emg signal at five scales, and the singular value and energy value of each wavelet coefficient are extracted as the eigenvectors.

After preprocessing the sEMG signals collected by the four motion modes of the legs, the feature scatter diagram distribution after feature extraction by wavelet decomposition method is shown in Fig. 6.

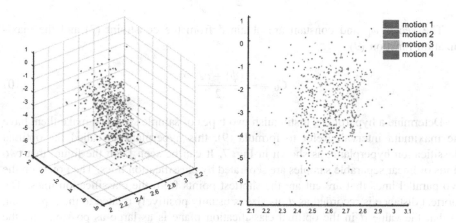

Fig. 6. Spatial distribution scatter of four kinds of action eigenvalues.

It can be seen from the feature scatter diagram that the feature values of different actions have a very light clustering effect, which can realize the feature representation of the four action modes and satisfy the desired pattern recognition rate.

4 Classification and Recognition

The method of support vector machines (SVM) is based on statistical learning theory. Suppose that the selected linear separable sample set is [23]:

$$(x_i, y_i)(i = 1, 2, \ldots, N, x_i \in R^n, y \in \{-1, 1\}) \tag{6}$$

According to the different y categories, it can be divided into positive sample subset X^+ and negative sample subset X^-. There is a unit vector $\phi \|\phi = 1\|$ and constant C, which makes (7) true.

$$\begin{cases} \langle X^+ \cdot \phi \rangle \rangle c \\ \langle X^- \cdot \phi \rangle \langle c \end{cases} \tag{7}$$

For any unit vector ϕ, determine two values

$$\begin{cases} c1(\phi) = min\langle X^+ \cdot \phi \rangle \\ c2(\phi) = max\langle X^- \cdot \phi \rangle \end{cases} \tag{8}$$

Find ϕ_0 to maximize the following:

$$r(\phi) = \frac{c1(\phi) - c2(\phi)}{2}, \|\phi\| = 1 \tag{9}$$

The vector ϕ_0 and constant are obtained from the constraint (7) and the maximization function (9)

$$C_0 = \frac{c1(\phi) + c2(\phi)}{2} \tag{10}$$

Determine a hyperplane, distinguish two types of sample sets, and make them have the maximum interval. Refer to formula (9), this hyperplane is called the optimal classification hyperplane, as shown in Fig. 7. It can be seen from the figure that two kinds of linear separable samples are separated by classification lines. The points on the two parallel lines that are cut are the shortest points from the classification lines. The shortest distance is recorded as d, its size is usually positively related to the separation, so that the value d in the optimal classification plane is as large as possible and the segmentation effect is better. The main goal of constructing the optimal hyperplane is found, and a vector and constant B are found to satisfy the following constraints

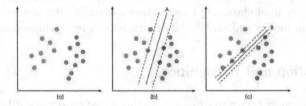

(a)   (b)   (c)

Fig. 7. SVM optimal classification surface.

The main goal of constructing the optimal hyperplane is found a vector W^* and constant b are found to satisfy the following constraints [24].

$$\begin{cases} \langle X^+ \cdot W^* \rangle + b^* \geq 1 \\ \langle X^- \cdot W^* \rangle + b^* \leq -1 \end{cases} \tag{11}$$

And the vector W^* has the minimum norm

$$min\rho(W) = \frac{1}{2} \|W^*\|^2 \tag{12}$$

$$f(X) = W^* \cdot X + b^* \tag{13}$$

Under the condition of (11), the relationship between the vector obtained by minimization (13) and the vector W^* forming the optimal hyperplane is

$$\phi0 = \frac{W^*}{\|W^*\|} \tag{14}$$

The interval $r(\phi0)$ between the optimal hyperplane and the classification vector is

$$r(\phi0) = sup\frac{1}{2}(c1(\phi0) - c2(\phi0)) = \frac{1}{\|W^*\|} \tag{15}$$

In order to find the optimal classification hyperplane, we need to solve the quadratic programming problem. Under the condition of linear constraint (11), we need to minimize the quadratic form, see Eq. (12). It can be solved by Lagrange multiplier method, and the equation is as follows:

$$L(W, a, b) = \frac{1}{2}\|W\|^2 - \sum_{i=1}^{N} ai\{yi\langle Xi \cdot W + b\rangle - 1\} \tag{16}$$

The weight coefficient vector of the optimal classification surface of the support vector, then its optimal classification surface function is:

$$W* = \sum_{i=0}^{N} ai^*yiXi \tag{17}$$

The wavelet decomposition method is used to extract the eigenvalues for SVM classification training, and the accuracy reaches 93.72%. Table 1 shows the accuracy after classification training (see Table 1).

Table 1. The accuracy after classification training.

	Motion 1	Motion 2	Motion 3	Motion 4
Motion 1	94	3	0	3
Motion 2	1	93	1	5
Motion 3	1	4	95	0
Motion 4	2	0	4	92

In order to analyze the performance of the algorithm adopted in this paper, the classification accuracy of several feature combinations mentioned in previous literatures is compared. Commonly used feature combinations include time-domain feature combinations: average absolute value (MAV), zero crossing rate (ZC), wavelength (WL), AR model coefficient, and the results are shown in Table 2.

Table 2. The accuracy after classification training.

Feature extraction method	Recognition accuracy
Average absolute value (MAV)	82
Zero crossing rate (ZC)	84
Wave length (WL)	81
AR model coefficient	72

5   Verification

The classification model obtained through training was applied to the lower extremity exoskeleton prototype shown in Fig. 8 to verify the rationality of the conclusions in this paper. According to the indicator light, the subjects made complete continuous movements one by one, each of which was completed 100 times and lasted for 5 s.

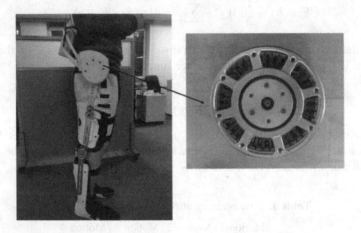

Fig. 8. Lower limb exoskeleton robot prototype.

The classifier model is used for real-time prediction, and the subject controls the exoskeleton to complete the corresponding actions. Figure 9 shows the effect of online real-time exoskeleton movement controlled by a subject. Since it takes some time for the subjects to complete the action according to the change of the indicator light during the experiment, the experimental results will have a slight error and the accuracy will be reduced. After all the subjects completed real-time control of the movement of the lower extremity exoskeleton, the average accuracy of the four groups of continuous movements was 95.14%, 92.51%, 94.30% and 91.78%, respectively. The accuracy statistics results are shown in Fig. 10.

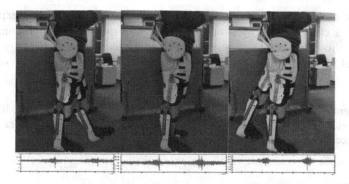

Fig. 9. Real-time control of the exoskeleton movement process.

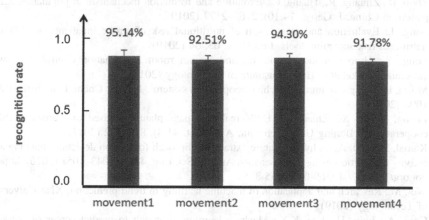

Fig. 10. Four groups of continuous motion accuracy comparison

6 Conclusion

In this paper, a new "time-frequency domain wavelet transform" method is proposed based on the strong temporal sequence of sEMG signal data. The method combines Angle and sEMG to segment the local motion data in the effective leg movement segment as sample data.

At the same time, wavelet analysis is used to extract feature values for parallel support vector machine training, and the Angle classification and recognition of continuous motion are realized. Experimental results show that the preprocessing method and classification recognition algorithm presented in this paper have better classification results. Been method application in the treatment of lower limb exoskeleton robot prototype, lower limb swing to the human body muscle in the process of electrical signals are analyzed in real time, the corresponding relationship between the Angle at the same time, the simulation analysis are carried out using MATLAB, based on the motion process of the lower limbs with multi-channel sEMG analysis, get lower limb joint Angle in the process of sports information, and through the prototype proves the rationality of the design of lower limb exoskeleton. Through verification experiments, this method is characterized by strong real-time performance and high control accuracy, which shows the application prospect of gait recognition technology based on sEMG signal in the field of exoskeleton, and lays a solid foundation for subsequent research.

Acknowledgements. Liaoning Provincial Department of Education key projects. Research on key technologies for collaborative walking robots based on fine electromyographic signal analysis; Project number: LZGD2019001.

References

1. Zhou, R., Zhuang, R., Huang, C.: Evolution and formation mechanism of population aging pattern in China. J. Geogr. **74**(10), 2163–2177 (2019)
2. Song, J.: Evaluation and comparison of nutritional risk of stroke inpatients by different nutritional risk screening tools. Jilin Univ. **03**, 64 (2019)
3. Long, Y.: Human motion prediction and human robot coordination control for lower extremity exoskeleton. Harbin Institute of Technology (2017)
4. Millot, P.: Designing human-machine cooperation systems. J. Phys. Chem. Lett. **3**(9), 1094–1098 (2016)
5. Zhang, S., Li, X., Zhang, P.: UAV real-time path planning based on human-machine cooperation. J. Beijing Univ. Aeronaut. Astronaut. **43**(4), 814–822 (2017)
6. Kamal, S., Jalal, A.: A hybrid feature extraction approach for human detection, tracking and activity recognition using depth sensors. Arab. J. Sci. Eng. **41**(3), 1043–1051 (2016). https://doi.org/10.1007/s13369-015-1955-8
7. Wei, M.: Research and application of machine learning in trend prediction. Xian University of Technology (2019)
8. Choi, A., Jung, H., Lee, K.Y.: Machine learning approach to predict center of pressure trajectories in a complete gait cycle: a feedforward neural network vs. LSTM network. Med. Biol. Eng. Comput. **57**(12), 2693–2703 (2019). https://doi.org/10.1007/s11517-019-02056-0
9. Yu, Y.: Ecognition of human lower limb motion and analysis of joint torque based on sEMG signal. Suzhou University (2016)
10. Li, Y., Jin, Z., Ji, S.: Design of a new 3-DOF hybrid mechanical arm. Sci. China Ser. E: Technol. Sci. **52**(12), 3592–3600 (2009)
11. Yuan, L., Hu, B., Wei, K.: Control Principle and MATLAB Simulation of Modern Permanent Magnet Synchronous Motor. Beijing University of Aeronautics and Astronautics Press, Beijing (2016)
12. Wang, Z., Zin, H., Zhou, S.: Design and performance analysis of humanoid robot four-DOF manipulator. Sci. Technol. Vis. **2020**(09), 137–139 (2020)
13. Liu, X., Li, M.: Effectiveness analysis of muscle fatigue in rehabilitation process based on surface EMG signal. J. Biomed. Eng. **36**(01), 6 (2019)
14. Zhang, Y., Jing, Y.: Analysis of descending and mapping of lower extremity surface EMG signals. J. Sens. Technol. **31**(07), 1046–1053 (2008)
15. Jiang, Y., Zou, R.: Analysis and recognition of lower limb spasm signal characteristics of surface EMG signals. Electron. Sci. Technol. **30**(11), 38–41 (2017)
16. Sun, G., Yan, Z.: Motion classification based on bispectral analysis of surface EMG signals. J. Beijing Univ. Technol. **43**(07), 1045–1050 (2017)

17. Liu, J.: Aanalysis of muscle fatigue based on spectral entropy of multi-channel surface EMG signals. J. Biomed. Eng. **33**(03), 431–435 (2016)
18. Shi, W.-T., Lyu, Z.-J., Tang, S.-T.: A bionic hand controlled by hand gesture recognition based on surface EMG signals: a preliminary study. Biocybern. Biomed. Eng. **38**(1), 126–135 (2018)
19. Huang, P., Yang, Q.: Feature extraction algorithm of sEMG based on amplitude cube and BP neural network. China Mech. Eng. **23**(11), 1332–1336 (2012)
20. Qiao, X., Hu, W.: Sample entropy and wavelet entropy analysis of EEG EMG signal. J. Test. Technol. **30**(4), 292–298 (2016)
21. Zou, X., Lei, M.: Pattern recognition of sEMG signal based on multi-scale maximum Lyapunov index. Chin. J. Biomed. Eng. **31**(1), 7–12 (2012)
22. Rajagopal, A., Dembia, C.L.: Full-body musculoskeletal model for muscle-driven simulation of human gait. IEEE Trans. Biomed. Eng. **63**(10), 2068–2079 (2016)
23. Jie, H., Lu, W.: Hand sEMG recognition based on SVM optimization based on artificial fish swarm algorithm. Sens. Microsyst. **35**(02), 23–25 (2016)
24. Lu, L., Liu, S.: Application analysis of sEMG gesture recognition based on nonlinear SVM fusion LDA. Laser J. **35**(08), 26–29 (2014)

Modelling for a 6DOF Anthropomorphic Manipulator

Zhongliang Liu[1], Yina Wang[1(✉)], Junyou Yang[1], and Jianghao Shi[2]

[1] Shenyang University of Technology, 111 Shenliao West Road,
Shenyang Economic and Technological Development Zone, Shenyang 11087,
Liaoning, People's Republic of China
liu.zhongliang@sut.edu.cn, wangyina0402@126.com
[2] Beihang University, No. 37 Xueyuan Road, Haidian District,
Beijing 100191, People's Republic of China

Abstract. Anthropomorphic robotics is a novel challenging domain of study and research. To collaborate with people, the anthropomorphic robot is required not only a humanoid structure but a human-like feature regarding the motion. But most anthropomorphic robots use the system which is mature algorithms integrated or closed to customers. To avoid this, remodeling is necessary and important for theoretic studying as well as precise controlling which not only strengthens the understanding of robot structure but also makes the research more targeted. In this paper, a geometric-analytical mix approach is proposed to solve the inverse kinematics for a 6-DOF arm of Seed-Noid R7F robot due to its dissatisfactory of Pieper's criteria, comparing with the general approach which doesn't consider the payload changing, a dynamics model with the payload scenario is built to respond the payload sensibility. At last, the proposed solution is verified using MATLAB when the payload is from 2 kg to 3 kg separately.

Index Terms: Life support robot · 6DOF · Anthropomorphic manipulator · Pieper's criteria · Kinematics and dynamics · Geometric-analytical mix approach

1 Introduction

With the advent of aging, the demand for assistive technology is increasing due to the caregiver shortage. To overcome this contradiction, the personal care robot has been developed to improve people's independence and quality of life (QOL) [1]. Dual-arm anthropomorphic robot, which is trending to widely application in not only industry but also habitual human life, is being developed to assist and interact with people, including simple conversations [2], autonomous navigation [3], item transportation [4], medical care [5]. Currently, China domestic robotic manufacturers and universities are focusing more on industrial robots and Automated Guided Vehicles relative fields. The AGVs and industrial robots with low accuracy have been mass produced by some key robotic OEMs. The rehabilitation robot is still in the initial stage, and the research challenges are mainly in the key fields of robot dynamics and dual-arm coordination. The robotic kinematics is a research hotpot, the treatment of kinematics of robot manipulators can be found in several classical robotics texts, such as [6–9]. Specific

© Springer Nature Singapore Pte Ltd. 2020
J. Qian et al. (Eds.): ICRRI 2020, CCIS 1335, pp. 64–82, 2020.
https://doi.org/10.1007/978-981-33-4929-2_5

texts are [10–12]. The inverse kinematic derives from the spherical wrist is utilized in [13–15]. Numerical methods for the solution of the inverse kinematics problem based on iterative algorithms are proposed in [16]. But, the literature for inverse kinematics with dissatisfactory of Pieper's solution are very scarce because there is no uniform solution. And this paper tries to propose a fast computing but a good precision approach aiming at the dissatisfactory of Pieper's criteria comparing with above mentioned methods. The derivation of the dynamic model for rigid manipulators can be found in several classical robotics texts, such as [6, 7, 9, 17–19]. The dynamics based on the Lagrange formulation are proposed in [20, 21]. The Newton–Euler formulation is proposed in [22], and a computationally efficient version for inverse dynamics can be found in [23]. But, comparing with the industrial robot, the anthropomorphic robot is sensitive to the payload of manipulator due to its precision and small workspace.

In this paper, we firstly analyze the open- architecture of the Seed-Noid R7F anthropomorphic robot, then we derive the D-H parameter, joints and links mass, coordinated system etc. information from URDF document of ROS. Based on these parameters, the forward kinematics of the 6-DOF left arm is modelled, and aiming at its dissatisfactory of Pieper's criteria, a geometric-analytical mix inverse kinematics approach is proposed, then a dynamics model with the payload scenario is established to respond its mass sensibility. At last, the proposed solution is verified using MATLAB when the payload is 2 kg and 3 kg.

2 Kinematics and Dynamics Model

2.1 Overview of the Robot

Seed-Noid R7F is an anthropomorphic robot which is designed as a life-sized support scenario cobot as shown in Fig. 1, and it's 141–170 cm in height (with a prismatic joint on its leg) and weights about 80 kg with two identical robotic arms. Moreover Seed-Noid R7F is equipped with an industrial motherboard, a mini PC-based controller with Linux system, Android touch-screen tablet PC, four omni-wheels under the subbase, six ultrasonic sensors (0.02–3 m) on the subbase, two 3D cameras (in the head and

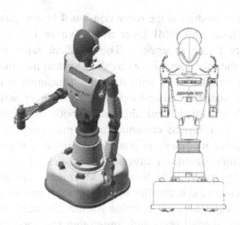

Fig. 1. Seed-Noid R7F anthropomorphic dual-arm robot

chest separately), two range sensors (0.06–4 m, $-120° \sim 120°$), one telecontrol board for controlling and communicating, two rechargeable DC 12 V/22 Ah lead batteries for the motors driving and the electronic board using.

2.2 System Architecture of the Robot

Seed-Noid R7F humanoid robot uses the open-architecture control system as the main platform based on Robot Operation System (ROS), which is a meta-operating system with the philosophy to promote code reuse against the repetitive code work among different programable robots [24]. As an open-source programing framework, ROS has been improved by hundreds of user-contributed ROS packages in the past few years. The main characteristic of this platform is that data are transferred between modules by using inter-process communications, which makes it easy to implement a centralized topology [25].

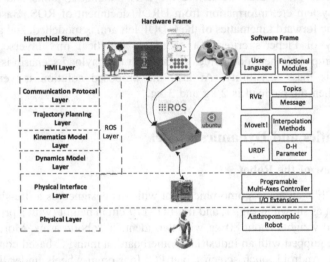

Fig. 2. System architecture of Seed-Noid R7F with ROS

The open-architecture system of the robot consists 4 layers: physical layer, physical interface layer, ROS layer and HMI layer as shown in Fig. 2. Each layer has its hardware and software frame separately. The physical layer includes the drivers, motors and sensors of the robot, and drives by the physical interface layer-an industrial programable multi-axes motherboard with IOs-which meanwhile receives information from the encoder and shapes a close-loop. The ROS layer with several advanced techniques such as RViz, MoveIt! and universal robotic description format (URDF), can be separated into four layers: communication protocol layer which provides a method to connect various modules via predefined protocol, trajectory planning layer which generate trajectory based on inverse kinematics of the robot and velocity planning mathematically by MoveIt! function, kinematics model layer which is the principal part for the kinematic controller for the forward and inverse kinematics, and dynamics model layer which is novel function for the advanced robot controlling. The Gigabyte Mini PC with ethernet ports and some other interfaces, which contributes to the ROS layer and plays an important role in the system architecture, is implemented on the basis of ROS. The parameters of kinematics and dynamics model, such as D-H parameters, link mass and initial matrix etc. are configured in the URDF.

2.3   Kinematics

The mapping between the joints of the robot arm and the end-effector in the Cartesian space is described kinematics matrix, The Denavit–Hartenberg convention was firstly introduced in [26]. A modified method is utilized in [17–19] where the modified D-H coordinate system of Seed-Noid R7F is established for each link of chains [27] in Table 1.

Table 1. Modified D-H parameters of the left-arm manipulator

Modified D-H parameters of a 6-DOF anthropopathic manipulator					
$^{i-1}T_i$	Link twist α_{i-1}	Link length a_{i-1}	Joint offset d_i	Joint angle θ_i	Offset
$^0T_1$	0	0	0.05978	θ_1	0
$^1T_2$	$-90°$	0.028046	0	θ_2	90°
$^2T_3$	$-90°$	0.01	0.2895	θ_3	90°
$^3T_4$	$-90°$	-0.035	0	θ_4	0
$^4T_5$	90°	0.035	0.2655	θ_5	90°
$^5T_6$	$-90°$	0	0	θ_6	0

- Here we move O_1, O_3 and O_5 to O'_1, O'_3 and O'_5 separately, and $d'_1 = 0.05978$, $d'_3 = d_3 + d_4 = 0.2895$ and $d'_5 = d_5 + d_6 = 0.2655$

The kinematic configuration of Seed-Noid R7F is shown in Fig. 3.

The transformation between two successive frames ({i} to {i − 1}) representing the preceding four movements is the product of the four matrices representing them, the transformation matrix based on modified D-H approach is as follows.

$$^{i-1}T_i = Rot(\hat{x}, \alpha_{i-1})Trans(\hat{x}, a_{i-1})Trans(\hat{z}, d_i)Rot(\hat{z}, \theta_i) =$$

$$\begin{bmatrix} \cos\theta_i & -\sin\theta_i & 0 & a_{i-1} \\ \sin\theta_i\cos\alpha_{i-1} & \cos\theta_i\cos\alpha_{i-1} & -\sin\alpha_{i-1} & -d_i\sin\alpha_{i-1} \\ \sin\theta_i\sin\alpha_{i-1} & \cos\theta_i\sin\alpha_{i-1} & \cos\alpha_{i-1} & -d_i\cos\alpha_{i-1} \\ 0 & 0 & 0 & 1 \end{bmatrix} \quad (1)$$

Then, the transformation matrix of each joint can be derived from expression (1) as

$$^0T_1 = \begin{bmatrix} \cos\theta_1 & -\sin\theta_1 & 0 & 0 \\ \sin\theta_1 & \cos\theta_1 & 0 & 0 \\ 0 & 0 & 1 & d_1 \\ 0 & 0 & 0 & 1 \end{bmatrix} \quad (2)$$

$$
{}^{0}T_{1} = \begin{bmatrix} \cos\theta_2 & -\sin\theta_2 & 0 & a_1 \\ 0 & 0 & 1 & 0 \\ -\sin\theta_2 & -\cos\theta_2 & 0 & 0 \\ 0 & 0 & 0 & 1 \end{bmatrix} \tag{3}
$$

$$
{}^{2}T_{3} = \begin{bmatrix} \cos\theta_3 & -\sin\theta_3 & 0 & a_2 \\ 0 & 0 & 1 & d_3 \\ -\sin\theta_3 & -\cos\theta_3 & 0 & 0 \\ 0 & 0 & 0 & 1 \end{bmatrix} \tag{4}
$$

$$
{}^{3}T_{4} = \begin{bmatrix} \cos\theta_4 & -\sin\theta_4 & 0 & a_3 \\ 0 & 0 & 1 & 0 \\ -\sin\theta_4 & -\cos\theta_4 & 0 & 0 \\ 0 & 0 & 0 & 1 \end{bmatrix} \tag{5}
$$

$$
{}^{4}T_{5} = \begin{bmatrix} \cos\theta_5 & -\sin\theta_5 & 0 & a_4 \\ 0 & 0 & -1 & -d_5 \\ -\sin\theta_5 & \cos\theta_5 & 0 & 0 \\ 0 & 0 & 0 & 1 \end{bmatrix} \tag{6}
$$

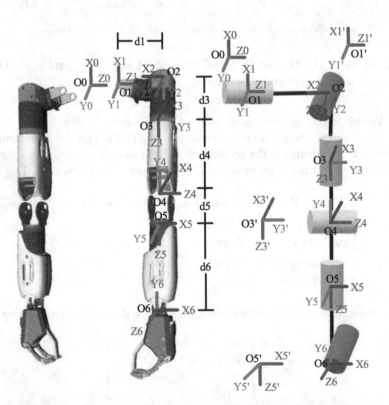

Fig. 3. Frame assignments of a 6DOF anthropomorphic manipulator

$$^5T_6 = \begin{bmatrix} \cos\theta_6 & -\sin\theta_6 & 0 & 0 \\ 0 & 0 & 1 & 0 \\ -\sin\theta_6 & -\cos\theta_6 & 0 & 0 \\ 0 & 0 & 0 & 1 \end{bmatrix} \tag{7}$$

The forward kinematics of this 6DOF arm is obtained by sequentially multiplying these transformation matrices. The total transformation of the anthropomorphic robot left arm can be obtained from

$$^0T_6 = \prod_{i=1}^{6} {}^{i-1}T_i = {}^0T_1 {}^1T_2 {}^2T_3 {}^3T_4 {}^4T_5 {}^5T_6 = \begin{bmatrix} n_x & o_x & a_x & p_x \\ n_y & o_y & a_y & p_y \\ n_z & o_z & a_z & p_z \\ 0 & 0 & 0 & 1 \end{bmatrix} \tag{8}$$

Where the right-hand side (RHS) of the Eq. (8) as follows.

$$n_x = c_1c_2c_3c_4c_5c_6 - c_1c_2c_3s_4s_6 - c_1c_2s_3s_5s_6 + c_1s_2s_4c_5c_6 + c_1s_2c_4s_6 + s_1s_3c_4c_5c_6 \\ - s_1s_3s_4s_6 + s_1c_3s_5c_6 \tag{9}$$

$$n_y = s_1c_2c_3c_4c_5c_6 - s_1c_2c_3s_4s_6 - s_1c_2s_3s_5s_6 + s_1s_2s_4c_5c_6 + s_1s_2c_4s_6 + c_1s_3c_4c_5c_6 \\ - c_1s_3s_4s_6 - c_1c_3s_5c_6 \tag{10}$$

$$n_z = -s_2c_3c_4c_5c_6 + s_2c_3s_4s_6 + s_2s_3s_5c_6 + c_2s_4c_5c_6 + c_2c_4s_6 \tag{11}$$

$$o_x = -c_1c_2c_3c_4c_5s_6 - c_1c_2c_3s_4c_6 + c_1c_2s_3s_5s_6 - c_1s_2s_4c_5s_6 + c_1s_2c_4c_6 - s_1s_3c_4c_5s_6 \\ - s_1s_3s_4c_6 - s_1c_3s_5s_6 \tag{12}$$

$$o_y = -s_1c_2c_3c_4c_5s_6 - s_1c_2c_3s_4c_6 + s_1c_2s_3s_5s_6 - s_1s_2s_4c_5s_6 + s_1s_2c_4c_6 + c_1s_3c_4c_5s_6 \\ - c_1s_3s_4c_6 + c_1c_3s_5s_6 \tag{13}$$

$$o_z = s_2c_3c_4c_5s_6 + s_2c_3s_4c_6 - s_2s_3s_5s_6 - c_2s_4c_5s_6 - c_2c_4c_6 \tag{14}$$

$$a_x = -c_1c_2c_3c_4s_5 - c_1c_2s_3c_5 - c_1s_2s_4s_5 - s_1s_3c_4s_5 - s_1c_3c_5 \tag{15}$$

$$a_y = -s_1c_2c_3c_4s_5 - s_1c_2s_3c_5 - s_1s_2s_4s_5 + c_1s_3c_4s_5 + c_1c_3c_5 \tag{16}$$

$$a_z = s_2c_3c_4s_5 + s_2s_3c_5 - c_2s_4s_5 \tag{17}$$

$$p_x = a_4c_1c_2c_3c_4 + d_5c_1c_2c_3s_4 + a_3c_1c_2c_3 + a_4c_1s_2s_4 - d_5c_1s_2c_4 + \\ a_2c_1c_2 - d_3c_1s_2 + a_4s_1c_3c_4 + d_5s_1s_3c_4 + d_5s_1s_3s_4 + a_3s_1s_3 + a_1c_1 \tag{18}$$

$$p_y = a_4 s_1 c_2 c_3 c_4 + d_5 s_1 c_2 c_3 s_4 + a_3 s_1 c_2 c_3 + a_4 s_1 s_2 s_4 - d_5 s_1 s_2 c_4 +$$
$$a_2 s_1 c_2 - d_3 s_1 s_2 - a_4 c_1 s_3 c_4 - d_5 c_1 s_3 s_4 + d_5 s_1 s_3 s_4 - a_3 c_1 s_3 + a_1 s_1 \qquad (19)$$

$$p_z = -a_4 s_2 c_3 c_4 - d_5 s_2 c_3 s_4 - a_3 s_2 c_3 + a_4 c_2 s_4 - d_5 c_2 c_4 - a_2 s_2 - d_3 c_2 + d_1 \qquad (20)$$

Where the notation s_i means $\sin\theta_i$ and c_i means $\cos\theta_i$, and vector $\mathbf{a} = (a_x, a_y, a_z)^T$ is defined as the direction of end-effector approaching the object, vector $\mathbf{o} = (o_x, o_y, o_z)^T$ is the direction of the end-effector from left to right. vector $\mathbf{n} = (n_x, n_y, n_z)^T$ is the normal vector and is perpendicular to the both. Vector $\mathbf{p} = (p_x, p_y, p_z)^T$ is the coordinates of the end-effector. Note that, the final expression of (9)–(20) should be verified by Matlab or Mathematica, otherwise, a small mistake will bring all the inverse kinematics effort collapse.

Similarly, angle of each joint of robotic arm is derived from the followings if the end-effector coordinate $(p_x, p_y, p_z)^T$ is known, it is named as inverse kinematics. Because Seed-Noid robot single arm doesn't satisfy Pieper's solution in which either three adjacent joint axes are parallel to one another or they intersect at a single point referring Fig. 3 [27], we can clearly find that the position vector is only relative to four variables θ_1, θ_2, θ_3 and θ_4 with three equations from the expression of $\mathbf{p} = (p_x, p_y, p_z)^T$, hence the closed-form joint solution doesn't exist in general solution. In this paper, a geometric-analytical mix approach is proposed, which not only have a fast computing but a good precision.

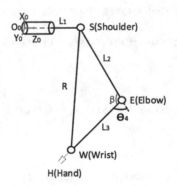

Fig. 4. Front view of the left-arm manipulator

From the geometric relationship in Fig. 4, it is clearly found that Point S is fixed at $(0, 0, d_1)^T$, point W is the given variables $(p_x, p_y, p_z)^T$, from the position vectors of the wrist and shoulder expressed in the frame, the lengths of L_1, L_2, L_3 and R is extracted from the given data to be as in

$$L_1 = d_1$$
$$L_2 = d_3$$
$$d_3 = d_5 \tag{21}$$
$$R^2 = (p_x - 0)^2 + (p_y - 0)^2 + (p_z - d_1)^2$$

Then we can easily obtain the angle β as in

$$\cos\beta = \frac{L_1^2 + L_2^2 - R^2}{2L_1L_2} = \frac{d_3^2 + d_5^2 - p_x^2 - p_y^2 - (p_z - d_1)^2}{2d_3d_5} \tag{22}$$

$$\sin\beta = \pm\frac{\sqrt{4d_3^2d_5^2 - d_3^2 - d_5^2 + p_x^2 + p_y^2 + (p_z - d_1)^2}}{2d_3d} \tag{23}$$

Hence

$$\beta = atan2\left(\pm\sqrt{4d_3^2d_5^2 - d_3^2 - d_5^2 + p_x^2 + p_y^2 + (p_z - d_1)^2}, d_3^2 + d_5^2 - p_x^2 - p_y^2 - (p_z - d_1)^2\right) \tag{24}$$

Therefore

$$\theta_4 = \pi - \beta = \pi - atan2\left(\pm\sqrt{4d_3^2d_5^2 - d_3^2 - d_5^2 + p_x^2 + p_y^2 + (p_z - d_1)^2}, d_3^2 + d_5^2 - p_x^2 - p_y^2 - (p_z - d_1)^2\right) \tag{25}$$

We can use both analytical and geometric approach to solve the θ_1 inverse kinematics, for the analytical solution, we can find $p_x^2 + p_y^2 + p_z^2$ is only the function of θ_2, θ_3 and θ_4 from the expression (18), (19) and (20), that is

$$p_x^2 + p_y^2 + p_z^2 = f(\theta_2, \theta_3, \theta_4) \tag{26}$$

Then, θ_2 and θ_3 can be solved when we substitute θ_4 with (25) into expression (20) and (26), but here we prefer the geometric approach to get the simplified expression of θ_1 and θ_2.

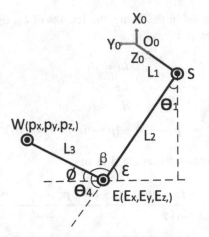

Fig. 5. Side view of the left-arm manipulator

We can find the wrist position, which is also the $\mathbf{p} = (p_x, p_y, p_z)^T$ vector of (8) in Fig. 5, is only relative with θ_2 and θ_4, vector $\mathbf{a} = (a_x, a_y, a_z)^T$ which is perpendicular to $\underset{WE}{\rightarrow}$, the constrains are as follows.

$$\varepsilon = \frac{\pi}{2} - \theta_1 \tag{27}$$

$$\beta = \pi - \theta_4 \tag{28}$$

$$\phi = \pi - \beta - \varepsilon = \theta_1 + \theta_4 - \frac{\pi}{2} \tag{29}$$

Therefore, the position of point $E = (E_x, E_y, E_z)^T$ can be yield by vector $\mathbf{p} = (p_x, p_y, p_z)^T$, that is

$$\vec{a} \cdot \vec{W} E = a_x(E_x = p_x) + a_y(E_y - p_y) + a_z(E_z - p_z) = 0 \tag{30}$$

$$(E_x - p_x)^2 + (E_y - p_y)^2 + (E_z - p_z)^2 = L_3^2 \tag{31}$$

And

$$E_x^2 + E_y^2 + (E_z - d_1)^2 = L_2^2 \tag{32}$$

Hence, we can obtain θ_1, θ_2 and θ_3,

$$\theta_1 = \text{atan2}(E_y, E_x) = \text{atan2}\left(L_2^2 - L_3^2 - p_x^2 - p_y^2, \pm\sqrt{4L_2^2L_3^2 - \left[(L_2^2 + L_3^2) - (L_x^2 + L_y^2)\right]^2}\right) - \text{atan2}(-p_x, p_y) - \theta_4 \tag{33}$$

$$\theta_2 = \text{atan2}\left(E_z - L_1, \pm\sqrt{L_2^2 - (L_1 - E_z)^2}\right) \tag{34}$$

$$\theta_3 = \arccos\left(\frac{a_4 c_2 s_4 - d_5 c_2 c_4 - a_2 s_2 - d_3 c_2 + d_1 - p_z}{a_4 s_s c_4 + d_5 s_2 s_4 + a_3 s_2}\right) \tag{35}$$

Finally, we can yield θ_5 from expression (17) and θ_6 from expression (11) and (14), they are

$$\theta_5 = \text{atan2}(s_2 s_3, c_2 c_4 - s_2 c_3 c_4) - \text{atan2}\left(a_z, \pm\sqrt{(s_2 s_3)^2 + (c_2 c_4 - s_2 c_3 c_4)^2 - a_z^2}\right) \tag{36}$$

$$\theta_6 = wrapToPI\left(\arcsin\left(\frac{bn_z - ao_z}{a^2 + b^2}\right)\right) \tag{37}$$

Where, we define $a \doteq -s_2 c_3 c_4 c_5 + s_2 s_3 s_5 + c_2 s_4 c_5$ and $b \doteq s_2 c_3 s_4 + c_2 c_4$.

2.4 Dynamics

Robot dynamics is the relationship between the forces/torque acting on a robot and the resulting motion of the robot. A common way to derive robot dynamics is to use Lagrange's equation of motion, which relates generalized coordinates with generalized forces via the kinetic and potential energies of a conservative system [28].

Lagrange mechanics is an "energy-based" approach to dynamics. The Lagrange is defined as followed:

$$L \doteq K - P \tag{38}$$

Where the Lagrange Equation as followed:

$$\tau_i = \frac{\partial}{\partial t}\left(\frac{\partial L}{\partial \dot{\theta}_i}\right) - \frac{\partial L}{\partial \theta_i} \tag{39}$$

Where τ_i is the summation of all external torques for a rotational motion, and θ_i is the system variable.

Let us consider the link i as the ith link of the Seed-Noid robot's left arm shown in Fig. 3, the arm tip will pick a load described by object M whose mass is m_L, we can yield the general dynamics equation as followed [28]:

$$T_i = \sum_{j=1}^{i} D_{ij}\ddot{q}_j + I_{i(act)}\ddot{q}_i + \sum_{j=1}^{i}\sum_{k=1}^{n} D_{ijk}\dot{q}_j\dot{q}_i + D_i \tag{40}$$

Where:

$$D_{ij} = \sum_{p=\max(i,j)}^{n} Trace\left(U_{pj}J_pU_{pi}^T\right) \tag{41}$$

$$D_{ijk} = \sum_{p=\max(i,j,k)}^{n} Trace\left(U_{pjk}J_pU_{pi}^T\right) \tag{42}$$

$$D_i = \sum_{p=i}^{n} -m_pg^T U_{pi}\bar{r}_p \tag{43}$$

and:

$$U_{pj} = \frac{\partial T_p^0}{\partial \theta_j} \tag{44}$$

$$U_{pjk} = \frac{\partial U_{pj}}{\partial \theta_k} = \frac{\partial\left(\frac{\partial T_p^0}{\partial \theta_j}\right)}{\partial \theta_k} \tag{45}$$

and J_i is the Pseudo Inertia Matrix whose exactly values are given in URDF documents of ROS.

$$J_i = \begin{bmatrix} \frac{1}{2}\left(-I_{xx}+I_{yy}+I_{zz}\right) & I_{ixy} & I_{ixz} & m_i\bar{x}_i \\ I_{ixy} & \frac{1}{2}\left(I_{xx}-I_{yy}+I_{zz}\right) & I_{ixy} & m_i\bar{y}_i \\ I_{ixz} & I_{iyz} & \frac{1}{2}\left(I_{xx}+I_{yy}-I_{zz}\right) & m_i\bar{z}_i \\ m_i\bar{x}_i & m_i\bar{y}_i & m_i\bar{z}_i & m_i \end{bmatrix} \tag{46}$$

Due to the hand link is a $\theta - r$ manipulator shown in Figure, the transformation matrix of hand center to wrist is

$$T_H^W = T_H^6 = \begin{bmatrix} 1 & 0 & 0 & 0 \\ 0 & 1 & 0 & -r \\ 0 & 0 & 1 & 0 \\ 0 & 0 & 0 & 1 \end{bmatrix} \tag{47}$$

Hence, the transformation matrix of hand center to arm base frame {0} is

$$T_H^0 = T_6^0 T_H^6 \tag{48}$$

Then we can get the velocity of the hand center as

$$v_H = \frac{d\left(T_6^0 T_H^6 r_H\right)}{dt} \tag{49}$$

The kinetic energy of the load M with the mass of m_L is

$$K_L = \frac{1}{2}m_L v_H^2 = \frac{1}{2}m_L trace\left(v_H v_H^T\right) \tag{50}$$

Similarly, the potential energy of the load M is

$$P_L = -m_L g^T T_H^0 r_H = -m_L[-g \quad 0 \quad 0 \quad 0]T_H^0 r_H \tag{51}$$

Where P_L is a scalar quantity, and the g^T is the gravity matrix referring frame {0}.

$$
\begin{aligned}
\frac{\partial T_i^{i-1}}{\partial \theta_i} &= \partial \left. \begin{bmatrix} \cos\theta_i & -\sin\theta_i & 0 & a_{i-1} \\ \sin\theta_i\cos\alpha_{i-1} & \cos\theta_i\cos\alpha_{i-1} & -\sin\alpha_{i-1} & -d_i\sin\alpha_{i-1} \\ \sin\theta_i\sin\alpha_{i-1} & \cos\theta_i\sin\alpha_{i-1} & \cos\alpha_{i-1} & d_i\cos\alpha_{i-1} \\ 0 & 0 & 0 & 1 \end{bmatrix} \right/ \partial\theta_i \\
&= \begin{bmatrix} \cos\theta_i & -\sin\theta_i & 0 & a_{i-1} \\ \sin\theta_i\cos\alpha_{i-1} & \cos\theta_i\cos\alpha_{i-1} & -\sin\alpha_{i-1} & -d_i\sin\alpha_{i-1} \\ \sin\theta_i\sin\alpha_{i-1} & \cos\theta_i\sin\alpha_{i-1} & \cos\alpha_{i-1} & d_i\cos\alpha_{i-1} \\ 0 & 0 & 0 & 1 \end{bmatrix}\begin{bmatrix} 0 & -1 & 0 & 0 \\ 1 & 0 & 0 & 0 \\ 0 & 0 & 0 & 0 \\ 0 & 0 & 0 & 0 \end{bmatrix} \\
&= \begin{bmatrix} -\sin\theta_i & -\cos\theta_i & 0 & 0 \\ \cos\theta_i\cos\alpha_{i-1} & -\sin\theta_i\cos\alpha_{i-1} & 0 & 0 \\ \cos\theta_i\sin\alpha_{i-1} & -\sin\theta_i\sin\alpha_{i-1} & 0 & 0 \\ 0 & 0 & 0 & 0 \end{bmatrix} = T_i^{i-1}Q_i
\end{aligned}
\tag{52}
$$

Where Q_i matrices are always constants for the revolute joint as shown in Eq. (52). Similarly, we can find

$$
\begin{aligned}
\frac{d}{dt}\frac{\partial T_i^{i-1}}{\partial\theta_i} &= \frac{\left(d \left. \partial \begin{bmatrix} \cos\theta_i & -\sin\theta_i & 0 & a_{i-1} \\ \sin\theta_i\cos\alpha_{i-1} & \cos\theta_i\cos\alpha_{i-1} & -\sin\alpha_{i-1} & -d_i\sin\alpha_{i-1} \\ \sin\theta_i\sin\alpha_{i-1} & \cos\theta_i\sin\alpha_{i-1} & \cos\alpha_{i-1} & d_i\cos\alpha_{i-1} \\ 0 & 0 & 0 & 1 \end{bmatrix} \right/ \partial\theta_i \right)}{dt} \\
&= \begin{bmatrix} \cos\theta_i & -\sin\theta_i & 0 & a_{i-1} \\ \sin\theta_i\cos\alpha_{i-1} & \cos\theta_i\cos\alpha_{i-1} & -\sin\alpha_{i-1} & -d_i\sin\alpha_{i-1} \\ \sin\theta_i\sin\alpha_{i-1} & \cos\theta_i\sin\alpha_{i-1} & \cos\alpha_{i-1} & d_i\cos\alpha_{i-1} \\ 0 & 0 & 0 & 1 \end{bmatrix}\begin{bmatrix} -1 & 0 & 0 & 0 \\ 0 & -1 & 0 & 0 \\ 0 & 0 & 0 & 0 \\ 0 & 0 & 0 & 0 \end{bmatrix} \\
&= \begin{bmatrix} -\cos\theta_i & -\sin\theta_i & 0 & 0 \\ -\sin\theta_i\cos\alpha_{i-1} & \cos\theta_i\cos\alpha_{i-1} & 0 & 0 \\ -\sin\theta_i\sin\alpha_{i-1} & \cos\theta_i\sin\alpha_{i-1} & 0 & 0 \\ 0 & 0 & 0 & 0 \end{bmatrix} = -T_i^{i-1}P_i
\end{aligned}
\tag{53}
$$

Therefore, the final equations of motion for a general anthropopathic 6 DOF arm with load m_L can be summarized as follows:

$$\tau_i = T_i + \left(\frac{d}{dt}\frac{\partial(K_L - P_L)}{\partial q_i} - \frac{\partial(K_L - P_L)}{\partial q_i}\right) = \sum_{j=1}^{i} D_{ij}\ddot{q}_j + I_{i(act)}\ddot{q}_i + \sum_{j=1}^{i}\sum_{k=1}^{n} D_{ijk}\dot{q}_j\dot{q}_k + D_i$$

$$+ \frac{d}{dt}\frac{\partial\left(\frac{1}{2}m_L trace\left(v_H v_H^T\right) + m_L[-g\quad 0\quad 0\quad 0]T_H^0\tau_H\right)}{\partial q_i}$$

$$- \frac{\partial\left(\frac{1}{2}m_L trace\left(v_H v_H^T\right) + m_L[-g\quad 0\quad 0\quad 0]T_H^0\tau_H\right)}{\partial q_i}$$

$$(54)$$

3 Simulation and Verification

In this paper, the forward and inverse kinematics are programed via Matlab .m document, the core program please refer Appendix A. In order to verify the geometric-analytical mix approach for the inverse kinematics, we use the closed-loop process for the validation of the computation accuracy [29], as shown in Fig. 7. To save the time, the Matlab Robotics Toolbox is utilized as well [30] as shown in Fig. 6.

Seed Noid R7F: : 6 axis, RRRRRR, modDH, slowRNE

j	theta	d	a	alpha	offset
1	q1	0.05978	0	0	0
2	q2	0	0.028046	-1.5708	0
3	q3	0.2895	0.01	-1.5708	0
4	q4	0	-0.035	-1.5708	0
5	q5	0.2655	0.035	1.5708	0
6	q6	0	0	-1.5708	0

Fig. 6. SeedNoid R7F forward kinematics based on Matlab robotics toolbox

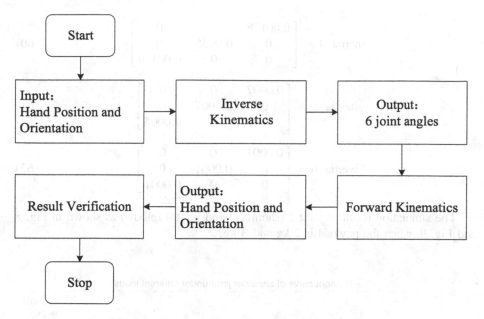

Fig. 7. Validation process for the inverse kinematics

For the modeling of dynamics, we substitute the parameter as following.

$$\text{mass\_center\_list} = \begin{bmatrix} 0 & 0 & -0.05978 \\ 0 & 0 & 0 \\ 0 & 0 & -0.1625 \\ 0 & 0 & 0 \\ 0 & 0 & -0.1955 \\ 0 & 0 & 0 \end{bmatrix} \tag{55}$$

$$\text{mass\_list} = \begin{bmatrix} 0.242 & 0.744 & 0.574 & 0.711 & 0.132 & 0.005 \end{bmatrix} \tag{56}$$

$$\text{inertia\_1} = \begin{bmatrix} 0.000263 & 0 & 0 \\ 0 & 0.000228 & 0 \\ 0 & 0 & 0.000263 \end{bmatrix} \tag{57}$$

$$\text{inertia\_2} = \begin{bmatrix} 0.000623 & 0 & 0 \\ 0 & 0.000228 & 0 \\ 0 & 0 & 0.000623 \end{bmatrix} \tag{58}$$

$$\text{inertia\_3} = \begin{bmatrix} 0.00214 & 0 & 0 \\ 0 & 0.00217 & 0 \\ 0 & 0 & 0.0035 \end{bmatrix} \tag{59}$$

$$inertia\_4 = \begin{bmatrix} 0.00178 & 0 & 0 \\ 0 & 0.0035 & 0 \\ 0 & 0 & 0.00179 \end{bmatrix} \quad (60)$$

$$inertia\_5 = \begin{bmatrix} 0.0007 & 0 & 0 \\ 0 & 0.0007 & 0 \\ 0 & 0 & 0.0004 \end{bmatrix} \quad (61)$$

$$inertia\_6 = \begin{bmatrix} 0.0001 & 0 & 0 \\ 0 & 0.0001 & 0 \\ 0 & 0 & 0.0001 \end{bmatrix} \quad (62)$$

The simulation result of joint 2 (shoulder) and joint 4 (elbow) as shown in Fig. 8 and Fig. 9. when the payload is 2 kg and 3 kg.

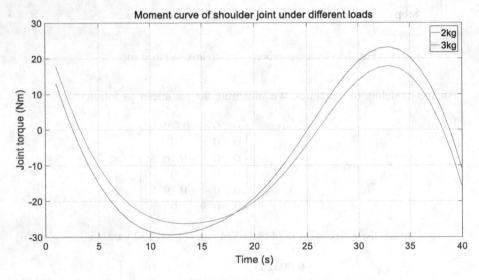

Fig. 8. Torque of joint 2 with Payload of 2 kg and 3 kg

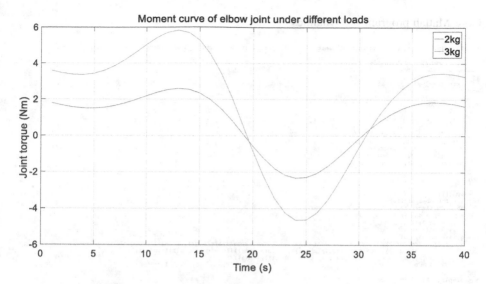

Fig. 9. Torque of joint 4 with Payload of 2 kg and 3 kg

4 Conclusion and Future Work

In this paper, we have demonstrated a geometric-analytical mix inverse kinematics approach for a 6-DOF left arm of Seed-Noid R7F robot due to its dissatisfactory of Pieper's criteria, then a dynamics model with the payload scenario is established to respond its mass sensibility. At last, the proposed solution is verified using MATLAB when the payload is 2 kg and 3 kg separately, the simulation result verifies the model's correct. This paper is the basic research of kinematics and dynamics of a single anthropomorphic manipulator, the further work is to provide the basic model for the coordinated control of the dual arms.

Appendix A

Core Matlab Program of Kinematics and Dynamics
Core Matlab program of Kinematics

```
>> clear; % clear workspace
>> clc; % clear command line
>>addpath(genpath('c:\Program Files\MATLAB\R2016a\toolbox\symbolic\symbolic'))£»%add MATLAB SYMBOLIC TOOLBOX
path
>>savepath£» % save path
>>startup_rvc£»% start up MATLAB Robotics Toolbox
>> syms q1 q2 q3 q4 q5 q6 d1 d2 d3 d4 d5 d6 a0 a1 a2 a3 a4 a5 alpha0 alpha1 alpha2 alpha3 alpha4 alpha5;
>> L(1)=Link([0,d1,a0, alpha0,0],'modified');
>> L(2)=Link([0,d2,a1, alpha1,0],'modified');
>> L(3)=Link([0,d3,a2, alpha2,0],'modified');
>> L(4)=Link([0,d4,a3, alpha3,0],'modified');
>> L(5)=Link([0,d5,a4, alpha4,0],'modified');
>> L(6)=Link([0,d6,a5, alpha5,0],'modified');
>> six_link=SerialLink(L,'name','Seed Noid R7F');
>> T=six_link.fkine([q1 q2 q3 q4 q5 q6])
```

Core Matlab program of Dynamics

```matlab
function [M,C,G,torque] = LagrangianDynamics(dh_list, mass_list, mass_center_list, inertia_tensor_list)

[rows, columns] = size(dh_list);
number_of_links = rows;
if columns ~= 4
    error('wrong DH parameters!')
end

for i = 1:rows
    % define q dq and ddq
    eval(['syms ','q',num2str(i),' real;']);
    % execute syms qi real;
    eval(['syms ','dq',num2str(i),' real;']);% execute syms dqi real;
    eval(['syms ','ddq',num2str(i),' real;']);%execute syms ddqi real;
    eval(['q(i)=','q',num2str(i),';']);%execute q(i)=qi;
    eval(['dq(i)=','dq',num2str(i),';']);%execute dq(i)=dqi;
    eval(['ddq(i)=','ddq',num2str(i),';']);%execute ddq(i)=dqi;

end

A = sym([]);% build null A matrix
for i = 1:number_of_links
    dh = dh_list(i,:);%dh=dh_list the ith row
    alpha(i) = dh(1); % alpaha(i)
    a(i) = dh(2);% a(i)
    d(i) = dh(3);% d(i)
    q(i) = dh(4);% q(i)
    A(:,:,i) = [cos(q(i)),   -sin(q(i))*cos(alpha(i)),  sin(alpha(i))*sin(q(i)),  a(i)*cos(q(i));
                sin(q(i)),  cos(q(i))*cos(alpha(i)), -sin(alpha(i))*cos(q(i)), a(i)*sin(q(i));
                0,          sin(alpha(i)), cos(alpha(i)),  d(i);
                0,          0,             0,             1];
end
A = simplify(A);% simplify A

% calculate echa matrix in frame(0)
A0 = sym([]);% build null matrix A0
for i = 1:number_of_links
    A0(:,:,i) = eye(4,4);% A0(:,:,i)
    for j = 1:i
        A0(:,:,i) = A0(:,:,i)*A(:,:,j);% A0(:,:,i) circulation
    end
end
A0 = simplify(A0);%

J = sym([]);
for i = 1:number_of_links
    m=mass_list(i);
    c=mass_center_list(i,:);
    I=inertia_tensor_list(:,:,i);
    x=c(1);
    y=c(2);
    z=c(3);
    J(:,:,i)=[(-I(1,1)+I(2,2)+I(3,3))/2 I(1,2) I(1,3) m*x;
        I(1,2) (I(1,1)-I(2,2)+I(3,3))/2 I(2,3) m*y;
        I(1,3) I(2,3) (I(1,1)+I(2,2)-I(3,3))/2 m*z;
        m*x m*y m*z m];
end

% Calculate M(q)
syms tr
for i = 1:number_of_links
    for j = i:number_of_links
        tr = 0;
        for k = max(i,j):number_of_links
            tr = tr + trace(eval(['diff(A0(:,:,k),q',num2str(i),')'])*J(:,:,k)*...
                eval(['diff(transpose(A0(:,:,k)),q',num2str(j),')']));
        end
        M(i,j) = simplify(tr);
        M(j,i) = M(i,j);
    end
end

% calculate C(q)
for i = 1:number_of_links
    for j = 1:number_of_links
        c = 0;
        for k = 1:number_of_links
            c = c + 1/2*(eval(['diff(M(i,j),q',num2str(k),')'])...
                + eval(['diff(M(i,k),q',num2str(j),')'])...
                - eval(['diff(M(j,k),q',num2str(i),')']))*eval(['dq',num2str(k)]));
        end
        C(i,j) = simplify(c);
    end
end

syms gc real
g = [0,0,-gc,0]';

% Calculate G(q)
for i = 1:number_of_links
    gi = 0;
    for j = 1:number_of_links
        gi = gi - mass_list(j)*g'...
            *eval(['diff(A0(:,:,j),q',num2str(i),')'])...
            *[mass_center_list(j,:),1]';
    end
    G(i) = simplify(gi);
end
G = G';

torque=M*ddq'+C*dq'+G;
end
```

References

1. Smarr, C.A., Fausset, C.B., Rogers, W.A.: Understanding the potential for robot assistance for older adults in the home environment. Technical report HFA-TR-1102School of Psychology, Human Factors and Aging Laboratory, Georgia Institute of Technology, Atlanta, GA (2011)
2. Srinivasan, V., Bethel, C.L., Murphy, R.R.: Evaluation of head gaze loosely synchronized with real-time synthetic speech for social robots. IEEE Trans. Hum.-Mach. Syst. **44**(6), 767–778 (2014)
3. Truong, X.T., Ngo, T.D.: Toward socially aware robot navigation in dynamic and crowded environments: a proactive social motion model. IEEE Trans. Autom. Sci. Eng. **14**(4), 1743–1760 (2017)
4. Wakita, Y., Tanaka, H., Matsumoto, Y.: Projection function and hand pointer for user-interface of daily service robot. In: Proceedings of the IEEE International Conference on Robotics and Biomimetics, pp. 2218–2224 (2017)
5. Azeta, J., Bolu, C., et al.: A review on humanoid robotics in healthcare. In: MATEC Web of Conferences, vol. 153, no. 5, p. 02004, January 2018
6. Paul, R.P.: Robot Manipulators: Mathematics, Programming, and Control. MIT Press, Cambridge (1981)
7. Asada, H., Slotine, J.-J.E.: Robot Analysis and Control. Wiley, New York (1986)
8. Sciavicco, L., Siciliano, B.: Modelling and Control of Robot Manipulators, 2nd edn. Springer, London (2000). https://doi.org/10.1007/978-1-4471-0449-0
9. Spong, M.W., Hutchinson, S., Vidyasagar, M.: Robot Modeling and Control. Wiley, New York (2006)
10. Bottema, O., Roth, B.: Theoretical Kinematics. North Holland, Amsterdam (1979)
11. Angeles, J.: Spatial Kinematic Chains: Analysis, Synthesis, Optimization. Springer, Berlin (1982). https://doi.org/10.1007/978-3-642-48819-1
12. McCarthy, J.M.: An Introduction to Theoretical Kinematics. MIT Press, Cambridge (1990)
13. Featherstone, R.: Position and velocity transformations between robot end effector coordinates and joint angles. Int. J. Robot. Res. **2**(2), 35–45 (1983)
14. Hollerbach, J.M.: A recursive lagrangian formulation of manipulator dynamics and a comparative study of dynamics formulation complexity. IEEE Trans. Syst. Man Cybern. **10**, 730–736 (1980)
15. Paul, R.P., Zhang, H.: Computationally efficient kinematics for manipulators with spherical wrists based on the homogeneous transformation representation. Int. J. Robot. Res. **5**(2), 32–44 (1986)
16. Tsai, L.W., Morgan, A.P.: Solving the kinematics of the most general six-and five-degree-of-freedom manipulators by continuation methods. ASME J. Mech. Transm. Autom. Des. **107**, 189–200 (1985)
17. Yoshikawa, T.: Foundations of Robotics. MIT Press, Boston (1990)
18. Craig, J.J.: Introduction to Robotics: Mechanics and Control, 3rd edn. Pearson Prentice Hall, Upper Saddle River (2004)
19. Khalil, W., Dombre, E.: Modeling, Identification and Control of Robots. Hermes Penton Ltd., London (2002)
20. Uicker, J.J.: Dynamic force analysis of spatial linkages. ASME J. Appl. Mech. **34**, 418–424 (1967)
21. Bejczy, A.K.: Robot arm dynamics and control, memo. TM 33-669, Jet Propulsion Laboratory, California Institute of Technology (1974)

22. Orin, D.E., McGhee, R.B., Vukobratovic, M., Hartoch, G.: Kinematic and kinetic analysis of open-chain linkages utilizing Newton-Euler methods. Math. Biosci. **43**, 107–130 (1979)
23. Luh, J.Y.S., Walker, M.W., Paul, R.P.C.: On-line computational scheme for mechanical manipulators. ASME J. Dyn. Syst. Measur. Control **102**, 69–76 (1980)
24. Xin, G., Mu, Y., Gao, Y.: Optimal trajectory planning for robotic manipulators using improved teaching-learning-based optimization algorithm. Ind. Robot Int. J. **43**(3), 308–316 (2016)
25. Quigle, M., Gerkey, B.P., Conley, K.: ROS: an open-source robot operating system. In: Conference: ICRA Workshop on Open Source Software, January 2009
26. Denavit, J., Hartenberg, R.S.: A kinematic notation for lower-pair mechanisms based on matrices. ASME J. Appl. Mech. **22**, 215–221 (1955)
27. Asfour, T., Dillmann, R.: Human-like motion of a humanoid robot arm based on a closed-form solution of the inverse kinematics problem. In: 2003 Proceedings of the IEEE/RSJ International Conference on Intelligent Robots and Systems, IROS 2003, vol. 2, November 2003
28. Niku, S.B.: Introduction to Robotics: Analysis, Systems, Applications, 2nd edn (2010)
29. Ho, T., Kang, C.G., Lee, S.: Efficient closed-form solution of inverse kinematics for a specific six-DOF arm. Int. J. Control Autom. Syst. **10**, 567–573 (2012). https://doi.org/10.1007/s12555-012-0313-9
30. Cork, P.: Robotics, Vision and Control: Fundamental Algorithms in MATLAB. Springer Tracts in Advanced Robotics, 2nd edn, vol. 118, pp. 205–211. https://doi.org/10.1007/978-3-319-54413-7

Machine Vision Application

Object Texture Transform
Based on Improved CycleGAN

Luyue Zhang[1], Bin Pan[1(✉)], Evgeny Cherkashin[2], Xiaoyang Zhang[1],
Xiaomeng Ruan[1], and Qinqin Li[1]

[1] Liaoning Shihua University, Fushun, China
panbin@lnpu.edu.cn
[2] ISDCT SB RAS, Ac. Lavrentieva Ave. 17, 630090 Novosibirsk, Russia

Abstract. Image style transfer is a kind of image-to-image conversion method, which has been widely used in the field of computer vision and is an important research direction. In the application of target deformation, the traditional algorithm has the disadvantages of unclear positioning and difficult separation from the background. In this paper, an improved cyclic generative antagonistic network (CycleGAN) is proposed. The model adds a layer of self-attention algorithm to the discriminator structure, and uses clues from all feature locations to establish global dependence, to obtain global geometric features of images, and to achieve long correlation. Compared with CycleGAN algorithm, the results show that the algorithm in this paper can learn to generate a style transfer image with clear texture and full rendering, and at the same time the image has no obvious loss in details. Compared with the original algorithm, the convergence speed in the learning process of the algorithm in this paper is obviously improved, the oscillation amplitude of loss function is small, and the number of steps required for network convergence is reduced. In the process of testing, the algorithm stylizes 1163 images, and the consumption time of each generation only increases by 17 s.

Keywords: Image translation · Image style transfer · Generative antagonistic network · Self-attention mechanism · Deep learning

1 Introduction

In the areas of image processing, computer graphics, and computer vision, many problems can be thought of as "converting" an input image into an output image. A lot of progress has been made in this direction. Convolutional neural network (CNN) can complete various image prediction tasks. By learning to minimize the loss, the loss function is used as the standard to improve the quality of generated results. For example, [1] uses CNN to minimize the difference between the generated image and the target image to learn the mapping function of parameters. However, this method is not a unsupervised learning method and requires a large amount of manpower to design an

The National Natural Science Foundation of China under Grant Nos. 61602228, 61572290; the Revitalization Talents Program of LiaoningProvince under Grant No. XLYC1807266.

© Springer Nature Singapore Pte Ltd. 2020
J. Qian et al. (Eds.): ICRRI 2020, CCIS 1335, pp. 85–97, 2020.
https://doi.org/10.1007/978-981-33-4929-2_6

efficient loss function. On this basis, generative antagonism network (GAN) [2] realizes the automatic learning and generation of loss function. GAN's method is to use discriminator model to complete the learning of the loss function to distinguish whether the image is real or not, and then train a generator model to minimize the loss. While traditional methods may require different loss functions to achieve different goals, GANs can be directly applied to a number of different tasks. In 2017, Zhu et al. designed CycleGAN [3] based on generative confrontation network (GAN), which realized the mismatch of training data. In training, only the images of the source and target fields need to be input, and no matching of the contents is required. CycleGAN's method is to train two generative adversary networks (GANs), in which the generator-discriminator model of each network converts one kind of image into another, instead of the traditional GAN matching training for a single type of image. At the same time, loss of cyclic consistency was added to obtain better data distribution capture ability than ordinary GAN and better image conversion effect. Target deformation is a special task in image transformation. CycleGAN is mainly used for image style transfer. Target deformation in CycleGAN can be regarded as the style transfer of an instance, which transforms a certain type of specific object in an image into another type of specific object. The particularity of target deformation is that it does not want to learn the image as a whole, but expects to separate the background in the image and only convert the required target type. The model of CycleGAN relies on convolution to model the dependence between different image regions. Small convolution kernel is difficult to find the long-distance dependence of image, and large convolution kernel has the problem of low computational efficiency. In this regard, based on the realization of target deformation by CycleGAN, this paper introduces the mechanism of self-attention [4]. The attention module uses the attention vector to determine how relevant an element is to other elements, and uses the feature weighted sum of all locations to calculate the response, where the weight is only calculated at a small computational cost, so as to better balance the relationship between the model's long-term correlation and computational statistical efficiency.

2   Related Work

2.1   Generative Adversarial Networks

Generative antagonistic network (GANs) has achieved many good results in the direction of image generation [6, 7]. This is a framework consisting of two subnetworks: discriminator (D) is designed to distinguish the generated image from the real image, while generator (G) is designed to generate the real image to trick the discriminator into not being able to tell which is the real image and which is the generated image. The two continue to cycle through this process, each updating its own parameters with each other's parameters to minimize losses. Through such alternate training, the difference between the real image and the generated image is continuously narrowed.

Now, GAN achieves Nash equilibrium mainly by optimizing the following loss functions:

$$\min_{G} \max_{D} E_{x \sim q_{data}(x)}[\lg D(x)] + E_{z \sim p(z)}[\lg(1 - D(G(z)))] \tag{2.1}$$

In order to generate more realistic images, a series of multi-level generation models have been developed gradually [8, 9]. [8] proposes a composite generative adversarial network, which decomposes complex components in images and generates different parts of images with multiple generators. The layered recursive GAN [9] learns to recursively generate image foreground and background. GAN has achieved great success in the application of image-to-image conversion. In the problem of conversion from wireframe to photo, from map to 3d real scene, from night to day and other images to images, [5, 10, 11] introduced cyclic consistency loss to solve the problem of unpaired images to image translation, and carried out experiments on the deformation of target objects (such as brown horse to zebra, apple to orange). Although GAN has made great progress and successful applications, there are still some problems that need to be optimized: GAN's network is extremely sensitive to super parameters, the training process is unstable, and it often fails to reach Nash equilibrium. The problems of model collapse and gradient disappearance occur frequently. Certain types of structures and geometries cannot be found in the image.

2.2 CycleGAN

CycleGAN is a Generative Adversarial Network model proposed by Zhu et al. It is a ring network composed of two symmetric GANs. In the training, CycleGAN only needs to take the image of the source field and the image of the target field as input, and there is no requirement on whether the image contents of the source field and the target field are matched. That is to say, CycleGAN can realize the mutual conversion between different styles of images in the case of unpaired training images, with good scalability and wider application. CycleGAN's two symmetric GANs share two generators and each has a discriminator, that is, two discriminators and two generators. Each one-way GAN has two loss function, so CycleGAN has four loss function. The total objective function of CycleGAN is:

$$L(G, F, DX, DY) = L_{gan}(G, DX, X, Y) + L_{gan}(F, DX, Y, X) + \lambda L_{cyc}(G, F) \tag{2.2}$$

The main method of CycleGAN is to train two generative adversary networks, each of which has a generator-discriminator model to transform one type of image into another. Cyclic consistency loss is used as a transmissibility method to monitor training and motivate CycleGAN cycling (Fig. 1).

The model gets the input image from the domain D_A, then passes the image to the first Generator G_{A2B}, and converts it to the image in the target domain D_B. This newly generated image is passed to the next G_{B2A}. The G_{B2A} converts the image *Generated_B* in the target domain D_B back to the original domain D_A. The image *Cyclic_A* converted back to the original domain must be similar to the original input image, a process that defines a mismatched mapping to a dataset that does not already exist.

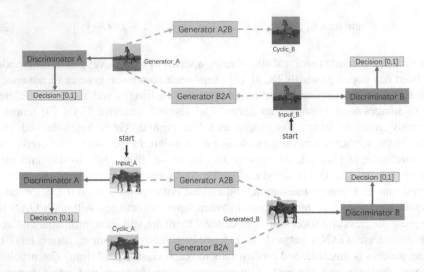

Fig. 1. CycleGAN's model structure, the above is the A2B generative adversarial nets, the following is B2A generative adversarial nets.

2.3   Self-attention Algorithm

In the traditional convolutional neural network, the size of each convolution kernel is limited, so each convolution operation can only cover a neighborhood around the pixel, and it is not easy to find the features with a long distance. Therefore, traditional GAN models often learn texture features, rather than specific structure and geometric features, such as human head and limbs.

Self-attention GAN is a GAN model proposed by Han Zhang et al. It adds a layer of self-attention algorithm to the generator and discriminator of the traditional GAN model. By directly calculating the relationship between any two pixels in the image, it can better learn the dependency relationship between global features. Another innovation of the model is the application of spectral normalization to the generator and discriminator, which stabilizes the training process and improves the quality of the generated images (Fig. 2).

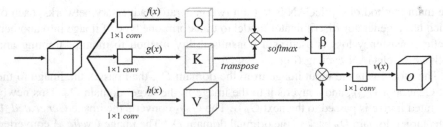

Fig. 2. The proposed self-attention module for the SAGAN. The $\otimes$ denotes matrix multiplication. The softmax operation is performed on each row.

As shown in the figure, two convolution kernels of 1 * 1 are used to carry out linear transformation and channel compression for the feature map of convolution, and then the two tensors are reshaped into matrix form. The attention map is obtained through the multiplication of transpose and the flexible maximum function. On the other side, the original feature map uses the convolution of 1 * 1 again for linear transformation, and multiplies the result with the attention force matrix to obtain the self-attention feature map. Finally, the self-attention feature map and the original convolution feature map are weighted and summed (the weight is learnable) to get the final output.

3 Model

CycleGAN converts the sample in A domain to the sample in B domain. The goal is to learn the mapping from A to B G_{A2B}, namely the generator in GAN. G converts the input image *Input_A* selected in A domain to the image *Generated_A* in B domain with G_{A2B}, and then uses GAN's D_B to determine whether it is a real image or a generated image. From this step, the loss function of GAN discriminator is constructed as follows:

$$
\begin{aligned}
L_G((G_{A2B}, D_B, A, B) &= E_{y \sim p_{data}(y)}[\log D_B(y)] \\
&+ E_{x \sim p_{data}(x)}[\log D_B(G_{A2B}(x))]
\end{aligned}
\tag{3.1}
$$

Then, CycleGAN converts the generated image in the B domain to the image in the A domain, learns to map G_{B2A}, and constructs the GAN discriminant loss function as follows:

$$
\begin{aligned}
L_G((G_{B2A}, D_A, A, B)) &= E_{x \sim p_{data}(x)}[\log D_A(y)] \\
&+ E_{y \sim p_{data}(y)}[\log D_A(G_{B2A}(y))]
\end{aligned}
\tag{3.2}
$$

In this way, CycleGAN simultaneously trains two mappings, one is the discriminator of generator G_{A2B}, the other is the discriminator of generator GB2A, where the input of G_{B2A} is *Generated_A* generated by the G_{A2B} of *Input_A*, the discriminator of training discriminates the authenticity of the real image and the image generated twice by the two generators, the expression is:

$$
G_{B2A}(x) \cong x
\tag{3.3}
$$

$$
G_{A2B}(G_{B2A}(y)) \cong y
\tag{3.4}
$$

Finally, the loss function generated by cyclic consistency is:

$$
\begin{aligned}
L_{cyc}(G_{A2B}, G_{B2A}, A, B) &= E_{x \sim A}[\||G_{B2A}(G_{A2B}(x))_x\||_1] \\
&+ E_{y \sim B}[\||G_{A2B}(G_{B2A}(y))_y\||_1]
\end{aligned}
\tag{3.5}
$$

The final total loss function is obtained: the generated antagonistic discriminator loss function from A to B, the generated antagonistic discriminator loss function from B to A, and the cyclic consistency loss function.

$$L(G_{A2B}, G_{B2A}, D_A, D_B) = L_G((G_{A2B}, D_B, A, B))$$
$$+ L_G((G_{B2A}, D_A, A, B)) + \lambda L_{cyc}((G_{A2B}, G_{B2A}, A, B))$$
(3.6)

In the improved model, a layer of attention algorithm is added to the training structure of the two discriminators. The original discriminator structure is composed of three convolution layers. First, the input image is cropped to a fixed size of 256 * 256. The first convolution process goes through the convolution kernel of 1 * 1 of stride 2 to obtain 64 * 128 * 128 feature map. The second convolution process then goes through the 2 * 2 convolution kernel of stride 2 to obtain the feature map of 256 * 32 * 32. Then enter the attention algorithm, after weight change, still output 256 * 32 * 32 feature map; The final output matrix is obtained by the convolution of 1 * 1 of stride 1 (Fig. 3).

Fig. 3. The improved discriminator structure adds a layer of self-attention algorithm between the second and third hidden layers.

The attention algorithm is divided into two steps. The first step is to find the similarity and then to normalize it. Firstly, the feature map $x \in \mathbb{R}^{C \times N}$ obtained from the previous convolution layer is transformed into two feature Spaces f and g to calculate the attention map, where $f(x) = W_f x, g(x) = W_g x$, C is the number of channels, and N is the number of feature positions of the features of the previous hidden layer.

To multiply matrices, convert $C \times N$ to $C \times (W \times H)$. After the operation of $f(x)$ and $g(x)$, the output is [C/8, N]. After the transpose of the result of $f(x)$ and the multiplication of $g(x)$, we get the mixing matrix s of size [N, N], which represents the correlation between each pixel and can be regarded as the correlation matrix. The operation of $h(x)$ is different, and the output is [C, N]. Then use softmax function to normalize the s matrix and get the matrix β.

$$\beta_{j,i} = \frac{\exp(s_{ij})}{\sum_{i=1}^{N} \exp(s_{ij})}$$
(3.7)

$$s_{j,i} = f(x_i)^T g(x_j)$$
(3.8)

$\beta_{j,i}$ represents the attention degree of the model to the ith position when synthesizing the jth region, that is, whether the model pays attention to the ith position when regenerating into the jth region can be regarded as an attention map. s_{ij} stands for similarity.

The second step is to output attention layer. Note that the output of the layer is $o = (o_1, o_2, \cdots, o_j, \cdots, o_N) \in \mathbb{R}^{C \times N}$, applying the obtained attention map to the output feature map of $h(x)$, multiply the generated $h(x_i)$ that has an impact on the jth pixel by its corresponding impact factor $\beta_{j,i}$, and sum them together, which can generate new j pixel according to the degree of influence. Convolve this result with another layer to obtain the result o of the feature map with added attention weight:

$$o_j = v(\sum_{i=1}^{N} \beta_{j,i} h(x_i)) \tag{3.9}$$

$$h(x_i) = W_h x_i \tag{3.10}$$

$$v(x_i) = W_v x_i \tag{3.11}$$

Finally, the output of the attention layer is multiplied by the scale parameter and added to the feature map of the initial input. The final output obtained is the result of the superposition of the attention mechanism of the original feature map:

$$y_i = \gamma o_i + x_i \tag{3.12}$$

4 Experiment

At the experimental stage, the GAN model in this paper is constructed by using Python3.6 and torch0.4. The training time is eight hours on a Linux platform configured with four NVIDIA GTX1020. The datasets of brown horse and zebra are provided by CycleGAN, and CycleGAN implements advanced unpaired object deformation, so the results generated in this paper are also compared with CycleGAN.

4.1 Visual Effect

In terms of the visual effect evaluation of the generated image samples of generative antagonistic network, the model performance of generative antagonistic network is described by evaluating the clarity, diversity and approximate degree of the generated image with the training set through quantitative indexes. In the aspect of image stylization, the clarity of stylized image is consistent with the texture background of the style image, and the clarity is difficult to accurately reflect the effect of image stylization. In terms of diversity and approximation degree, stylized image is different from content image and style image, but it integrates the characteristics of both, and cannot be measured by diversity and approximation degree index. In the evaluation of the visual effect of image stylization, it is mainly based on people's subjective feelings to evaluate the quality of the visual effect. This paper uses the degree of style rendering, texture detail clarity to evaluate the effect of style extraction and fusion;

The information retention degree of the content image is evaluated by the detail of the image content so as to obtain the subjective evaluation standard of the image stylization visual effect. The degree of style rendering mainly refers to the information of image color, overall effect and brightness of image texture, texture detail refers to the degree of image clarity, and image content detail refers to the degree of retention of content information of content image. Under the condition of training rounds of 200 and learning rate of 0.0002, the image stylization effect of CycleGAN algorithm and the improved algorithm on horse2zebra data set is shown in Fig. 4 and 5.

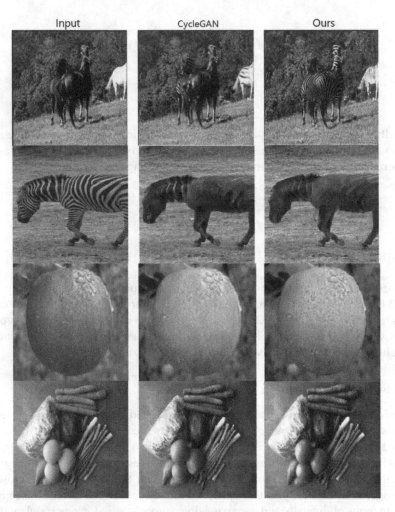

Fig. 4. Comparison between the improved model and CycleGAN.

In Fig. 4, the first two lines are the results of horse and zebra generation in B → A (zebra to horse) and A → B (horse to zebra), respectively. The last two lines are the results of apple and orange in B → A (orange to apple) and A → B (apple to orange),

Fig. 5. Comparison between the improved model and CycleGAN.

respectively. The first column is the input image, the second column is the result of CycleGAN generation, and the third column is the result of this paper. It can be seen that the improved algorithm is superior to CycleGAN algorithm in target recognition. The results show that the improved algorithm is more accurate and the texture effect is more integrated. In the first row of results, CycleGAN confused the background with the target, resulting in incomplete and incomplete transformation of zebra, while the improved model for horse transformation was more integrated; in the second row of results, CycleGAN mistook a large part of the background area as a horse target, and converted it into black and white stripes. The improved algorithm greatly reduced the misjudgment area, basically successfully stripped the background area, and did not send The zebra is generated in the correct position and keeps the background consistent. In the third row of results, CycleGAN did not convert some orange objects, and mistakenly converted the leaves into Apple textures. The improved algorithm transformed all orange objects in the image, and successfully separated the background without error conversion when the background leaves and the target had occlusion relationship. In the fourth row of results, CycleGAN algorithm was used In the shadow part of apple, confusion is generated, which is converted into fuzzy background leaf texture. Although a small part of the area is misjudged as background, the target and background are clearly distinguished and the error is greatly reduced.

In Fig. 5, the first two lines are the results of horse and zebra generation in B → A (zebra to horse) and A → B (horse to zebra), respectively, while the last two lines are the results of apple and orange generation in B → A (orange to Apple) and A → B (apple to orange), respectively. The first column is the input image, the second column is the result of CycleGAN generation, and the third column is the result of this paper. It can be seen that the improved algorithm is superior to CycleGAN algorithm in target recognition. The results show that the improved algorithm is more accurate and the texture effect is more integrated. In the first row of results, CycleGAN confused the background with the target, resulting in incomplete and incomplete transformation of zebra, while the improved model for horse transformation was more integrated; in the second row of results, CycleGAN mistook a large part of the background area as a horse target, and converted it into black and white stripes. The improved algorithm greatly reduced the misjudgment area, basically successfully stripped the background area, and did not send The zebra is generated in the correct position and keeps the background consistent. In the third row of results, CycleGAN did not convert some orange objects, and mistakenly converted the leaves into Apple textures. The improved algorithm transformed all orange objects in the image, and successfully separated the background without error conversion when the background leaves and the target had occlusion relationship. In the fourth row of results, CycleGAN algorithm was used In the shadow part of apple, confusion is generated, which is converted into fuzzy background leaf texture. Although a small part of the area is misjudged as background, the target and background are clearly distinguished and the error is greatly reduced.

Through the analysis of Fig. 4 and 5, we can see that the stylized image rendering using the improved algorithm in this paper is sufficient, can accurately identify the target contour area, carry out the overall texture conversion, and can retain the details of the original image content, with good visual effect.

4.2 Stability and Speed of Convergence

Under the condition that the training algebra is 140 and the learning rate is 0.0002, the CycleGAN algorithm and the improved algorithm in this paper show the changes of the loss function in the training process on horse2zebra data set as shown in Fig. 6 and Fig. 7.

It can be seen from Fig. 6 and Fig. 7 that the CycleGAN algorithm has a large loss function in the early stage of training, and then slowly converges. In the second half of the training process, there are still high pulses, resulting in a large loss function. In this paper, the improved algorithm converges rapidly at the beginning of training, and then shows stable convergence. During the training, the pulse is smaller, and the oscillation

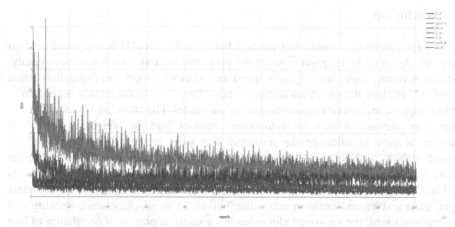

Fig. 6. CycleGAN's loss function changes on the horse2zebra dataset.

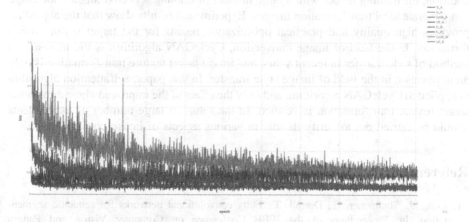

Fig. 7. Improved algorithm's loss function changes on the horse2zebra dataset.

amplitude is smaller than that of CycleGAN algorithm, showing fast convergence speed and strong stability, and the final convergence degree is better than that of CycleGAN algorithm. Through the analysis of Fig. 6 and Fig. 7, it can be seen that while the algorithm in this paper has stronger stability, the number of training steps required for network convergence is also less, and the final convergence degree of loss function is also higher than that of CycleGAN.

5 Conclusion

In this paper, an image conversion method based on CycleGAN is proposed, and an improved algorithm is designed to solve the problems such as poor generation quality, unstable network, slow convergence speed and difficult target discrimination when CycleGAN realizes the target texture conversion task. The self-attention algorithm is used to improve the discriminator structure of the model. The attention vector is used to judge how relevant an element is to other elements, and the weighted sum of all positions is used to calculate the response. From the perspective of visual effect, network stability and convergence speed, the experiment was carried out on a target texture transformation data set, and the following conclusions were obtained: in terms of visual effect, the improved algorithm can generate transform images with complete target, clear texture and uniform color distribution. In terms of network stability and convergence speed, the improved algorithm has a small amplitude of oscillation of loss function and no violent pulse during training, and requires fewer steps for the algorithm to converge, with faster convergence speed and higher final convergence degree. In addition, the training model with a large number of training sets covering a wide range can generate ideal transformation images. Experimental results show that the algorithm provides high quality and practical optimization results for the target texture transformation. In the field of image conversion, CycleGAN algorithm is the mainstream method of style transfer in recent years, and image target texture transformation is only an application in the field of image style transfer. In this paper, selfattention algorithm is applied to CycleGAN algorithm, and only the effect of the improved algorithm on the target texture transformation is verified. In the future, a large number of experiments should be carried out to verify its role in various aspects of image conversion.

References

1. Long, J., Shelhamer, E., Darrell, T.: Fully convolutional networks for semantic segmentation. In: Proceedings of the IEEE Conference on Computer Vision and Pattern Recognition, pp. 3431–3440 (2015)
2. Goodfellow, I., Pouget-Abadie, J., Mirza, M., et al.: Generative adversarial nets. In: Advances in Neural Information Processing Systems, NIPS 2014, Montréal, 8–13 December 2014, vol. 27, pp. 2672–2680 (2014)
3. Zhu, J.Y., Park, T., Isola, P., et al.: Unpaired image-to-image translation using cycle consistent adversarial networks. In: Proceedings of 2017 IEEE International Conference on Computer Vision, pp. 2242–2251 (2017)
4. Zhang, H., Goodfellow, I., Metaxas, D., et al.: Self-attention generative adversarial networks. arXiv arXiv:1805.08318 (2018)
5. Yi, Z., Zhang, H., Tan, P., Gong, M.: DualGAN: unsupervised dual learning for image-to-image translation. In: The IEEE International Conference on Computer Vision (ICCV) (2017)
6. Denton, E.L., Chintala, S., Fergus, R., et al.: Deep generative image models using a Laplacian pyramid of adversarial networks. In: Advances in Neural Information Processing Systems, pp. 1486–1494 (2015)

7. Radford, A., Metz, L., Chintala, S.: Unsupervised representation learning with deep convolutional generative adversarial networks. arXiv preprint arXiv:1511.06434 (2015)
8. Kwak, H., Zhang, B.T.: Generating images part by part with composite generative adversarial networks. arXiv preprint arXiv:1607.05387 (2016)
9. Yang, J., Kannan, A., Batra, D., Parikh, D.: LR-GAN: layered recursive generative adversarial networks for image generation. In: 5th International Conference on Learning Representations (ICLR) (2017)
10. Zhu, J.Y., Park, T., Isola, P., Efros, A.A.: Unpaired image-to-image translation using cycle-consistent adversarial networks. In: The IEEE International Conference on Computer Vision (ICCV) (2017)
11. Kim, T., Cha, M., Kim, H., Lee, J.K., Kim, J.: Learning to discover cross domain relations with generative adversarial networks. In: Proceedings of the 34th International Conference on Machine Learning, pp. 1857–1865 (2017)

Research on Human Interaction Recognition Algorithm Based on Interest Point of Depth Information Fusion

Yangyang Wang[1]([✉]), Xiaofei Ji[1], and Zhuangzhuang Jin[2]

[1] School of Automation, Shenyang Aerospace University,
Shenyang 110136, China
wyy2004101@163.com
[2] Liaoning Shihua University, Fushun 113001, China

Abstract. The traditional feature description methods of human interaction action based on RGB video are greatly affected by illumination change, object occlusion and environmental change. In this paper, depth and color information of the image are fused in the processing of feature extraction, and a novel human interaction recognition algorithm based on interest point of depth information fusion is proposed. Firstly, the spatio-temporal interest points (STIP) are extracted from the video, and the spatio-temporal interest points are processed hierarchically by the corresponding depth information. Secondly, three-dimensional scale invariant feature transform (3D SIFT) are used to obtain the feature description of hierarchical interest points, the model of the bag of words is constructed by using K-means clustering and Gaussian mixture clustering, and the representation of the training video is obtained based on dictionary projection. Finally, the support vector machine is used to classify different layers of video features respectively, then human interaction recognition is achieved by decision-level fusion of recognition probability for different layers of a video. The experimental results show that the accuracy of the human interaction recognition is improved by combing the depth information, and the proposed algorithm achieves 88.75% recognition accuracy in the SBU Kinect interaction dataset. The validity of the proposed method is proved.

Keywords: Interaction action recognition · Spatio-temporal interest points · Bag of words · Depth information · Decision-level fusion

1 Introduction

Human interaction recognition has recently been an active research in the field of computer vision [1, 2]. It has many applications such as public violent conflict monitoring, virtual reality and context-based video understanding [3]. Compared to single action recognition, human interaction recognition is more complex. Because more types of limb movements are involved, the coordination and arrangement between the limbs is also more diverse. Therefore, how to effectively extract action feature and establish a reasonable interaction model are the two main problems for human interaction recognition and understanding [4, 5].

© Springer Nature Singapore Pte Ltd. 2020
J. Qian et al. (Eds.): ICRRI 2020, CCIS 1335, pp. 98–109, 2020.
https://doi.org/10.1007/978-981-33-4929-2_7

To solve the above two problems, numerous human interaction representation and recognition methods are proposed. In the past ten years, the methods mainly focus on RGB videos. For example, Li et al. [6] proposed a hybrid feature to describe and recognize human interactions. Their hybrid feature consists of global motion and shape information, S-T correlation of local spatio-temporal interest points (STIP). A representation of 3D XYT spatio-temporal volume based on the co-occurrence of visual words has been proposed for interactions [7]. These methods considered co-occurrence relationship between the persons involved in the interaction. But only two-dimensional spatio-temporal information are provided in RGB videos, the correct interaction recognition rate is still relatively low in complex background.

Recently, the devices which can simultaneously capture visual RGB images and depth images are appeared, for example Kinect sensor produced by Microsoft. The emergence of these devices has significantly reduced the cost of deep information acquisition. Therefore, deep information has been widely introduced into interaction recognition. Song et al. [8] proposed an end-to-end spatial and temporal attention model based on human skeleton data. They built their model based on recurrent neural networks (RNNs) using long short-term memory (LSTM) method. The model learning was realized by regularized cross-entropy loss. Yun et al. [9] described features using geometric relational features based on the distance between all pairs of joints. The experiments proved that their proposed feature outperforms plane feature and velocity feature. Ji et al. [10, 11] firstly described the interactive body part with joint distance and joint motion. And then they built contrast mining model for human interaction recognition. The comparison proves that the methods based on depth images obtained higher recognition accuracy than the methods based on RGB images. However, there still exist some bottleneck problems, such as the extraction accuracy of skeleton data captured by sensors, the matching accuracy between the body joints.

Therefore, some scholars have tried to combine RGB data and depth data of videos to improve the accuracy of the interaction recognition. Zhu et al. [12] tested the performance of STIP features combined with depth map through a lot of experiments. Firstly they chose three STIP detectors to extract STIP features. Secondly, they built visual dictionary by K-means clustering to describe the features of human interaction. Lastly, they adopted SVM to recognize the interactions. Their method used the correspondence between the STIPs in RGB videos and the depth sequences, and removed the irrelevant interest points through human skeleton information. The experimental results showed that through the above two processes the validity of STIP features was improved, and a better recognition result was obtained. However, their method was poor in real-time. Because the method needed to compute the joint distance between the current pair-wise joints, the joint difference between current posture and the previous one, and the joint difference between current posture and the original one. All of these took a lot of time. Ijjina et al. [13] combined motion information and deep learning for interaction recognition. They built temporal templates to represent actions. Motion history image (MHI) and motion energy image (MEI) are computed as the weighted. The motion representation of RGB and depth video was given as input to a convolutional neural network (CNN) to extract ConvNet features. Finally, they adopted extreme learning machines (ELM) classifiers for recognition. Baradel et al. [14] proposed a two-stream method. They made use of multi-modal video data such as articulated pose and RGB frames to describe actions. In order to associate the convolutional layer with the abstract meaning layer, the joints of body are ordered on the basis of the topology of the human body. And they processed the raw RGB steam using a spatio-temporal soft-attention mechanism of the pose network. Each time the image locations were received as the input of LSTM network. The final fusion of LSTM features were finally done at the logical layer using a temporal attention mechanism. Lin et al. [15] proposed a deep model with radius–margin bound. They considered the temporal variations difference of the same type of actions which completed by different people. Therefore, they decomposed the actions into several sub-actions through latent variables. Each sub-action corresponded to a segment of the video, and it was mapped to a sub-network of the deep architecture. Then they regularized radius–margin and combined it with the deep feature learning. The above methods which combined RGB data and depth data effectively improved the generalization performance of classification. However, these methods took a lot of time to train the deep learning network. And the correlation information between the skeleton joint points was difficult to obtain.

Our proposed method is partially motivated by these works. We make full use of information complementarity of RGB image and depth image, and their respective advantages. In our proposed method, depth information and STIP features are fused for human interaction recognition. The main contributions of our work are two aspects. First, we propose a novel feature representation. The depth information and the color information of an image are fused in the process of feature extraction. Second, the STIP features are hierarchically represented. Through experiments we have proved that the discrimination of our descriptors is improved without increasing computational complexity. Our proposed method is simple to implement and finally achieves satisfactory recognition results.

2 The Framework of Proposed Method

A structure chart of our method is shown in Fig. 1. The main processes are of two parts: training phase and recognition phase.

2.1 Training Phase of Interest Point of Depth Information Fusion

Interest Point Detection and Laying. The descriptors based on the STIPs have been successful for RGB single action recognition [16]. The reasons are that the STIPs are robust to background clutter, viewpoint changes and occlusion [17]. And the STIPs can accurately locate the areas with obvious motion variance.

Therefore our method firstly detects the STIPs on training RGB video. Then we use the position information matching principle of depth image and RGB image to deal with the STIPs. The matched STIPs are divided into L layers according to the gray value of the depth information.

Feature Representation. The STIPs of L layers are described using 3-dimensional scale invariant feature transform (3D SIFT). For all of the training videos, the STIPs in

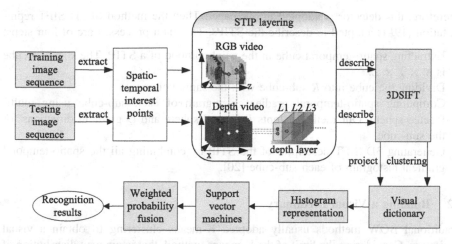

Fig. 1. The structure chart of our proposed method.

the same layer are clustered using K-means clustering and Gaussian mixture clustering method. The cluster centers of L layers are regarded as visual words.

For each training video, the 3D SIFT descriptors of L layers are projected into the visual dictionary of corresponding layer. Then we obtain a histogram representation of bag of words (BOW) for each layer, and send them to the support vector machine for training.

2.2 Recognition Phase of Interest Point of Depth Information Fusion

For testing image sequence, similar to training videos, STIP detection and hierarchical division are firstly performed. The next step is the STIP layering representation. The representation is called as hierarchical STIP descriptor. We classify the hierarchical STIP descriptor of the test video using the support vector machine classifier. The similarity probability of the action template is obtained. Finally the recognition results of L layer are fused through weighted probability. The fusion result is the final interaction recognition result.

3 Hierarchical STIP Representation with Depth Information

Compared traditional STIP representation based on BOW, our proposed method joined depth information. The STIPs are layered according to the gray value of the depth information.

3.1 STIP Detection and Description

Dollars et al. [18] detected STIPs through Gabor filter and Gaussian filter. They calculated the response value of a function based on the combination of Gabor and Gaussian filter. This method has been proved effective for action recognition.

Therefore, this detector is adopted in our paper. Then the method of 3D SIFT representation [19] is adopted to describe the STIPs. The main processes are of four steps:

- Extracting spatio-temporal cube in the neighborhood of a STIP. The size of a cube is $X \times Y \times Z$.
- Dividing the cube into R sub-cube with the same volume.
- Computing spatio-temporal gradient histogram of each sub-cube using multi-faceted sphere. Our method adopts P faceted sphere and Q gradient directions for the sub-cube.
- Generating 3D SIFT descriptor of the STIP by combining all the spatio-temporal gradient histogram of each sub-cube [20].

3.2 Building a Visual Dictionary

Traditional BOW methods usually adopted K-means clustering to obtain a visual dictionary. Considering the limit of the K-means method, that is, uneven distribution of the dictionary, our method uses Gaussian mixture clustering to process the hierarchical STIP features. Each layer of the STIP corresponds to a visual dictionary.

Gaussian mixture model is a clustering algorithm based on probability method [21]. It assumes that all data samples are generated from a multivariate Gaussian distribution of a given parameter. The Gaussian mixture model is defined as:

$$p(x) = \sum_{k=1}^{K} \pi_k N\left(x; \mu_k, \sum_k\right) \tag{1}$$

where π_k is weight factor, K is the number of the model, μ_k is mean, $\sum_k$ is standard deviation. Any Gaussian distribution $N(x; \mu_k, \sum_k)$ is called as a component of this model. Each component is a clustering center.

$$N\left(x; \mu_k, \sum_k\right) = \frac{1}{\sqrt{2\pi|\sum|}} \exp\left(-\frac{1}{2}(x-\mu)^T \sum^{-1}(x-\mu)\right) \tag{2}$$

This paper uses the maximum likelihood method for parameter estimation [22]. In this method the probability value is generally low. When the value of N is large, the result of the multiplication is very small, and it is easy to cause floating-point underflow problems. Therefore, formula (3) is adopted to solve this problem:

$$\max \sum_{i=1}^{N} \log\left(\sum_{k=1}^{K} \pi_k N\left(x; \mu_k, \sum_k\right)\right) \tag{3}$$

Formula (3) is solved using expectation maximization (EM) algorithm [23]. The solution process is as follows:

1. E-step: Calculation the probability of x which generated by each component. For each data x_i, its probability generated by the kth component is:

$$\gamma(i,k) = \frac{\pi_k N\left(x_i|\mu_k, \sum_k\right)}{\sum_{j=1}^{K} \pi_j N\left(x_i|\mu_j, \sum_j\right)} \tag{4}$$

2. M-step: The maximum likelihood estimate is obtained by maximizing μ_k and $\sum_k$ in formula (4):

$$\mu_k = \frac{1}{N_k} \sum_{i=1}^{N} \gamma(i,k) x_i \tag{5}$$

$$\sum_k = \frac{1}{N_k} \sum_{i=1}^{N} \gamma(i,k)(x_i - \mu_k)(x_i - \mu_k)^T \tag{6}$$

$$N_k = \sum_{i=1}^{N} \gamma(i,k) \tag{7}$$

3. Repeating the iteration in the above two steps until the value of the likelihood function converges.

3.3 Hierarchical Feature Representation

Firstly, according to the STIPs of the moving human body and the STIPs around the human body, the depth image is hierarchically processed based on gray level. Therefore, the position correspondence between depth information and color information is built. In this way, hierarchical information is mapped to the construction of STIP features in RGB image.

Secondly, the STIP features of each layer are clustered. M feature vector is obtained. That is, each clustering center is regarded as a word in the visual dictionary. Each feature vector is represented by a nearest word.

Finally, the frequency of the words in the video is computed to generate the histogram representation. The hierarchical feature representation is as follows:

$$L_1^Z = [Z_1^l, Z_1^u], L_2^Z = [Z_2^l, Z_2^u], \cdots, L_m^Z = [Z_m^l, Z_m^u] \tag{8}$$

$$h_m = \frac{1}{M} \sum_{Z(x_i) \in L_m^Z} v_i, \forall m = 1, 2, \cdots, M \tag{9}$$

Where l is the depth layer of multi-channel, h_m is the histogram representation of the mth channel.

4   Experiments and Results Analysis

4.1   Dataset and Experimental Settings

SBU Kinect interaction dataset [8] is used in our paper. This dataset for human interaction recognition has been acquired with a Kinect sensor. It contains depth images, RGB images, and skeleton images. Eight types of interactions are performed by 7 people in the same environment. These interactions consist of approaching, departing, kicking, punching, pushing, hugging, shaking hands and exchanging. These examples are shown in Fig. 2. This dataset is a challenging benchmark due to the similarity and non-periodicity of motions.

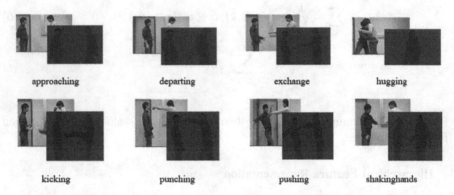

Fig. 2.   The examples of SBU Kinect interaction dataset.

The experiments were completed on the Matlab 2012 software platform with a frequency of 3.60 GHz, 8 GB of memory and 64-bit win7 operating system.

This paper adopts SVM to respectively obtain recognition probability of RGB feature and depth feature representations. The final result is the fusion of the two probabilities.

4.2   The Recognition Results of Different Gray Values

In our experiments, 10 videos are chosen for testing, the others for training. The parameters are chosen as follows: The size of the cube in the neighborhood of the STIP is $12 \times 12 \times 12$. And then the cube is divided into 2 sub-cubes using multi-faceted sphere. The faceted sphere and gradient directions are respectively 4 and 32. Therefore the feature dimension of a 3D SIFT descriptor of the STIP is 256. In the process of clustering, the first layer is to cluster each video. The number of the cluster centers is 50. The second layer is to cluster 8 kinds of videos, the number of the cluster centers is 25. The depth image is robust to illumination and environmental change. Therefore, in our paper the depth information is introduced to assist RGB image to cluster the STIPs. The recognition rate of different gray values is shown as Table 1. From Table 1 it can be seen that the discrimination of our representation is improved.

Table 1. The recognition accuracy of different gray values.

Experiments	Gray values of depth information	Recognition accuracy based on K-means clustering (%)	Fusion Result (%)	Recognition accuracy based on Gaussian clustering (%)	Fusion Result (%)
Experiment 1	$0 \leq x \leq 255$	76.25	/	77.50	/
Experiment 2	$0 \leq x <50$	65.00	88.75	77.50	88.75
	$50 \leq x \leq 255$	82.50		80.00	
Experiment 3	$0 \leq x <39$	61.25	78.75	65.00	80.00
	$39 \leq x <55$	62.50		61.25	
	$55 \leq x \leq 255$	68.75		77.50	

Experiment 1 shows the results of traditional STIP method without hierarchy. Experiment 2 and 3 show the results of our proposed hierarchical STIP representation method based on depth information. And the result of each layer is fused with weighted probability. In experiment 2, the basis for the hierarchical division is whether the points are inside the human body. In experiment 3, the STIPs are layered based on three types of locations, that is, inside the human body, on the human edge, and outside the human body. The recognition accuracy based on K-means clustering of experiment 2 and experiment 3 is 88.75% and 77.75%, respectively 12.5% and 2.5% higher than the rate of experiment 1 as shown in Table 1. The results prove that hierarchical STIP representation according to the gray value of the depth information is helpful to improve the recognition accuracy.

4.3   The Recognition Results of Decision-Level Fusion

Each grayscale range of experiment 2 and experiment 3 in Table 1 corresponds to a STIP layer. The recognition rate of each layer is respectively computed. Finally, decision-level fusion is adopted. The optimal weights are 0.29 and 0.71 in experiment 2. The optimal weights are 0.31, 0.29 and 0.40 in experiment 3. Figure 3 and Fig. 4 respectively give the confusion matrix after decision-level fusion of experiment 2 and experiment 3. It can also be concluded from Table 1 that the effect of building a dictionary with Gaussian mixture clustering is better than that of K-means clustering. Therefore, both the confusion matrix are computed based on Gaussian mixture clustering.

It can be observed in Fig. 3 the recognition accuracy gets improvement 11.25% on Gaussian mixture clustering compared with experiment 1. In Fig. 4 the accuracy also gets improvement 2.5%. The interaction "departing" is wrongly recognized into "approaching". The proportions of misrecognition are 40% and 50% respectively for experiment 2 and experiment 3. The reason may be that the two actions involve similar limb movements. The difference is that the order of limb separation and intersection is different. This increases recognition difficulty.

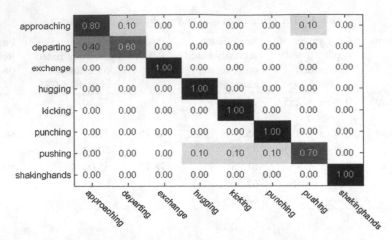

Fig. 3. The confusion matrix after decision-level fusion of experiment 2.

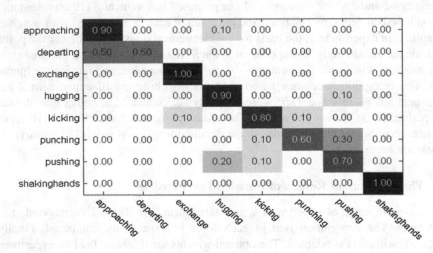

Fig. 4. The confusion matrix after decision-level fusion of experiment 3.

4.4   The Recognition Results Compared with Other Methods

Table 2 compares our recognition results with other methods for SBU Kinect interaction dataset.

As can be seen from Table 2, the recognition accuracy of our two layers STIPs based on depth information (experiment 2) is higher than the literature [9] and [11]. Our recognition accuracy is 88.75%, respectively 8.45% and 1.85% higher than the rate of literature [9] and [11] as shown in Table 2. The accuracy is comparable to that of the literature [10]. The accuracy of the literature [8, 13–15] is slightly higher than that of our paper. However, the above methods adopted deep learning, for example, literature [8] and [14] used LSTM, literature [15] used LCNN. The algorithms for obtaining the relationship of the human skeleton joints are complicated. A large number of samples

Table 2. The recognition accuracy of different methods.

Literature	Method	Accuracy (%)
Our Proposed Method	Two Layers STIPs + BOW + Depth Information	88.75
	Three Layers STIPs + BOW + Depth Information	78.75
Literature [9]	Raw Skeleton	49.70
	Joint Features	80.30
Literature [11]	Raw Skeleton + Contrast Mining	79.40
	Joint Features + Contrast Mining	86.90
Literature [10]	Joint Features + BOW	82.50
	Joint Features + CFDM	89.40
Literature [14]	LSTM	72.00
	Joint Features + LSTM	90.50
Literature [13]	MHI + MEI + CNN	90.98
Literature [8]	SA-LSTM	88.00
	TA-LSTM	89.00
	STA-LSTM	91.50
Literature [15]	Softmax + LCNN	92.40
	SVM + LCNN	92.80

are required for learning. Time complexity is high. And their system operations require higher hardware configuration. Our paper directly extracts RGB features according to depth information. The processing speed is only 15 fps, faster than those of deep learning methods.

5 Conclusion

Our paper makes full use of the complementarity of RGB images and depth images, proposes a human interaction recognition algorithm based on interest point of depth information fusion. Our method considers the discrimination of the STIP descriptor to local limbs in different gray levels. In the process of feature extract, depth information and RGB information are combined to match the position of STIPs. And then according to the gray value of the depth image, the matched STIPs are divided into L layers to obtain hierarchical feature representation. The experiments for SBU Kinect interaction dataset prove that our proposed method shows satisfying results. Our method does not need to use Kinect device to estimate the skeleton model. The processing speed is 15 fps. Therefore this method is simple and easy to implement. It provides a new solution for human interaction recognition.

Acknowledgements. This work is supported by the Scientific Research Youth Project of Education Department of Liaoning Province, China (No. L201745).

References

1. Slimani, K.N.E.H., Benezeth, Y., Souami, F.: Human interaction recognition based on the co-occurrence of visual words. In: Proceedings of the IEEE Conference on Computer Vision and Pattern Recognition Workshops, pp. 461–466. IEEE, Columbus, USA (2014)
2. Mukherjee, S., Biswas, S.K., Mukherjee, D.P.: Recognizing interaction between human performers using "key pose doublet". In: Proceedings of the 19th ACM International Conference on Multimedia, pp. 1329–1332. ACM, Scottsdale, USA (2011)
3. Ji, X., Qin, L., Zuo, X.: Human interaction recognition based on the co-occurring visual matrix sequence. In: Proceedings of the 12th International Conference on Intelligent Robotics and Applications, pp. 489–501. Springer, Shenyang, China (2019)
4. Kantorov, V., Laptev, I.: Efficient feature extraction encoding and classification for action recognition. In: Proceedings of the IEEE Conference on Computer Vision and Pattern Recognition, pp. 2593–2600. IEEE, Columbus, USA (2014)
5. Zhang, X., Cui, J., Tian, L.: Local spatio-temporal feature based voting framework for complex human activity detection and localization. In: Proceedings of the 1st Asian Conference on Pattern Recognition, pp. 12–16. IEEE, Beijing, China (2011)
6. Li, N., Cheng, X., Guo, H., Wu, Z.: A hybrid method for human interaction recognition using spatio-temporal interest points. In: Proceedings of the 22nd International Conference on Pattern Recognition, pp. 2513–2518. IEEE, Stockholm, Sweden (2014)
7. Slimani, K., Benezeth, Y., Souami, F.: Human interaction recognition based on the co-occurrence of visual words. In: Proceedings of the IEEE Conference on Computer Vision and Pattern Recognition Workshops, pp. 455–460. IEEE, Columbus, USA (2014)
8. Song, S., Lan, C., Xing, J., Zeng, W., Liu, J.: An end-to-end spatio-temporal attention model for human action recognition from skeleton data. In: Proceedings of the 31st AAAI Conference on Artificial Intelligence, pp. 4263–4270. AAAI, California, USA (2016)
9. Yun, K., Honorio, J., Chattopadhyay, D., Berg, T.L., Samaras, D.: Two person interaction detection using body-pose features and multiple instance learning. In: Proceedings of the IEEE Conference on Computer Vision and Pattern Recognition Workshops, pp. 28–35. IEEE, Providence, USA (2012)
10. Ji, Y., Cheng, H., Zheng, Y., Li, H.: Learning contrastive feature distribution model for interaction recognition. J. Vis. Commun. Image Rep. **33**, 340–349 (2015)
11. Ji, Y., Ye, G., Cheng, H.: Interactive body part contrast mining for human interaction recognition. In: IEEE International Conference on Multimedia and Expo Workshops, pp. 1–6. IEEE, Chengdu, China (2014)
12. Zhu, Y., Chen, W., Guo, G.: Evaluating spatiotemporal interest point features for depth-based action recognition. Image Vis. Comput. **32**(8), 453–464 (2014)
13. Ijjina, E.P., Chalavadi, K.M.: Human action recognition in RGBD videos using motion sequence information and deep learning. Pattern Recogn. **72**, 504–516 (2017)
14. Baradel, F., Wolf, C., Mille, J.: Pose-conditioned spatio-temporal attention for human action recognition. In: Proceedings of the IEEE Conference on Computer Vision and Pattern Recognition, pp. 1–10. IEEE, Honolulu, USA (2017)
15. Lin, L., Wang, K., Zuo, W., Wang, M., Luo, J., Zhang, L.: A deep structured model with radius–margin bound for 3D human activity recognition. Int. J. Comput. Vis. **118**(2), 256–273 (2016)
16. Ji, X., Zhou, L., Wu, Q.: A novel action recognition method based on improved spatio-temporal features and AdaBoost-SVM classifiers. Int. J. Hybrid Inf. Technol. **8**(5), 165–176 (2015)

17. Li, C., Su, B., Liu, Y., Wang, H., Wang, J.: Human action recognition using spatio-temporal descriptor. In: Proceedings of the 6th International Congress on Image and Signal Processing, pp. 107–111. IEEE, Hangzhou, China (2013)
18. Dollar, P., Rabaud, V., Cottell, G., Belongie, S.: Behavior recognition via sparse spatio-temporal features. In: Proceedings of the 2nd Joint IEEE International Workshop on Visual Surveillance and Performance Evaluation of Tracking and Surveillance, pp. 65–72. IEEE, Beijing, China (2005)
19. Ngoc, L.Q., Viet, V.H., Son, T.T., Hoang, P.M.: A robust approach for action recognition based on spatio-temporal features in RGB-D sequences. Int. J. Adv. Comput. Sci. Appl. 7(5), 166–177 (2016)
20. Ji, X., Wu, Q., Ju, Z., Wang, Y.: Study of human action recognition based on improved spatio-temporal features. Int. J. Autom. Comput. 5(11), 500–509 (2014)
21. Zhou, Y., Rangarajan, A., Gader, P.D.: A Gaussian mixture model representation of endmember variability in hyperspectralunmixing. IEEE Trans. Image Proces. 27(5), 2242–2256 (2018)
22. Gebru, I.D., Alameda-pineda, X., Forbes, F., Horaud, R.: EM algorithms for weighted-data clustering with application to audio-visual scene analysis. IEEE Trans. Pattern Anal Mach. Intell. 38(12), 2402–2415 (2016)
23. Watanabe, H., Muramatsu, S., Kikuchi, H.: Interval calculation of EM algorithm for GMM parameter estimation. In: Proceedings of IEEE International Symposium on Circuits and Systems: Nano-Bio Circuit Fabrics and Systems, pp. 2686–2689. IEEE, Grenoble, France (2010)

Pose Estimation for Planar Target Based on Monocular Visual Information

Yuqi Zhou[1], Hongwei Gao[1,2(✉)], Jian Sun[1], and Yueqiu Jiang[1]

[1] College of Automation and Electrical Engineering,
Shenyang Ligong University, Shenyang 110159, China
30963915@qq.com
[2] State Key Laboratory of Robotics, Shenyang Institute of Automation,
Chinese Academy of Sciences, Shenyang 110016, China

Abstract. Target pose estimation is a key technology in the field of artificial intelligence. This paper focuses on the pose estimation method based on monocular vision. Firstly, a cooperative target composed of five circular patterns is designed to increase the measurement accuracy. To solve the shortcomings of traditional image processing algorithms, we add two constraints, area threshold and ellipse roundness, to achieve more accurate feature extraction and segmentation. Then, the least square method is used to fit ellipses, and an ellipse sorting method is designed. Combined with the established models, the pose estimation of the target can be realized via the PNP problem. A large number of experiments prove the effectiveness and feasibility of the proposed method. Experiments show that the mean error for the attitude angle and position are less than 0.4° and 0.5 mm respectively. In a word, this method has the advantages of simple process, reasonable cost and high precision, and can be applied to engineering practice.

Keywords: Monocular vision · Pose estimation · Image processing · Recognition

1 Introduction

Pose estimation always has an essential task in different fields. Compared with the high cost of high-precision instruments, the measurement method based on the vision system is more prevalent [1–4].

According to the number of cameras, there are many kinds of vision system that can be used to estimate the target pose. Three-eye and four-eye visual techniques were employed to achieve the target measurement in Refs. [5, 6]. However, this kind of multi-camera measurement system is complicated in the system calibration, and the overall cost is high, and the flexibility is poor. The application of the binocular vision also has many disadvantages, such as complex system structure, high algorithm complexity and many limitations [7]. Therefore, a method by using the monocular vision system seems to be more practical [8, 9]. Schlobohm et al. [10] achieved pose information measurement based on contour features, which can overcome the lack of features on the object. However, this method may fail to detect changes in the depth

© Springer Nature Singapore Pte Ltd. 2020
J. Qian et al. (Eds.): ICRRI 2020, CCIS 1335, pp. 110–123, 2020.
https://doi.org/10.1007/978-981-33-4929-2_8

and angle, so it has many limitations and cannot be widely used. According to establish 2D-3D correspondences, the target pose information can be calculated by using a Perspective-n-Point (PnP) algorithm. Zhang et al. [11] proposed a new method to determine the pose of target by using circular markers. Long et al. [12] studied algorithms for the multipoint point camera pose estimation, and the feasibility was proved in a large number of experiments. In Ref. [13], Huang et al. used a dual circle with the different center to perform pose estimation. It is worth noting that these methods use different numbers of circles as the main components of the target pattern. Therefore, the two issues should also be considered in the pose estimation: (1) number of feature points; (2) accurate extraction of feature information [14].

In this paper, we propose a method based on the monocular vision to estimate the target pose information. In order to improve the accuracy of pose estimation, we design a particular recognition target with five circles. According to the targets used in this paper, we improve the traditional image preprocessing and feature extraction methods to ensure that the feature information can be accurately screened, such as adding constraints including roundness and area. Then, the 2D coordinates of the center point are obtained by the ellipse fitting method, and the accurate pose information can be calculated by combining with the solution of PnP problem. Finally, the feasibility and accuracy of the proposed method are proved by experiments.

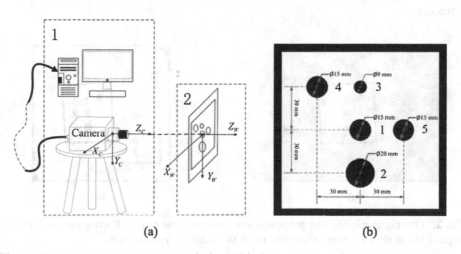

(a)   (b)

Fig. 1. The system structure and target design: (a) is the system structure; (b) illustrates the key parameters of the target.

2 Image Feature Detection and Recognition

The whole composition of this paper can be simply divided into two parts, as shown in Fig. 1(a): (1) image acquisition and processing part, which includes industrial computer, camera, data transfer lines and corresponding image processing programs; (2) target part, which is composed of the target designed in this paper and the object to be measured. When creating the target in this paper, it is necessary to follow the principles of the regular layout, simple pattern and easy installation. The target structure and key parameters can be shown in Fig. 1(b).

2.1   Image Preprocessing

Image preprocessing can enhance the recognition of features and blur the noise caused by uncontrollable factors such as the hardware manufacturing process. At the same time, it can avoid the interference of useless information produced by the complex background.

In this paper, median filtering is our first choice. Median filter can eliminate independent noises, and it can also preserve edges while removing noises [15].

In a laboratory environment with less scene variation, the image segmentation algorithm based on a fixed threshold can meet the requirements. We prefer that the proposed method can adapt to more complex environment, however, so the Otsu segmentation algorithm is used in this paper. In Fig. 2(b), we can see the edge information is roughly detected.

As can be seen in Fig. 2(b), there are still many interference areas, which are difficult to be eliminated by a single image preprocessing method such as morphological algorithm or connected area extraction. Therefore, after the connected region extraction, we add the constraints of area and roundness to complete the task of feature extraction.

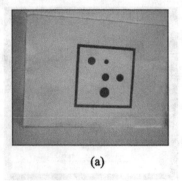
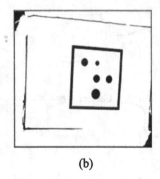

(a)   (b)

Fig. 2. Filtering processing and image segmentation: (a) is the median filtering operation of the original image after the gray processing; (b) is the result of segmentation.

The area threshold can help us to extract the target pattern initially. We count the pixel areas contained in all the outermost contour of the image, and then complete the screening by setting high and low thresholds, which are obtained by experiments in the range of camera height.

Then, we calculate the minimum circumscribed rectangles of all contours in the outer contour that meets the requirements, and then complete the more accurate feature extraction through ellipse roundness.

$$\mu = \begin{cases} \frac{H}{W}, & \text{if } H < W \\ \frac{W}{H}, & \text{if } H > W \end{cases} \quad (1)$$

On the equation, W is the width of the minimum circumscribed rectangle, and H is the height. μ is the ellipse roundness. In addition, we set this threshold to 0.67 and remove the corresponding contour when μ is below 0.67. By completing the above two constraint steps, the feature information can be accurately and completely extracted, as shown in Fig. 3(a). In order to facilitate the following ellipse fitting, we also perform edge detection based on canny algorithm [16], as shown in Fig. 3(b).

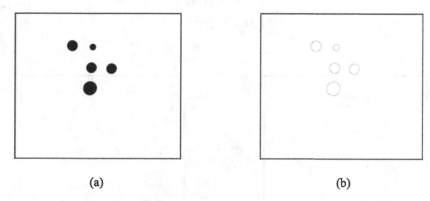

(a)   (b)

Fig. 3. Results of adding constraints and edge detection: (a) is the feature extraction operation; (b) represents Canny algorithm operation.

2.2 Ellipse Fitting and Ellipse Center Acquisition

After the above operations, we use a non-iterative method proposed by Ref. [17] to achieve the ellipses fitting and parameters solution.

$$\frac{[(x_i - x_{Cet})\cos\alpha + (y_i - y_{Cet})\sin\alpha]^2}{r_1^2} + \frac{[(x_{Ctt} - x_i)\sin\alpha + (y_i - y_{Cet})\sin\alpha]^2}{r_2^2} = 1 \quad (2)$$

where (x_i, y_i) is the coordinate value of the edge on the ellipse, r_1 and r_2 are the semi-major axis and semi-minor axis respectively. α is the angle between the major axis and X axis, and (x_{Cet}, y_{Cet}) is the center point. Equation (2) is usually rewritten as second order polynomial form.

$$ax_i^2 + bx_i y_i^2 + cy_i^2 + dx_i + ey_i + f = 0 \quad (3)$$

The constraint $4ac - b^2 = 1$ can be used to an ellipse-specificity of the solution. In Eq. (3), we define $Z = [a, b, c, d, e, f]^T$ and $X = [x_i^2, x_i y_i^2, y_i^2, x_i, y_i, 1]^T$. The algebraic distance from the point (x_i, y_i) to the ellipse can be recorded as $F_1(Z, X_i)$ [18].

Therefore, a least squares fitting based on the algebraic criterion can be calculated as follows [19].

$$\min_{Z} = \sum_{i=1}^{n} (F_1(Z, X_i))^2 = \sum_{i=1}^{n} \left(ax_i^2 + bx_iy_i^2 + cy_i^2 + dx_i + ey_i + f \right)^2 \qquad (4)$$

After determining the all coefficient of the elliptic equation [20], we can calculate the central coordinates $\{C_i = (x_{Ceti}, y_{Ceti})\}_{i=1}^{5}$ of all ellipses in the feature image by using Eq. (3) [21]. Figure 4(a) shows the results of these operations.

$$\begin{cases} x_{Ceti} = \frac{2cd - be}{b^2 - 4c} \\ y_{Ceti} = \frac{2e - bd}{b^2 - 4c} \end{cases} \qquad (5)$$

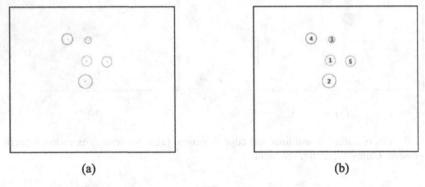

| (a) | (b) |

Fig. 4. The result of ellipse information processing: (a) represents ellipse fitting and center point extraction; (b) adopts the sorting method of this paper

In the process of pose estimation, it is required to provide the accurate and ordered coordinates of the center of the ellipse, and also to provide the world coordinates of the corresponding center point. Therefore, the operation for storing the central coordinates in a certain order is essential. In this paper, we design a feature ellipses sorting method according to the target we used, as shown in Fig. 4(b).

1. We construct the minimum enclosing rectangle according to the center coordinates of all ellipses. Then, the mass center coordinate of the rectangle can be obtained by using Eq. (5), and record it as C_0.

$$\begin{cases} X_{MC} = \frac{1}{n} \sum_{i=0}^{n} x_{Recti} \\ Y_{MC} = \frac{1}{n} \sum_{i=0}^{n} y_{Recti} \end{cases} \qquad (6)$$

where (X_{MC}, Y_{MC}) is the mass center, (x_{Recti}, y_{Recti}) is the vertex coordinate of the smallest bounding rectangle, and we set $n = 4$. Then, the distance from the center point of each ellipse to C_0 can be calculated, and the ellipse corresponding to the center point of the minimum distance is determined as the first ellipse.

3. Area value is another important constraint in our method. We count the pixel area occupied by the remaining ellipse, and mark the ellipse corresponding to the maximum area value as the second one. In contrast, the ellipse with the smallest area is regarded as the third.
4. The center point C_2 is used as a reference position. In the distance from the center of all ellipses to C_2, the ellipse corresponding to the maximum distance is considered as the fourth ellipse.

3   Target Pose Estimation

3.1   Camera Model and Calibration

The camera model is associated with the transformation between coordinate systems. In this paper, we establish the coordinate relationship, as shown in Fig. 5.

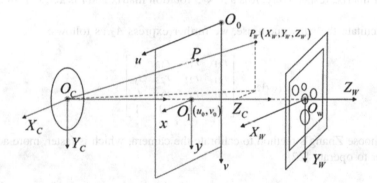

Fig. 5. The coordinate relations involved in this paper: $O_W X_W Y_W Z_W$ is the world coordinate system, $O_c X_c Y_c Z_c$ is the camera coordinate system, $O_1 xy$ is the image coordinate system, and $O_0 uv$ is the pixel coordinate system

As shown in Fig. 5, the image coordinate system is converted to the pixel coordinate system by using Eq. (7).

$$\begin{bmatrix} u \\ v \\ 1 \end{bmatrix} = \begin{bmatrix} 1/dx & 0 & u_0 \\ 0 & 1/dy & v_0 \\ 0 & 0 & 1 \end{bmatrix} \begin{bmatrix} x \\ y \\ 1 \end{bmatrix} \tag{7}$$

(X_C, Y_C, Z_C) is the representation of the space point P_W in the camera coordinate. According to the similar triangle principle, the relationship between the point in the camera coordinate system and the image coordinate system can be expressed as follows:

$$\begin{cases} x = f\frac{X_C}{Z_C} \\ y = f\frac{Y_C}{Z_C} \end{cases} \tag{8}$$

where f is the camera focus length, and the relationship between point P_W in the world coordinate system and the projected point P in image coordinate system can be expressed by Eq. (9)

$$Z_c \begin{bmatrix} u \\ v \\ 1 \end{bmatrix} = A_1 A_2 \begin{bmatrix} X_w \\ Y_w \\ Z_w \\ 1 \end{bmatrix} = \begin{bmatrix} \frac{f}{dx} & 0 & u_0 & 0 \\ 0 & \frac{f}{dy} & v_0 & 0 \\ 0 & 0 & 1 & 0 \end{bmatrix} \begin{bmatrix} R & t \\ 0^T & 1 \end{bmatrix} \begin{bmatrix} X_w \\ Y_w \\ Z_w \\ 1 \end{bmatrix} \tag{9}$$

where $\frac{f}{dx} = f_x$ and $\frac{f}{dy} = f_y$ are the scale factor on the axial direction, and (u_0, v_0) is the principle point. A_1 and A_2 are the camera's internal parameter matrix and external parameter matrix, respectively. R is a 3×3 rotation matrix and t is a 3×1 translation vector.

To facilitate the subsequent use, we further express A_2 as follows:

$$A_2 = \begin{bmatrix} r_{11} & r_{12} & r_{13} & t_x \\ r_{21} & r_{22} & r_{23} & t_y \\ r_{31} & r_{32} & r_{33} & t_z \\ 0 & 0 & 0 & 1 \end{bmatrix} \tag{10}$$

We choose Zhang's method to calibrate the camera, which is faster, more accurate and easier to operate [22].

3.2   Pose Estimation Method

In the above chapters, we have completed the filtering and extraction of feature points on the target and obtained the pixel coordinates of feature points. At the same time, the coordinates of feature points in the world coordinate system are known. Therefore, we can calculate the pose of the target with feature points. As the number of features increases, the accuracy of the computed results will increase. In this paper, we set five feature points in the target, which can not only ensure the flexibility of the target but also improve the overall anti-interference and accuracy.

$$\begin{cases} X_{Ci} = r_{11}X_{Wi} + r_{12}Y_{Wi} + r_{13}Z_{Wi} + t_x \\ Y_{Ci} = r_{21}X_{Wi} + r_{22}Y_{Wi} + r_{23}Z_{Wi} + t_y \\ Z_{Ci} = r_{31}X_{Wi} + r_{32}Y_{Wi} + r_{33}Z_{Wi} + t_z \end{cases} \tag{11}$$

where (X_{Ci}, Y_{Ci}, Z_{Ci}) and (X_{Wi}, Y_{Wi}, Z_{Wi}) are the coordinates of the feature point $P_i (i = 5)$ in the camera coordinate system and the world coordinate system, respectively. r_{11} to r_{33} are the constituent elements of the rotation matrix R in Eq. (10), and t_x, t_y and t_z are the translation vector t in Eq. (10).

By using Eq. (8), we can obtain the coordinates of the feature point P_i on the normalized plane of focal length, and these relationships can be used in Eq. (9) to get the system of equations like Eq. (12).

$$\begin{cases} r_{11}X_{Wi} + r_{12}Y_{Wi} - r_{31}X_{1Ci}X_{Wi} - r_{32}X_{1C_i}Y_{Wi} - X_{1Ci}t_z + t_x = 0 \\ r_{21}X_{Wi} + r_{22}Y_{Wi} - r_{31}Y_{1Ci}X_{Wi} - r_{32}Y_{1Ci}Y_{Wi} - Y_{1Ci}t_z + t_y = 0 \end{cases} \tag{12}$$

where (X_{1Ci}, Y_{1Ci}) represents the coordinates of P_i on the normalized plane. It is worth noting that we take a certain feature point as the origin of the world coordinate system, and these feature points P_i belong to a same plane. Therefore, we set $Z_{Wi} = 0$ and use it in subsequent operations.

When the number of points with known three-dimensional coordinates is n, the number of equations is $2n$. In this paper, we set $n = 5$, and Eq. (12) can be further expressed as follows:

$$S_1 H_1 + S_2 H_2 = \begin{bmatrix} X_{W1} & 0 & -X_{1C1}X_{W1} \\ 0 & X_{W1} & -Y_{1C1}X_{W1} \\ \vdots & \vdots & \vdots \\ X_{Wn} & 0 & -X_{1Cn}X_{Wn} \\ 0 & X_{Wn} & -Y_{1Cn}X_{Wn} \end{bmatrix} \begin{bmatrix} r_{11} \\ r_{21} \\ r_{31} \end{bmatrix} + \begin{bmatrix} Y_{W1} & 0 & -X_{1C1}Y_{W1} & 1 & 0 & -X_{1C1} \\ 0 & Y_{W1} & -Y_{1C1}Y_{W1} & 0 & 1 & -Y_{1C1} \\ \vdots & \vdots & \vdots & \vdots & \vdots & \vdots \\ Y_{Wn} & 0 & -X_{1Cn}Y_{Wn} & 1 & 0 & -X_{1Cn} \\ 0 & Y_{Wn} & -Y_{1Cn}Y_{Wn} & 0 & 1 & -Y_{1Cn} \end{bmatrix} \begin{bmatrix} r_{12} \\ r_{22} \\ t_x \\ t_y \\ t_z \end{bmatrix} \tag{13}$$

where S_1 is a 10×3 matrix, and S_2 is a 10×6 matrix. H_1 can provide the constraint condition for solving Eq. (13), which is $\|r_{11}^2 + r_{21}^2 + r_{31}^2\| = 1$.

The solution of Eq. (13) can be transformed into an optimized solution process by using an optimization function F_2.

$$\min F_2 = \|S_1 H_1 + S_2 H_2\|^2 + \delta\left(1 - \|r_{11}^2 + r_{21}^2 + r_{31}^2\|^2\right) \tag{14}$$

where δ is a random value, and H_1 and H_2 can be determined by Eq. (15).

$$\begin{cases} \left[S_1^T S_1 - S_1^T S_2 (S_2^T S_2)^{-1} S_2^T S_1\right] H_1 = \delta H_1 \\ H_2 = -(S_2^T S_2)^{-1} S_2^T S_1 H_1 \end{cases} \tag{15}$$

where H_1 is the eigenvector corresponding to the minimum eigenvalue of the matrix $\left[S_1^T S_1 - S_1^T S_2 (S_2^T S_2)^{-1} S_2^T S_1\right]$.

After the calculation of H_1 and H_2, the external parameter matrix $^c B_W$ can be obtained, which represents the camera coordinate system relative to the world coordinate system. Moreover, the third column of matrix $^c B_W$ can be obtained by computing the cross product of the first column and the second column, and $^c B_W$ is a rough description of pose information. $^c B_W$ consists of the rotation matrix and translation vector. To improve accuracy, the RLS (Recursive Least Square) method is used to realize the optimization [23]. In addition, the attitude can be expressed by three rotation angles: roll angle α, pitch angle β and yaw angle γ.

4   Experimental Results

To verify the validity of the method presented in this paper, we built an experimental platform. We choose the CMOS color camera produced by Basler, which has a maximum resolution of 2448 × 2048, and the frame rate is 23 fps. Besides, we use the 12 mm fixed focus lens. During camera calibration, the calibration board has the size of 400 × 300 mm with 12 × 9 array with 15 mm of each chessboard, and its accuracy is ±0.01 mm. All the programs and data are processed by a computer with an Intel Core i7 4710MQ CPU and 16 GB RAM. The software environment of this paper is Windows 7 system (Fig. 6).

(a)   (b)

Fig. 6. Monocular camera: (a) is the 12 mm fixed focus lens; (b) is the CMOS camera

Fig. 7. Calibration images.

4.1   Camera Calibration Experiment

We use the visual system to collect at least 21 images with the calibration plate in different positions, and part of the collected images is shown in Fig. 7. According to these images, we can obtain the camera parameters as shown in Table 1. $(k_1, k_2, k_3, p_1, p_2)$ are the lens distortion coefficients [24]. It is worth noticing that the principle point (u_0, v_0) is close to half the maximum resolution of the camera.

Table 1. Results of Camera calibration

(u_0, v_0)	(f_x, f_y)	(k_1, k_2, k_3)	(p_1, p_2)
(1222.68, 1000.14)	(3739.7, 3739.63)	(−0.0964, 0.3263, 0.3086)	(−0.0004, −0.0011)

In Fig. 8, the reprojection error is displayed in color-coded form of the cross, and this error can be used as the evaluation standard of the camera calibration. It can be seen that the error values are densely distributed at (0, 0), and this result also proves the accuracy of our calibration. At the same time, we can know that the mean calibration error is 0.0763 mm, and the maximum error is only 0.0911 mm. They are all less than 0.1 mm. Therefore, the calibration results can meet our requirements.

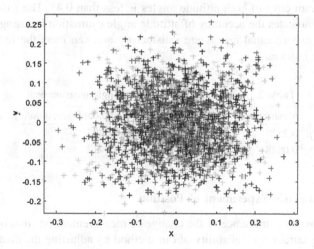

Fig. 8. Reprojection error.

4.2   Error Analysis Experiment of Attitude Angle

The distance between the camera and the target is 500 mm in the attitude measurement experiment. In addition, we set the initial attitude: $\alpha = -178.36°$, $\beta = 3.55°$ and $\gamma = 1.93°$. Considering the existing experimental conditions, only attitude angle β and γ are analyzed in this paper. The angle ranges of γ and β are $[-30°, 30°]$ and $[-10°, 16°]$, respectively. When measuring the attitude angle, it is necessary to ensure that only one axis changes its angle.

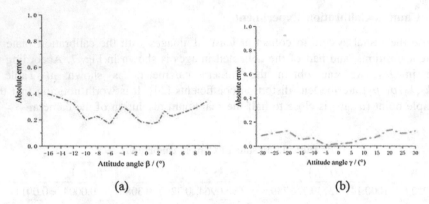

Fig. 9. Attitude angle solution error: (a) attitude angle β; (b) attitude angle γ.

In Fig. 9, we can see the error of β is proportional to the change of angle. With the increase of rotation angle, the target has greater distortion in the imaging plane, which can directly lead to more deviation in (x_{Cet}, y_{Cet}) and influence the measurement of β. By comparison, the error of attitude angle γ is small and the change is gentle.

Table 2 is the error analysis of attitude angle, from which we can see that the mean errors of attitude angles β and γ are 0.2579 and 0.0828 respectively. It should be noted that the maximum error of both attitude angles is less than 0.4°. The smaller standard deviation also indicates the accuracy of attitude angle estimation by using our method. Therefore, the experimental results are satisfactory and can meet the requirements of most industrial indexes.

Table 2. Error analysis of attitude Angle measurement

	Mean error/(°)	Maximum error/(°)	Standard deviation/(°)
β	0.2579	0.3969	0.0703
γ	0.0828	0.1276	0.0412

4.3   Error Analysis Experiment of Position

In order to analyze the influence of the change of measurement position on the system, we verify the accuracy and reliability of our method by adjusting the distance between the camera and the target. In this study, the measurement distance is the primary factor that can affect the accuracy of target position estimation.

In this paper, the camera moves along the Z_c axis in the range of 400 mm to 800 mm, and there is no axial movement occurs during the experiment. In addition, we set 510 mm as the reference point, and the sampling interval distance within the moving range is 10 mm. In order to avoid the randomness of the data, each sampling point is measured three times, and the average value is taken as the final measurement value.

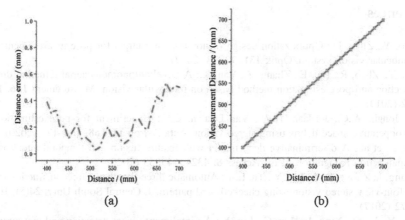

(a)   (b)

Fig. 10. Position solution error: (a) is the error of working distance; (b) represents the comparison between the measured value and actual value.

Figure 10 shows that the error tends to become larger with the increase or decrease of the measurement distance, and the probability of the maximum error appearing at the boundary of the measurement range is higher. The maximum error is only 0.4996 mm. Moreover, the mean error is only 0.2360 mm, and the standard deviation is 0.1463 mm. This level of error can meet our requirements for the ranging and localization.

5 Conclusion

In this paper, we propose a method based on the monocular vision to estimate the target pose information. Based on the monocular vision theory, the coordinate system model of the measurement system can be established. After the target image processing method proposed in this paper, the non-iterative ellipse fitting method is used to achieve the ellipse fitting and extract the center point of the ellipse. According to the PnP problem, the pose of the target can be estimated accurately.

The innovations and contributions of this paper are as follows: (a) we improve the traditional image processing method and add two constraints on the basis of Otsu method to better extract the feature information, and we study the sorting method of feature points in the target. (b) in order to improve the accuracy of pose estimation, we design a corresponding cooperative target. In addition, we build the monocular vision platform to verify the proposed method. Through the analysis of the experimental results, our method is accurate and effective, and the errors of the position and attitude angle are small, which can meet the needs of high precision. In a word, this method can be applied in practice, which is beneficial to the intelligent and unmanned industrial production. However, there are still some shortcomings in this paper. We pay more attention to the improvement of image processing algorithms. In the future research work, we will add more comparison of pose estimation experiments.

References

1. Guo, Y., Zhao, D.: Optimization design of non-coplanar target for pose measurement with monocular vision system. Optik **131**, 72–78 (2017)
2. Ma, Y., Zhao, R., Liu, E., Zhang, Z., Yan, K.: A novel autonomous aerial refueling drogue detection and pose estimation method based on monocular vision. Measurement **136**, 132–142 (2019)
3. De Jongh, W.C., Jordaan, H.W., Van Daalen, C.E.: Experiment for pose estimation of uncooperative space debris using stereo vision. Acta Astronaut. **168**, 164–173 (2020)
4. Yu, J., et al.: A discriminative deep model with feature fusion and temporal attention for human action recognition. IEEE Access **8**, 43243–43255 (2020)
5. Xiong, J., Zhong, S.-d., Liu, Y., Tu, L.-f.: Automatic three-dimensional reconstruction based on four-view stereo vision using checkerboard pattern. J. Central South Univ. **24**(5), 1063–1072 (2017)
6. Ma, Y., Li, Q., Xing, J., Huo, G., Liu, Y.: An intelligent object detection and measurement system based on trinocular vision. IEEE Trans. Circ. Syst. Video Technol. **30**(3), 711–724 (2020)
7. Yin, Y., Altmann, B., Pape, C., Reithmeier, E.: Machine-vision-guided rotation axis alignment for an optomechanical derotator. Optics Lasers Eng. **121**, 456–463 (2019)
8. Cui, J.-s., Huo, J., Yang, M., Wang, Y.-k.: Research on the rigid body posture measurement using monocular vision by coplanar feature points. Optik **126**(24), 5423–5429 (2015)
9. Sharma, S., D'Amico, S.: Comparative assessment of techniques for initial pose estimation using monocular vision. Acta Astronautica **123**, 435–445 (2016)
10. Schlobohm, J., Pösch, A., Reithmeier, E., Rosenhahn, B.: Improving contour based pose estimation for fast 3D measurement of free form objects. Measurement **92**, 79–82 (2016)
11. Zhang, S.N., Wang, B.S., Yuan, W.Q.: A method of determining the pose of planar target in vision detection system. Adv. Mater. Res. **659**, 156–161 (2013)
12. Long, Q., Zhongdan, L.: Linear N-point camera pose determination. IEEE Trans. Pattern Analysis Mach. Intell. **21**(8), 774–780 (1999)
13. Huang, B., Sun, Y., Zhu, Y., Xiong, Z., Liu, J.: Vision pose estimation from planar dual circles in a single image. Optik **127**(10), 4275–4280 (2016)
14. Safaee-Rad, R., Tchoukanov, I., Smith, K.C., Benhabib, B.: Three-dimensional location estimation of circular features for machine vision. IEEE Trans. Robot. Autom. **8**(5), 624–640 (1992)
15. Abid Hasan, S.M., Ko, K.: Depth edge detection by image-based smoothing and morphological operations. J. Comput. Des. Eng. **3**(3), 191–197 (2016)
16. Arunkumar, P., Shantharajah, S.P., Geetha, M.: Improved canny detection algorithm for processing and segmenting text from the images. Cluster Comput. **22**(3), 7015–7021 (2018). https://doi.org/10.1007/s10586-018-2056-8
17. Fitzgibbon, A., Pilu, M., Fisher, R.B.: Direct least square fitting of ellipses. IEEE Trans. Pattern Anal. Mach. Intell. **21**(5), 476–480 (1999)
18. Liang, J., et al.: Robust ellipse fitting based on sparse combination of data points. IEEE Trans. Image Process. **22**(6), 2207–2218 (2013)
19. Yu, J., Kulkarni, S.R., Poor, H.V.: Robust ellipse and spheroid fitting. Pattern Recogn. Lett. **33**(5), 492–499 (2012)

20. Zhang, G., Jayas, D.S., White, N.D.G.: Separation of touching grain kernels in an image by ellipse fitting algorithm. Biosys. Eng. **92**(2), 135–142 (2005)
21. Wu, W.-Y., Wang, M.-J.J.: Elliptical object detection by using its geometric properties. Pattern Recogn. **26**(10), 1499–1509 (1993)
22. Zhang, Z.: A flexible new technique for camera calibration. IEEE Trans. Pattern Anal. Mach. Intell. **22**(11), 1330–1334 (2000)
23. Choi, M., Oh, J.J., Choi, S.B.: Linearized recursive least squares methods for real-time identification of tire-road friction coefficient. IEEE Trans. Veh. Technol. **62**(7), 2906–2918 (2013)
24. Ye, N., Yang, B., Zhou, H., Zhang, L.: A calibration trilogy of monocular-vision-based aircraft boresight system. Measurement **117**, 133–143 (2018)

Context-Aware Convolutional Neural Network for Single Image Super-Resolution

Xiaotian Wang, Yan Zhang[✉], and Ziwei Lu

Liaoning Shihua University, Fushun 113001, China
976802661@qq.com

Abstract. Super-resolution is an algorithm for reconstructing high-resolution image from low-resolution image. It is one of the more active research topics in the field of computer vision. In recent years, convolutional neural network has made much progress in single image super-resolution. However, the super-resolution algorithms based on convolution network is still difficult to accurately reconstruct the image. In order to improve the reconstruction performance, it is essential to consider the channel information and spatial information of convolution network. We propose a model based on convolutional neural network and attention mechanism, which utilize channel attention modules and spatial attention modules to enhance the flow of information. The convolution network which combines the channel and spatial context information can enhance the network performance. Our model has better ability to utilize context features to more effectively reconstruct the image. Experimental results on several widely used datasets show that our model achieves better reconstruction performance than other model.

Keywords: Super-resolution · Attention mechanism · Deep learning

1 Introduction

Super-resolution (SR) is an algorithm for reconstructing high-resolution (HR) image from low-resolution (LR) image. High-resolution images provide more visual information than low-resolution images. However, limited to hardware and imaging conditions, it's hard to obtain high-resolution images. Super-resolution algorithms infer high-resolution images from low-resolution images. It is one of the more active research topics in the field of computer vision. The reconstructing the image should be close to the real image, which is a challenging task.

Classic super-resolution reconstruction methods such as bicubic and nearest neighbor are still in use today due to their low computational complexity and cost. However, the results of these methods still lack sufficient high-frequency details. In recent years, with the rapid development of deep learning technology, super-resolution reconstruction tasks based on convolutional neural network have become a research hotspot. Deep learning methods usually consist of 2 parts: feature extraction and upscaling. They showed better performance than the classical SR methods.

Convolutional neural networks have stronger learning capabilities than shallower models, and can more accurately reconstruct images. Early deep learning methods

© Springer Nature Singapore Pte Ltd. 2020
J. Qian et al. (Eds.): ICRRI 2020, CCIS 1335, pp. 124–134, 2020.
https://doi.org/10.1007/978-981-33-4929-2_9

usually up-scale LR images to HR image size using bicubic interpolation. The upscaled LR image is used as the input to the neural network to reconstruct the image. But upsampling may damage the key information of the original LR image. The original LR image may also not be suitable for upsampling. So, it has a negative impact on the feature extraction of convolutional neural networks. In this paper, we propose a new super-resolution model based on convolutional neural networks. We combine attention mechanism with convolution network.

2   Related Work

Image super-resolution algorithms can generally be divided into the following three categories: interpolation-based [1, 2], reconstruction based [3, 4], and learning-based methods [5, 6]. Linear, bicubic, and Lanczos filtering are fast to compute, but usually produce smooth textures. Convolutional sparse coding method [7] is proposed to reconstruct the real texture while avoiding edge artifacts. To further improve computational efficiency, cluster the LR feature space into numerous subspaces and learn simple mapping functions for each subspace.

This paper focuses on learning-based methods. Its basic idea is to establish a non-linear mapping from LR images to HR images. In recent years, super-resolution methods based on deep learning have achieved excellent performance. SRCNN [8] was proposed by Dong et al. A neural network with 3 convolutional layers learns the end-to-end mapping between LR and HR images. The LR image needs to be interpolated to the HR image size. VDSR [9] has been proposed by Kim et al. It has a deeper network structure and uses residual learning to better restore high-frequency details of the image. RD-IDN [10] employ dense skip connections to improve the learning ability of network.

Attention model refers to human visual attention mechanism. People often only focus on a part of the scene or picture to get important information. The key of attention model is to focus only on the meaningful input and allocate the limited resources more reasonably. Woo et al. propose an effective attention module for convolutional neural networks, which uses channel attention and spatial attention modules to filter features. It can enhance features that are beneficial to network performance and make the network perform better.

3   The Proposed Method

3.1   Network Architecture

Convolutional neural network learns the mapping of LR image and HR image to reconstruct the image. Due to the excellent performance of neural network, super-resolution can be accomplished through a single neural network:

$$I_{HR} = F(I_{LR}) \tag{1}$$

Where I_{HR} and I_{LR} denoted as HR and LR image, and function $F(\cdot)$ indicates the neural network to accomplish super-resolution task.

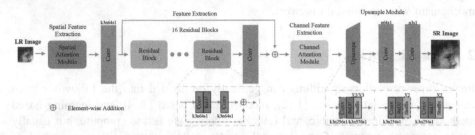

Fig. 1. Network architecture of our proposed model with convolutional kernel size (k), number of feature maps (n) and stride (s).

As shown in Fig. 1, our proposed method consists of four parts: spatial feature extraction, feature extraction, channel feature extraction and upsample module. Firstly, we use spatial attention module to extract shallow feature F_C with spatial context information:

$$F_S = S_I(I_{LR}) \tag{2}$$

where function S_I represent operations of spatial attention module the input layer. I_{LR} denotes LR image. F_S is then enhanced by one convolutional layer (Conv):

$$F_I = H_I(F_S) \tag{3}$$

Where H_I denotes convolution operation of the shallow feature F_S. Then, we construct n residual blocks to extract feature map. In this paper we define n = 16. Each residual block contains 2 convolutional layers. The first residual block is employed to generate feature F_{R1} based on F_I:

$$F_{R_1} = R_1(F_I) + F_I \tag{4}$$

Where function $R_1(\cdot)$ represent operation of residual block in the first residual block. F_I represents short skip connection. The short connections allow better flow of low-frequency features. We construct n residual block:

$$F_{R_n} = R_n(F_{R_{n-1}}) + F_{R_{n-1}} \tag{5}$$

Where $F_{R_{n-1}}$ represent short skip connection. After processing of n residual block. Then, we use channel attention module to extract feature F_S with channel context information.

$$F_C = S_C(H_c(F_{R_n}) + F_I) \tag{5}$$

where function S_C represent operations of channel attention module. $H_c(\cdot)$ denotes convolution operation of the last residual block. Respectively, the last term of F_I represents long skip connection. Finally, an upsampling operation and two Conv are used to obtain SR image:

$$I_H = H_{c2}(H_{c1}(H_U(F_C))) \tag{6}$$

Where function H_{c1}, H_{c2} represent operation of Conv. H_U denotes operations of upsampling. In upsampling layer, for $\times 2/\times 3$ super-resolution, we use PixelShuffle $\times 2/\times 3$. In the $\times 4$ model, the upsampling layer is changed to the cascading of two Shuffle $\times 2$.

3.2   Spatial Feature Extraction

The edge or texture area usually contains more high-frequency information, while the smooth area contains more low-frequency information. The spatial attention module makes the network focus on more important areas and helps the network to distinguish between different areas.

First input the LR image into the convolutional layer, which could be represented as:

$$F_s = \text{Conv}(F_{LR}) \tag{7}$$

Where F_{LR} represents LR image. Four spatial attention blocks are performed on input feature map. Firstly, each spatial attention block uses 2 convolutional layers to extract features. Then, we use 2 convolutional layers and sigmoid function to generate the spatial attention weight. The last convolutional layer is used to squeeze global spatial information into one channel. Spatial attention weight thus could be computed as:

$$S(F_i) = \text{sigmoid}(\text{Conv}_1(\text{relu}(\text{Conv}_2(F_s)))) \tag{8}$$

The $Conv_1(\cdot)$ and $Conv_2(\cdot)$ output feature map with $1 \times m \times n$ and $C \times m \times n$

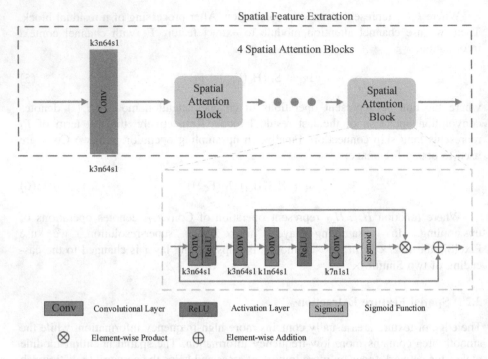

Fig. 2. Channel attention module, which contains 1 convolution layer and 4 channel attention blocks.

3.3   Channel Feature Extraction

Not all channel-wise features are important for recovering high-frequency details. We utilize the channel attention module to exploit cross-channel relationship.

As shown in Fig. 2, Conv is performed on input feature map. Then, we construct 4 channel attention blocks to offers an intuitive descriptor on inherent context property among different feature channels. In each channel attention block, we construct two Conv to extract feature map F_i:

$$F_c = \text{Conv}(\text{relu}(\text{Conv}(F_i))) \tag{9}$$

Where $Conv(\cdot)\cdot$ and $relu(\cdot)$ represent operation of convolution layer and relu activation function. Then we utilize global pooling to output feature map F_{avg} with size $C \times 1 \times 1$:

$$F_{avg} = \text{Avg}(Fc) \tag{10}$$

Channel Attention Module

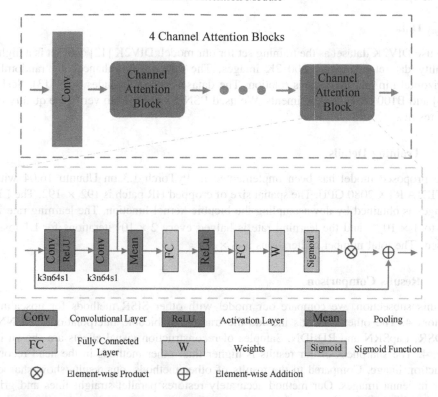

Fig. 3. Channel attention module, which contains 1 convolution layer and 4 channel attention blocks.

Where $Avg(\cdot)$ denotes global mean pooling operation. Then F_{avg} fed into two hidden layers. The first hidden layer performs dimension reduction to aggregate information among channels. Finally, the sigmoid function is used to output the channel weights:

$$C(F_i) = sigmoid\big(W_1\big(relu\big(W_0\big(F_{avg}\big)\big)\big)\big) \qquad (11)$$

Where $sigmoid(\cdot)$ refer to sigmoid activation function. W_0 and W_1 are learnable parameter matrices and defined with size $\frac{C}{r} \times C$ and $C \times \frac{C}{r}$, r is a pre-defined dimension reduction parameter and set it as 16 by experiments.

4   Experiment

4.1   Data

We use DIV2K dataset as the training set for our model. DIV2K [12] data set is a high-quality dataset containing 800 2K images. The data set is enhanced by randomly horizontal flip and 90 degrees rotation. The dataset we use includes Set5 [13], Set14 [14] and B100 [15] for experiments. We used PSNR and SSIM to verify the quality of the results.

4.2   Training Details

The proposed model has been implemented in PyTorch 1.3 on Ubuntu 16.04 with NVIDIA RTX 2080 GPU. The spatial size of cropped HR patch is 192×192. The LR images is obtained by downsampling the bicubic kernel function. The learning rate is set to 1×10^{-4}, and the learning rate is halved every 2×10^5 iterations for L1 loss-based. The total number of iterations is 8×10^5.

4.3   Results Comparison

In this subsection, we compare our model with other SISR methods for upscaling factors 4. The other methods include the traditional bicubic interpolation, SRCNN, VDSR, LapSRN and RD-IDN. Samples of reconstruction visual effects are shown in Fig. 4. The sharpness of our results is higher than other methods in the head reconstruction image. Compared to the results of other methods, our results show sharper hair in lenna images. Our method accurately restores parallel straight lines and grid patterns like windows.

As shown in Table 1, we compared the quantitative comparative results of various A algorithms kinds of SISR algorithms for $2 \times$, $3 \times$ and $4 \times$ SISR tasks. Comparing the PSNR and SSIM results, our model shows better performance than other models.

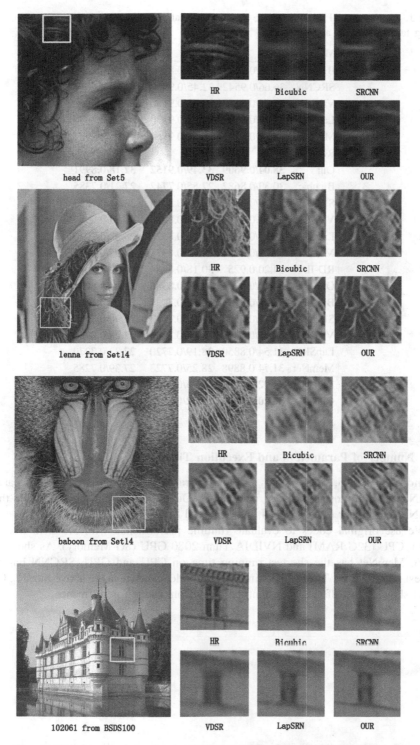

Fig. 4. Visual comparisons for 4 × SISR on Set5, Set14 and BSDS100 dataset.

Table 1. The average PSNR/SSIM are used to evaluate the super-resolution algorithm. The scaling factors are 2, 3, and 4.

Scale	Methods	Set5	Set14	BSDS100
2 ×	Bicubic	33.66/0.9299	30.24/0.87688	29.56/0.8431
	SRCNN	36.66/0.9542	32.45/0.9067	31.36/0.8879
	VDSR	37.53/0.9590	33.05/0.9130	31.90/0.8960
	LapSRN	37.52/0.9591	33.08/0.9130	31.08/0.8950
	MemNet	37.79/0.9601	33.32/0.9154	32.05/0.8984
	RD-IDN	–	–	–
	Our	**38.04/0.9609**	**33.49/0.9152**	**32.18/0.8991**
3 ×	Bicubic	30.39/0.8682	27.55/0.7742	27.21/0.7385
	SRCNN	32.75/0.9090	29.30/0.8215	28.41/0.7863
	VDSR	33.67/0.9210	29.78/0.8320	28.83/0.7990
	LapSRN	33.82/0.9227	29.87/0.8320	28.82/0.7980
	MemNet	34.12/0.9254	30.04/0.8371	28.97/0.8025
	RD-IDN	34.20/0.925	30.18/0.838	28.98/0.801
	Our	**34.40/0.9276**	**30.24/0.8386**	**29.08/0.8035**
4 ×	Bicubic	28.42/0.8104	26.00/0.7027	25.96/0.6675
	SRCNN	30.48/0.8628	27.50/0.7513	26.90/0.7101
	VDSR	31.35/0.8830	28.02/0.7680	27.29/0.0726
	LapSRN	31.54/0.8850	28.19/0.7720	27.32/0.7270
	MemNet	31.74/0.8898	28.25/0.7723	27.39/0.7285
	RD-IDN	31.92/0.893	28.35/0.776	27.42/0.731
	Our	**32.20/0.8952**	**28.62/0.7826**	**27.59/ 0.7358**

4.4   Number of Parameters and Execution Time

As shown in Fig. 5, SRCNN has the least parameters and the worst performance. MemNet has the most number of parameters. Our model parameters are lower than MemNet and higher than RD-IDN, but achieved better performance.

We use original codes to evaluate runtime performance with 3.6 GHz Intel i7 9900 k CPU (32G RAM) and NVIDIA Titan 2080 GPU (8G Memory). As shown in Fig. 6, MemNet has the shortest running time on CPU and GPU. SRCNN has the shortest running time. The running time of our model is between the above models. Our model shows trade-offs between runtime and reconstruction performance.

PSNR(dB)

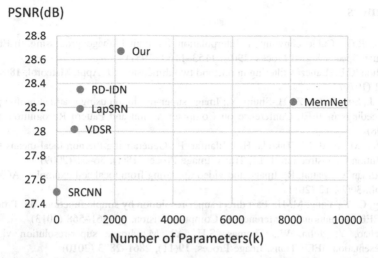

Fig. 5. Comparison of number of parameters. The results are evaluated on Set14 dataset, and the scale factor is set as ×4.

PSNR(dB)

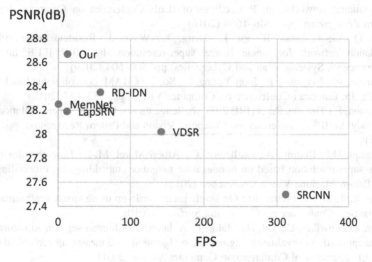

Fig. 6. Comparisons comparison of runtime (FPS) on Set14 dataset, and the scale factor is set as ×4.

5 Conclusion

In this paper, we proposed an image super-resolution model based on convolutional neural networks. According to the experimental results, our model is better than comparison method. Our reconstructed image can more accurately compound the original image. The validity of our method is confirmed by experimental results.

References

1. Keys, R.G.: Cubic convolution interpolation for digital image processing. IEEE Trans. Acoust. Speech Signal Proces. **29**(6), 1153–1160 (1981)
2. Duchon, C.E.: Lanczos filtering in one and two dimensions. J. Appl. Meteorol. **18**(8), 1016–1022 (1979)
3. Sun, J., Sun, J., Xu, Z., H.-Shum, Y.: Image super-resolution using gradient profile prior. In: Proceedings of IEEE Conference on Computer Vision and Pattern Recognition. pp. 1–8. (2008)
4. Protter, M., Elad, M., Takeda, H., Milanfar, P.: Generalizing the non-local-means to super-resolution reconstruction. IEEE Trans. Image Proces. **18**(1), 36–51 (2009)
5. Freedman, G., Fattal, R.: Image and video upscaling from local self examples. ACM Trans. Graph. 30(2), 12 (2011)
6. Yang, C.-Y., Yang, M.-H.: Fast direct super-resolution by simple functions. In: Proceedings of IEEE International Conference on Computer Vision, pp. 561–568 (2013)
7. Jianchao, Y., John, W., Thomas, S.H., Yi, M.: Image super-resolution via sparse representation. IEEE Trans. Image Proces. **19**(11), 2861–2873 (2010)
8. Dong, C., Loy, C.C., He, K., Tang, X.: Learning a deep convolutional network for image super-resolution. In Proceedings of European Conference on Computer Vision. pp. 184–199. (2014)
9. Kim, J., Lee, J.K., Lee, K.M.: Accurate image super-resolution using very deep convolutional networks. In: Proceedings of IEEE Conference on Computer Vision and Pattern Recognition. pp. 1646–1654 (2016)
10. Chen, Q., Li, J., Duan, B., Pu, L., Deng, X., Wang, J.: Residual dense information distillation network for single image super-resolution. In: 2019 IEEE International Conference on Systems, Man and Cybernetics. pp. 500–504 (2019)
11. Sanghyun, W., Jongchan, P., Joon-Young, L., So, K.: CBAM: convolutional block attention module. In: European Conference on Computer Vision. pp. 132–148 (2018)
12. Agustsson, E., Timofte, R.: NTIRE 2017 challenge on single image super-resolution: dataset and study. In: IEEE Conference on Computer Vision and Pattern Recognition. pp. 126–135 (2017)
13. Bevilacqua, M., Roumy, A., Guillemot, C., Alberi Morel, M.-L.: Low-complexity single-image super-resolution based on nonnegative neighbor embedding. In: Proceedings of the 23rd British Machine Vision Conference (2012)
14. Zeyde, R., Elad, M., Protter, M.: On single image scale-up using sparse-representations. In: International Conference on Curves and Surfaces (2010)
15. Martin, D., Fowlkes, C., Tal, D., Malik, J.: A database of human segmented natural images and its application to evaluating segmentation algorithms and measuring ecological statistics. In: IEEE International Conferenceon Computer Vision (2001)
16. Lai, W., Huang, J., Ahuja, N., Yang, M.: Deep laplacian pyramid networks for fast and accurate super-resolution. In: 2017 IEEE Conference on Computer Vision and Pattern Recognition, CVPR 2017. pp. 5835–5843 (2017)

Electric Drive and Power System Fault Diagnosis

Online Fault Diagnosis of High-Voltage Vacuum Circuit Breaker Based on Deep Convolutional Long Short-Term Memory Network

Deyin Xu, Lin Luo$^{(\boxtimes)}$, and Qiao Wang

School of Information and Control Engineering, Liaoning Shihua University,
Fushun 113001, Liaoning, People's Republic of China
luolin@lnpu.edu.cn

Abstract. The information of the mechanical conditions is contained in the vibration signals generated from the actuator control system of high-voltage vacuum circuit breaker. Analysis of the vibration signals makes noninvasive detection reliable and effective. The traditional fault detection methods are difficult to achieve impressive results and implement the monitoring in a real-time manner. This paper proposes a convolutional sequential deep network-based fault detection and diagnosis method. The convolutional neural network is designed to extract the features of the fault conditions. To avoid the vanishing gradient in the dynamic modeling, a long short-term memory method is employed to process the features. The experiments on a high-voltage vacuum circuit breaker demonstrate that the proposed method is competitive with the widely used support vector machine.

Keywords: High-voltage vacuum circuit breaker · Deep learning · Fault diagnosis · Long short-term memory

1 Introduction

With the development of China's economy, the requirement for the stable operation of the power system is also constantly improving. High voltage circuit breaker plays an important role in power system, which should not only guarantee the normal power supply of the power system, but also cut off the electrical equipment when it breaks down. Therefore, the normal operation of the high-voltage circuit breaker is the basis to ensure the stable operation of the power system. According to statistical analysis, more than 2/3 of the faults of high-voltage circuit breaker are caused by its poor mechanical performance, which is mainly caused by the abnormal mechanism operating. It is necessary to identify the fault of high voltage circuit breaker in real time [1–3].

High voltage circuit breaker contains rich vibration signals in the process of mechanical operation. Vibration signals' characteristic value is stable and

© Springer Nature Singapore Pte Ltd. 2020
J. Qian et al. (Eds.): ICRRI 2020, CCIS 1335, pp. 137–149, 2020.
https://doi.org/10.1007/978-981-33-4929-2_10

reliable. Effective feature vector can be extracted from the vibration signal, so as to accurately determine whether the circuit breaker has fault and the type of fault. On the other hand, the opening and closing coil current of the circuit breaker can reflect the working condition of the controlled mechanism in the operating process, and monitor the current signal [4]. Taking the current signal and vibration signal of mechanical operation as input signals, from which the effective characteristic factors can be extracted, and the defective parts and positions can be further analyzed, so as to monitor and diagnose the state of the circuit breaker quickly and effectively [5]. Due to the convenience of data acquisition, non-invasive fault diagnosis method based on vibration signal has gradually become the mainstream of research. The VMD is used to preprocess the vibration signal and obtain the required modal components, then SVM model is used to diagnose the modal components and determine the fault type [6,7]. The Hilbert value can be obtained through the empirical mode decomposition (EMD) to extract the feature vector of the vibration signal, then the feature vectors are delivered to the rough set theory, so as to distinguish the different types of the circuit breaker faults [8]. The signal is decomposed and the fault type is determined by calculating the logarithm energy entropy of each node.

In recent years, the development of deep network such as Convolutional Neural Networks (CNN) and Long Short-term Memory (LSTM) have promoted many applications of artificial intelligence in power system [9–11]. CNN is composed of input layer, convolution layer, pooling layer and full connection layer. The mechanical vibration signal of the high voltage circuit breaker contains multi-layer time series. These time series are taken as samples to form the index image data map, which are delivered to the input layer. In the process of data collection, the fitting-difference method is used for each sampling point synchronously, so that the characteristic vector of mechanical vibration signal of high-voltage circuit breaker can be obtained through CNN network. By comparing and analyzing the similar factors between the unknown state and the typical normal working state, we can identify the type of the unknown operating state and realize the diagnosis of the mechanical state of the high-voltage circuit breaker [12–15]. LSTM can effectively solve the problem of the gradient disappearance and gradient explosion in traditional neural network through the joint action of input gate, output gate and forgotten gate [16–19].

In this paper, a hybrid network model combining convolution neural network and long-short term memory network is proposed. This method combines the feature extraction of vibration signal in mechanical operation with mechanical fault diagnosis, thus an end-to-end fault diagnosis strategy is obtained. The problem of missing fault information caused by fault signal feature extraction and diagnosis strategy independently has been solved, and this method improves the diagnosis accuracy of high-voltage circuit breaker state diagnosis. Experiments on high voltage vacuum circuit breakers verify the possibility of this method.

2   Convolution Neural Network Model

Traditional mechanical fault recognition method based on shallow neural network has some problems, such as low accuracy, slow training speed and low recognition efficiency. The deep neural network' structure is more hierarchical than the shallow neural network to obtain the deeper mechanical state information in the vibration signal and diagnose the mechanical state of electrical equipment in time. Convolutional neural network in deep neural network can be used to deal with large number of complex vibration signals. In the past years, many scholars have benefited from the advantages of convolutional neural network, such as strong local sensing ability and common weights, to speed up the training process. In recent years, convolutional neural network is used to diagnose mechanical faults, which has a high performance in simulation results and the accuracy is much higher than other neural network models [20]. It has been proved that the convolutional neural network can keep the data information stable without considering scale, displacement and distortion. Convolutional neural network can accelerate the training process by weight sharing, and the most important feature information in the sample can be retained by using the maximum pooling operation in the pooling layer. After the last pooling operation, the 1D convolutional neural network transforms the extracted vectors into 1D vectors. The classifier can diagnose 1D vectors quickly [21].

In this paper, the convolution neural network is used to extract the characteristics of vibration signals of circuit breakers. Firstly, convolution operation is carried out by preset step sliding filter. Convolution operation can traverse the whole sample to obtain fault information in the sample. Then the fault information is passed to the pooling layer. Nonlinear activation in the pool layer is helpful to reduce the number of data points and avoid over fitting, so as to speed up the training. It also acts as a smooth process from which unnecessary noise can be eliminated. Finally, the obtained local information is transferred to the full connection layer to connect all previous feature maps and obtain the complete feature vector.

CNN considers the characteristics of the sample in the dimension but ignores the characteristics of the sample in the time series, so it may cause the loss of the feature information of the sample sequence. LSTM adopts an adaptive method to extract the feature vectors of vibration signals to avoid information loss and the extracted feature vectors contain time series characteristics. The hybrid network model, which combines convolutional neural network and long -short term memory network (CNN-LSTM) is proposed in this paper. In the process of mechanical state diagnosis of circuit breakers, the multi-dimensional characteristics of the original samples and the time series characteristics are both considered, which improve the diagnosis accuracy of the mechanical state.

3   CNN-LSTM Model

The traditional fault diagnosis model is independent of feature extraction and modeling. The feature expression ability depends on the preset parameters, and will directly affect the subsequent modeling process. This paper proposes an end-to-end deep learning model which combines feature extraction with sequence classification by convolution neural network and long-short term memory model.

3.1   Convolution Layer

In the input layer, the mechanical vibration signal of the high voltage circuit breaker contains a multivariate time series with N samples and D dimensions, which can be defined as $\mathbf{X} = (\mathbf{x}_1, \mathbf{x}_2, \cdots, \mathbf{x}_T), \mathbf{x}_t \in \mathbb{R}^D$, where t is the time step index ranging from 1 to T. Consider L convolutional layers, spatial correlations with the types of faults can be captured by a set of 1D filters, and the layer index is denoted as $l \in \{1, \cdots, L\}$. The filters for each layer are parameterized by the tensor $\mathcal{A}^{(l)} \in \mathbb{R}^{D_l \times d \times D_{l-1}}$, where d is filter duration, and D_l and D_{l-1} are the length of feature vectors on the current and the previous layer, respectively. For the l-th layer, the i-th component of the activation $\hat{E}_{i,t}^{(l)}$ is constructed by the incoming normalized activation matrix $\hat{E}_t^{(l-1)}$,

$$
\begin{aligned}
\hat{E}_{i,t}^{(l)} &= f\left(\mathcal{A}_{i,\cdot,\cdot}^{(l)} * E_{\cdot,t+d}^{(l-1)} + b_i^{(l)}\right), \forall t \in [1, T] \\
&= f\left(\sum_{t'=1}^{d} \left\langle \mathcal{A}_{i,t',\cdot}^{(l)}, E_{\cdot,t+d-t'}^{(l-1)}\right\rangle * E_{\cdot,t+d}^{(l-1)} + b_i^{(l)}\right)
\end{aligned}
\tag{1}
$$

where operators $*$ and $\langle \cdot \rangle$ are convolution and dot product, and $f(\cdot)$ is a nonlinear function, such as rectified linear unit (ReLU).

The pooling layers for subsampling the output of the convolutional layer is helpful to effectively obtain the activation during a long period. The objective of a pooling layer is to ensemble similar local features into the integration. One of the most common pooling is to utilize a max operation to the result of each filter at time step,

$$
m = \max_i \left\{ \hat{E}_{i,t}^{(l)} \right\}
\tag{2}
$$

The normalized activation matrix has the form of,

$$
E_t^{(l)} = \frac{1}{m + \varepsilon} \hat{E}_t^{(l)}
\tag{3}
$$

where ε is a small constant. According to Eq. (2), the highest value is extracted to capture the most important feature. With the pooling layers, the shifts and the distortions in a signal would be reduced by eliminating the non-maximal value. The structure of the convolutional layers extracting the spatial feature is shown in Fig. 1.

Two reasons justify the use of the max pooling operation. First, by eliminating the non-maximal valued, it reduces the computation of the upper layers. Second, it provides a form of translation invariance. The obtained feature vectors are then fed to the sequential layer. To capture the long distance dependence, LSTM is introduced into the sequential layer for vector composition.

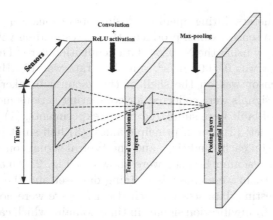

Fig. 1. Structure of convolutional layer for extracting the spatial feature.

3.2 Long-Short Term Memory Model

The deep neural network model proposed by S. Hochreiter is used to solve the long-distance dependence problem that the traditional neural network cannot deal with [22]. LSTM consists of repeatedly connected subnets, also known as memory blocks. There are three kinds of gate structures in memory unit: forgotten gate, input gate and output gate, which are used to provide continuous operation of reading, writing and resetting unit [23]. In this paper, we use a tree structure to solve the problem that LSTM relies too much on the training sequence.

For each parent node, two child nodes exist, namely, c_n and h_n, which represent the node hidden state and memory unit, respectively. The transfer equation of node n is expressed as follows:

$$\begin{bmatrix} C_n \\ f_n^l \\ f_n^r \end{bmatrix} = \begin{bmatrix} \tanh \\ \sigma \\ \sigma \end{bmatrix} T_{a,b} \begin{bmatrix} X_j \\ h_n^l \\ h_n^r \end{bmatrix} \tag{4}$$

where X_j stands for the input vector, superscripts l and r represent the left and right child nodes, respectively, σ is the sigmoid function, and $T_{a,b}$ denotes the linear change, where a and b are the network parameters to be learned. Then, we connect the output end of LSTM to the softmax classifier and obtain the classification result of the working condition recognition.

4   Experiment and Results

4.1   Acquisition of Vibration Signal of High Voltage Circuit Breaker

In the paper, a ZW32-12FG/630-20 vacuum breaker was taken as a mechanical testing system, which was equipped with spring operating mechanism. Vibration acceleration parameters during opening/closing operations were measured with YD-39 acceleration sensor, which produced the corresponding voltage. The voltage were digitized by using the NI 9234 data acquisition card. The measurement range of the sensor was 0–5000 m/s^2, sampling rate was 10 kHz. The position of acceleration sensor was on the shell of the vacuum breaker, on which the vertical vibration signals were collected for comparing various monitoring methods. Three types of fault were simulated in the experiments: (1) jam of tripping closing electromagnet; (2) jam of principle axles; (3) half shaft jam. We carried out 7 experiments on each condition, and the typical signals on the normal and fault status of the vacuum breaker were shown in Fig. 2. Among these seven datasets, three of them were chose as training data-set and the remaining ones were selected as testing data-sets for each status. There were no obvious differences and changes from vibration signals in time domain, which can be illustrated through Fig. 2a –Fig. 2d. It is a remarkable fact that the feature analysis on the weak vibration signal is particularly important for the state identification.

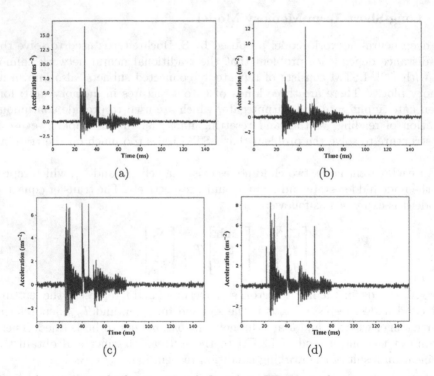

Fig. 2. Acquisition of vibration signals based on YD-39 acceleration sensor in time domain, (a) Normal condition; (b) tripping closing electromagnet jam; (c) principle axles jam; (d) half shaft jam

4.2   Model Parameter Setting and Evaluation Method

The CNN designed in this paper has a convolution layer and a pooling layer and the number of convolution filters are 64. The convolution kernel size is set to 1 and the step size is 1. The size of MaxPooling1D in the pooling layer is 1, The activation function in CNN adopts the ReLu function. The output dimension of vibration signal is reduced by convolution operation and pooling operation, then the sample is processed by flatten operation as the whole feature extraction of the network. The number of nodes in the network layer used in the LSTM model are 100, and sigmoid function is used as the activation function. ROC curve can be used to describe the sensitivity and specificity of the diagnosis model. Draw the curve with the false positive rate and the true rate as the horizontal and vertical coordinates, and calculate the area under the curve(AUC). The larger the area is, the higher the accuracy of diagnosis model has; the closer the curve is to the upper left corner, the more sensitive of the model is. If the ratio of normal state signal samples to fault state signal samples is large, the precision-recall curve (PR) can be used to further determine the classification performance of the model. Take recall as abscissa and precision as ordinate. When the recall is increasing, the precision is always high, that is the closer the curve is to the upper right corner, the faster the curve climbs. In this paper, SVM model is used as a comparison method. The fault diagnosis of the same vibration signal samples of high voltage vacuum circuit breaker is carried out by the CNN-LSTM method and SVM model. The results show in Receiver Operating Characteristic Curve (ROC), Precision-Recall Curve (PRC). Table 1 shows the settings of four mechanical vibration states.

Table 1. Setting of four mechanical vibration states

Mechanical vibration states	Status number
Normal condition	Class 0
Tripping closing electromagnet jam	Class 1
Principle axles jam	Class 2
Half shaft jam	Class 3

4.3   Experimental Analysis

Experiment 1: the diagnosis comparison of vibration signals between CNN-LSTM and SVM in the normal state and in the tripping closing electromagnet jam.

Firstly, in Fig. 3(a), the ROC curve of CNN-LSTM model is closer to the upper left corner of the coordinate axis, so the diagnosis accuracy is higher. The ROC curve identified as the tripping closing electromagnet jam state completely envelops the ROC curve in normal state, so the fault signal and normal signal can be classified accurately. Secondly, the curves of normal state and fault state in Fig. 3(a) remain stable with the change of threshold, while the curves of normal state in Fig. 3(b) fluctuate greatly during this period. CNN-LSTM model can be used to better predict the fault trend of tripping closing electromagnet jam. Finally, from Fig. 3(a), it can be seen that the AUC values of the normal state and the tripping closing electromagnet jam of CNN-LSTM are respectively 0.99 and 1.0. The AUC values corresponding to SVM model in Fig. 3(b) are 0.52 and 0.52. The AUC value is more higher in CNN-LSTM model.

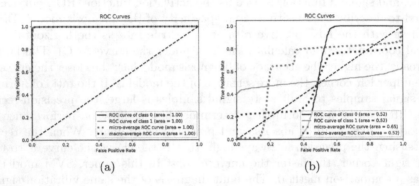

(a)   (b)

Fig. 3. ROC curve of (a) CNN-LSTM and (b) SVM over fault mode 1.

To further analyze the ability of the model, we use the precision recall (PR) curve which are shown in Fig. 4 to evaluate the model. As shown in Fig. 4(a), the PR curve of CNN-LSTM hybrid model is closer to the upper right corner of the coordinate axis than that in Fig. 4(b). It can be concluded that CNN-LSTM model has high precision no matter how the recall rate changes in the training process. In addition, the curves of normal state and tripping closing electromagnet jam state in Fig. 4(a) are always smooth with the change of threshold value in the diagnosis process, and the precision still changes smoothly with the increase of recall rate. The AUC values of the normal state and the tripping closing electromagnet jam of CNN-LSTM are respectively 0.992 and 0.997, however the AUC values corresponding to SVM model are 0.575 and 0.629. The AUC value of CNN-LSTM model is much higher than SVM model. From ROC curve and PR curve, it can be seen that CNN-LSTM model has higher sensitivity and accuracy for the condition diagnosis of breaker in normal state and in the tripping closing electromagnet jam state.

Experiment 2: the diagnosis comparison of vibration signals between CNN-LSTM and SVM in the normal state and in the principle axles jam.

From Fig. 5, it can be seen that the ROC curve of CNN-LSTM model in normal state and the principle axles jam state change smoothly with the change of threshold value, while the curve of SVM model shown in Fig. 5(b) fluctuate greatly during this period. CNN-LSTM model is better to predict the fault trend of circuit breakers. AUC values in normal state and in the principle axles jam

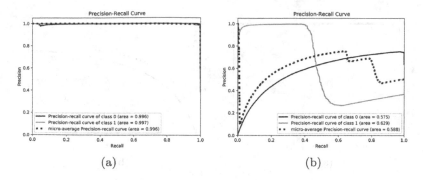

Fig. 4. Precision-recall curve of (a) CNN-LSTM and (b) SVM over fault mode 1.

state are 1.00 and 1.00 respectively, the AUC values corresponding to SVM model are 0.54 and 0.54. The AUC value of CNN-LSTM model is much higher than the value of SVM model.

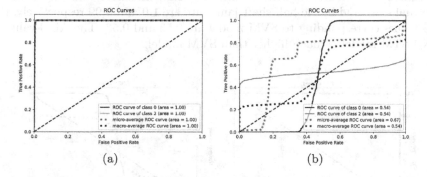

Fig. 5. ROC curve of (a) CNN-LSTM and (b) SVM over fault mode 2.

As shown in Fig. 6(a), the PR curve of CNN-LSTM model in the normal state and in the principle axles jam state are always smooth with the change of threshold value in the diagnosis process, and the precision is still stable with the increase of recall rate. AUC values in normal state and in the principle axles jam state are 0.997 and 0.997 respectively, the AUC values corresponding to SVM model are 0.585 and 0.647. The AUC value of CNN-LSTM model is higher. When monitoring the fault state and normal state, the mechanical state monitoring model of CNN-LSTM is better than SVM model in classifying the principle axles jam state and normal state. Therefore, CNN-LSTM model is more conducive to monitor the mechanical state of the high-voltage circuit breaker.

Experiment 3: The diagnosis comparison of vibration signals between CNN-LSTM and SVM in the normal state and in the half shaft jam.

First of all, from Fig. 7(a), it can be seen that the ROC curve of CNN-LSTM model is closer to the upper left corner of the coordinate axis. Secondly,

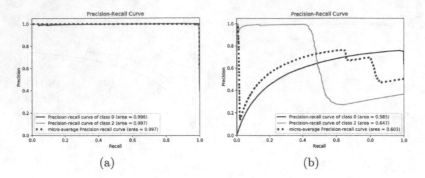

Fig. 6. Precision-recall curve of (a) CNN-LSTM and (b) SVM over fault mode 2.

when CNN-LSTM model is used to diagnosis the half shaft jam of the circuit breaker, the true positive ratio (TPR) is stable close to 1 with the increase of the threshold, while the TPR value of SVM model for fault diagnosis in Fig. 7(b) is maintained between 0.4–0.6 with the increase of the threshold value. AUC values in normal state and in the half shaft jam state are 1.0 and 0.99 respectively, the AUC values corresponding to SVM model are 0.53 and 0.53. The AUC value of CNN-LSTM model is much higher than SVM model.

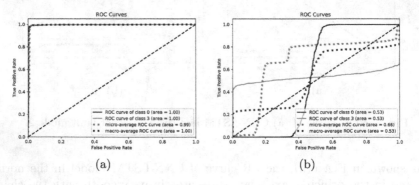

Fig. 7. ROC curve of (a) CNN-LSTM and (b) SVM over fault mode 3.

It can be seen from Fig. 8 that the PR curve of CNN-LSTM model is closer to the upper right corner of the coordinate axis than that of SVM model. In Fig. 8(a), the curves of normal state and the half shaft jam state are kept smooth, and the precision still changes smoothly with the increase of recall rate. Finally, AUC values in normal state and in the half shaft jam state are 0.997 and 0.992 respectively, the AUC values corresponding to SVM model are 0.583 and 0.642. Above all, it can be conclusion that CNN-LSTM model has higher accuracy than SVM model in the diagnosis of circuit breaker fault state.

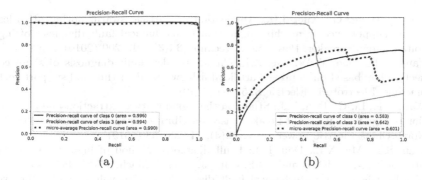

Fig. 8. Precision-recall curve of (a) CNN-LSTM and (b) SVM over fault mode 3.

5 Summary

Fault diagnosis of circuit breaker which is based on traditional neural network has some problems, such as complex process, slow training speed and low accuracy. In view of the above problems, this paper proposes CNN-LSTM model to detect the mechanical vibration signal of high-voltage circuit breaker. CNN-LSTM full play the advantage of CNN model in extracting potential feature vectors, so as to get a large number of effective fault information quickly. LSTM network model can better fit the time sequence and complex nonlinear relationship of mechanical vibration signals, so as to effectively identify the mechanical vibration state.

The results of three experiments show that CNN-LSTM model can extract effective feature vectors more quickly and diagnosis fault state more accurately than SVM model. CNN-LSTM model can diagnose feature vectors more accurately. With the increase of the number of samples, CNN-LSTM model can still maintain a high accuracy of fault diagnosis, SVM model has a downward trend of accuracy in the same situation. Therefore, CNN-LSTM model which are compared with the traditional neural network, can diagnose the mechanical fault of high-voltage circuit breaker more accurately in real time.

Acknowledgement. This work was supported by National Natural Science Foundation of China (Grants No. 61703191), the Foundation of Liaoning Educational Committee (No. L2017LQN028) and the Scientific Research Foundation of Liaoning Shihua University (No. 2017XJJ-012).

References

1. Wang, J., Xia, J.F., Liu, X.H.: Research on mechanical condition monitoring and fault diagnosis for DC circuit breaker in traction substation. Power Syst. Protect. Control **48**(1), 33–40 (2020)
2. Yang, L., Zhu, Y.L.: High voltage circuit breaker fault diagnosis of probabilistic neural network. Power Syst. Protect. Control **43**(10), 62–67 (2015)

3. Jia, R., Hong, G., Xue, J.H.: Application of particle swarm optimization-least square support vector machine algorithm in mechanical fault diagnosis of high-voltage circuit breaker. Power Syst. Technol. **34**(3), 197–200 (2010)
4. Yang, Y.F., Jiao, W.W., He, Z.J.: Circuit breaker fault diagnosis of shore connection box based on improved artificial fish swarm algorithm and support vector machine. Electrotech. Electric **8**, 57–64 (2019)
5. Wan, S.T., Li, C., Dou, L.J.: Study on the signal feature extraction and classification of high voltage circuit breaker based on vibration signal and current signal. J. North China Electric Power Univ. **46**(4), 31+38+53 (2019)
6. Wan, S.T., Ma, X.L., Dou, L.J.: Fault diagnosis method of high voltage circuit breaker based on VMD and MES. Chin. J. Constr. Mach. **17**(5), 444–449 (2019)
7. Lin, L., Chen, Z.Y.: Mechanical fault diagnosis of high voltage circuit breakers based on rough set neural networks and vibration signals. Trans. China Electrotech. Soc. **35**(S1), 277–283 (2020)
8. Li, B.B., Ke, Y.G., Tian, Y.: Mechanical condition detection of high voltage circuit breaker based on wavelet packet theory. J. Hefei Univ. Technol. **42**(7), 924–929 (2019)
9. Wen, C.L., Lu, F.Y.: Review on deep learning based fault diagnosis. J. Electron. Inf. Technol. **42**(1), 234–248 (2020)
10. Wang, Y., Sun, J.F., Xiao, X.Y.: Cable incipient fault classification and identification based on optimized convolution neural network. Power Syst. Protect. Control **48**(7), 10–18 (2020)
11. Yang, Z.W., Liu, H., Bi, T.S.: PMU bad data detection method based on long short-term memory network. Power Syst. Protect. Control **48**(7), 1–9 (2020)
12. Zhang, P., Yang, T., Liu, Y.N.: Feature extraction and prediction of QAR data based on CNN-LSTM. Appl. Res. Comput. **36**(10), 2958–2961 (2019)
13. Li, C., Zhang, D.H., Li, D.W.: Study on power supply capacity of distribution network based on CNN-LSTM. Foreign Electron. Measur. Technol. **38**(9), 16–21 (2019)
14. Al-Masni, M., Kim, D., Kim, T.: Multiple skin lesions diagnostics via integrated deep convolutional networks for segmentation and classification. Comput. Methods Programs Biomed. **190**, 105351 (2020)
15. Do, H., Guo, Y., Yoon, A.: Accuracy, uncertainty, and adaptability of automatic myocardial ASL segmentation using deep CNN. Magn. Reson. Med. **83**(5), 1863–1874 (2020)
16. Bogaerts, T., Masegosa, A., Angarita, J.: A graph CNN-LSTM neural network for short and long-term traffic forecasting based on trajectory data. Transp. Res. Part C Emerg. Technol. **112**, 62–77 (2020)
17. Song, R., Xiao, Z., Lin, J.: CIES: cloud-based intelligent evaluation service for video homework using CNN-LSTM network. J. Cloud Comput. Adv. Syst. Appl. **9**(1), 7 (2020). https://doi.org/10.1186/s13677-020-0156-5
18. Du, X.L., Chen, Z.G.: Fault diagnosis of bearing based on wavelet convolutional auto-encoder and LSTM network. J. Mech. Electric. Eng. **36**(7), 663–668 (2019)
19. Du, X.L., Chen, Z.G., Xu, X.: Fracture truck fault diagnosis based on wavelet, WAE and LSTM. China Petroleum Mach. **47**(10), 88–93+106 (2019)
20. Xu, Y.W., Wang, Z.Y., Li, L.C.: CNN-based fault detection algorithm for water meter. J. Fuzhou Univ. (Nat. Sci. Edn.) **48**(3), 1–5 (2020)
21. Li, X.W., Liu, S.Y., Gao, K.L.: Power system transient stability assessment based on bidirectional long short term memory network and convolutional neural network. Sci. Technol. Eng. **20**(7), 2733–2739 (2020)

22. Hochreiter, S., Schmidhunber, J.: Long short-term memory. Neural Comput. **9**(8), 1735–1780 (1997)
23. Ming, T.T., Wang, K., Tian, D.D.: Estimation on state of charge of lithium battery based on LSTM neural network. Guangdong Electric Power **33**(3), 26–33 (2020)

Online Fault Diagnosis of Power Transformer with Temporal and Spatial Features

Ting Liu[1], Lin Luo[2(✉)], Qiao Wang[2], and Shuai Chen[2]

[1] School of Electronic & Automation, City Institute, Dalian University
of Technology, Dalian 116021, Liaoning, People's Republic of China
[2] School of Information and Control Engineering, Liaoning Shihua University,
Fushun 113001, Liaoning, People's Republic of China
luolin@lnpu.edu.cn

Abstract. To address the feature extraction and online modeling problems on the traditional fault diagnosis based on the dissolved gas analysis, this paper proposes the spatiotemporal features-based online method for the power transformer fault diagnosis. The method combines the convolutional neural network with the long short-term memory. The features related to the spatial domain are extracted by the convolutional and pooling layer. Furthermore, a recurrent neural network with a gating structure is introduced to obtain the temporal feature. As a result, online end-to-end deep network is constructed with a sequential layer bridging the temporal and spatial features. The outstanding performance of this method is demonstrated and compared to the existing models in the data of dissolved gas analysis. The result shows that the prediction accuracy of the proposed method with insensitive time-steps is 10% – 30% higher than that of dynamic support vector machine and convolutional neural network.

Keywords: Dissolved gas analysis · Deep learning · Convolutional neural network · Long short-term memory

1 Introduction

Power transformer is one of the key equipments in power system, its operation state directly influence the safe and stable operation of the whole power system [1]. Under actual operation conditions such as electrical stress, mechanical stress and thermal pressure, will lead to the aging of internal insulation of power transformer. Moreover, the latent defects of the equipment will lead to the decrease of electrical strength of internal insulation, which cause the mechanical failure. As the rapid development of electrical technology, a variety of insulation defect diagnosis technologies emerge, mainly including: (1) Insulating oil characteristic test including: breakdown voltage [2] and dissolved gas analysis(DGA) [3]. (2) Dielectric response test including: frequency domain dielectric spectroscopy [4] and Polarization/depolarization current test [5]. (3) frequency response analysis

[6]. (4) partial discharge test [7]. DGA has been widely used in transformer fault diagnosis because of its advantages of intuitionistic observation results and easy operation.

DGA mainly detects the characteristic gas concentration produced by the cracking of transformer oil and insulating paper cellulose including: H_2, C_2H_6, C_2H_4, C_2H_2, Co and Co_2. Under normal working conditions, the transformer insulation system will release a certain amount of characteristic gas, but the characteristic gas concentration out of limit will occur because of local over-heating and large electrical stress point produced by discharge or arc. In recent years, DGA have been proposed as diagnostic criteria at home and abroad such as: IEEE C57-104–2019 [8], DL/T 722–2014 [9]. At the same time, the diagnosis methods based on DGA gas ratio are also proposed: Rogers ratio method, IEC three-ratio method. However, most of these methods need to be combined with the field experience. When facing different external factors such as aging degree, fault duration, voltage surge, will appear serious diagnostic errors.

With the rapid development of information processing and artificial intelligence technology, domestic and foreign scholars began to introduce intelligent diagnosis technology into the research of DGA insulation diagnosis of transformer, such as: artificial neural network(ANN) [10], fuzzy logic [11], wavelet transform [12], support vector machine(SVM) [13]. Although these methods have achieved good results, they also have some shortcomings. First of all, the existing methods usually separate the feature extraction and modeling process, but rarely combine the feature extraction and modeling process together to complete the task, thus decreasing certain degree of accuracy. Secondly, although ANN method has strong non-linear modeling ability, the error propagation form is easy to produce over fitting problem, which restricts its practicability. As a special kind of single hidden layer feedforward neural network, SVM can solve the problems of non-linear, high-dimensional and small samples. However, due to the lack of effective online strategy, its ability to process real-time data is limited to some extent.

In recent years, the development of deep network promotes the application of artificial intelligence technology in DGA insulation diagnosis of transformer such as convolutional neural networks(CNN) and long-short term memory(LSTM) [14–17]. CNN is composed of input layer, convolution layer, pooling layer and full connection layer. By comparing and analyzing the similar factors of the identification set of unknown state and normal state, CNN can identify the type of unknown operation state and realize the DGA insulation diagnosis of transformer. To solve the problem of real-time modeling of sequence data, recurrent neural network(RNN). RNN applies the preorder information to the postorder computation and carries out recursive modeling in the sequence evolution direction. LSTM uses input gate, output gate and forgetting gate to deal with sequence data, which can effectively solve the problem of gradient disappearance and gradient explosion when RNN processes long sequence data.

Based on the DGA and deep learning technology, which combined the convolutional neural network and long short memory model, this paper proposes an end-to-end deep network to process online DGA data. The structure of this paper is as follows: Sect. 1 briefly describes the current research situation at home and abroad; Sect. 2 introduces the process and main steps of CNN algorithm. In Sect. 3, the model of transformer fault diagnosis based on convolution characteristic and long-short term memory model is integrated. In Sect. 4, the accuracy of transformer on-line fault diagnosis between CNN-LSTM model, SVM model and CNN model is compared on the actual DGA data to verify the effectiveness and accuracy of proposed method. Section 5 is the conclusion and summary .

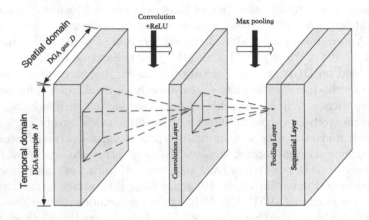

Fig. 1. Structure of CNN for extracting the spatial feature of DGA.

2   Convolution Neural Network Model

In order to extract information from spatial structure of data, LeCun [18] proposes CNN model, which uses convolution operation to replace matrix product operation in traditional network, and reduces over fitting by compressing data and parameters. Because the adjacent data points are closely related in space, the neurons in the network can extract the hidden features from the data only by focusing on the local information. Therefore, this paper will use convolution neural network to extract the relevant characteristics of transformer fault. DGA real-time data can be expressed as an ordered real-time series: $\mathbf{X} = (\mathbf{x}_1, \mathbf{x}_2, \cdots, \mathbf{x}_T), \mathbf{x}_t \in \mathbb{R}^D, t \in [1, T]$. D is the quantity of characteristic gas. Suppose the number of convolutions in CNN is L. We can use 1D filter to capture fault sensitive spatial features in DGA data, and express the filters on the convolution layer $l \in \{1, \cdots, L\}$ as a tensor form $A^{(l)} \in R^{D_l \times d \times D_{l-1}}$, D_l and D_{l-1} are the length of feature vectors on the current convolution layer and the previous layer respectively. The feature map of the previous layer is convoluted with a learnable convolution kernel, and the output of the activation function forms the neuron of this layer, thus forming the feature extraction layer, which is expressed as,

$$\hat{E}_{i,t}^{(l)} = f\left(\mathcal{A}_{i,\cdot,\cdot}^{(l)} * E_{\cdot,t+d}^{(l-1)} + b_i^{(l)}\right), \forall t \in [1,T]$$

$$= f\left(\sum_{t'=1}^{d}\left\langle \mathcal{A}_{i,t',\cdot}^{(l)}, E_{\cdot,t+d-t'}^{(l-1)}\right\rangle * E_{\cdot,t+d}^{(l-1)} + b_i^{(l)}\right) \qquad (1)$$

where $\hat{E}_t^{(l)} \in R^{D_l}$ is the denormalization activation on current layer. $E^{(l-1)} \in R^{D_{l-1} \times T_{l-1}}$ is the normalized activation matrix on the previous layer. The operators $*$ and $\langle\cdot\rangle$ are convolution and dot product, respectively. $f(\cdot)$ is a nonlinear function such as rectified linear unit(ReLU). In order to effectively calculate the activation value in a long period of time, CNN needs to use the pooling layer to sample the output of the convolution layer. The pooling operation generally uses the max operation of the filter with the step size,

$$m = \max_i \left\{ \hat{E}_{i,t}^{(l)} \right\} \qquad (2)$$

In this way, we normalize the activation vector after pooling operation, then

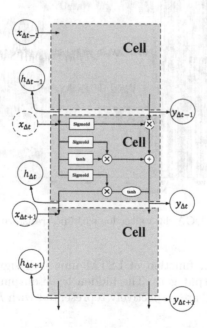

Fig. 2. Structure of LSTM for extracting the temporal feature of DGA.

$$E_t^{(l)} = \frac{1}{m+\varepsilon}\hat{E}_t^{(l)} \qquad (3)$$

where ε is any decimal. The maximum pooling operation has two functions. Firstly, the spatial resolution of the network is effectively reduced by eliminating the non maximum eigenvalues, so as to eliminate the tiny offset and distortion of the signal. Secondly, the maximum pooling runtime has translation invariance. The CNN network structure which is used to extract the spatial characteristics of DGA data is shown in Fig. 1.

3   LSTM Model

In order to capture the further temporal correlation of feature sequences, we input the spatial feature vectors extracted from the pooling layer into the sequential layer. The LSTM model is introduced into the sequence layer for vector synthesis. The whole LSTM network structure is used to extract DGA timing features. LSTM structure is shown in Fig. 2.

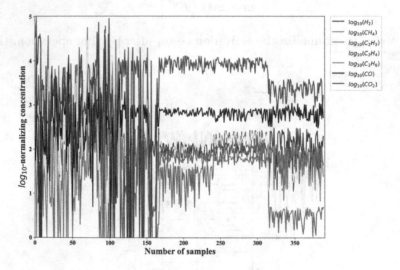

Fig. 3. DGA data after logarithmic transforming.

The state transition function of LSTM unit is designed with input gate, forgetting gate and output gate. The hidden layer output vector and memory unit of LSTM in step T are respectively represented with h_t and c_t. The LSTM unit can be described as:

$$\begin{bmatrix} z_t^o \\ z_t^i \\ z_t \\ z_t^f \end{bmatrix} = W_h h_{t-1} + W_E E_t + b \tag{4}$$

$$c_t = \sigma(z_t^f) \odot c_{t-1} + \sigma(z_t^i) \odot \tan(z_t) \tag{5}$$

$$h_t = \sigma(z_t^o) \odot \tan(c_t) \qquad (6)$$

where $W_h \in R^{4D_h \times D}, W_E \in R^{4D_h \times D}, b \in R^{4D_h}, \sigma(\cdot)$ is sigmoid function, $\tan(\cdot)$ is hyperbolic tangent function, operator $\odot$ is vector inner product, D_h is the number of nodes in the LSTM unit. As can be seen from Eq. (5), forgetting gate $\sigma(z_t^f)$ is used to control the next iteration Information; Input gate $\sigma(z_t^i)$ determines the input information of the current feature data; It can be seen from Eq. (6) that output gate $\sigma(z_t^o)$ is used to control the information obtained in the current unit. Because sigmoid function is used in each gate, these gate structures can selectively forget or retain the related sequence information in the feature sequence. In addition, it can be seen from Eqs. (5) and (6) that the hyperbolic tangent function is used in the state and output of the LSTM unit, because the function can be approximately regarded as a linear function near the zero point, which is convenient for subsequent processing.

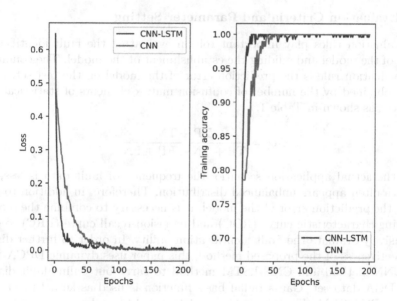

Fig. 4. Training accuracy and loss curves of CNN and CNN-LSTM.

4   Case Study

4.1   Fault Samples and Data Preprocessing

Through the data mining of DGA, we can find out many factors that cause transformer fault, such as high energy arc discharge, medium temperature overheating (300 – 700 °C), corona discharge, partial discharge, but the core problem is still to judge whether a long-term running transformer will be in danger of failure state. In this paper, 389 sets of transformer oil chromatogram data under normal working conditions and faults are collected. The characteristic gas is selected as: H_2, CH_4, C_2H_6, C_2H_4, C_2H_2, CO and CO_2. We marked the data that had been running for at least 5 years before the failure as normal conditions and the others as fault conditions. Under normal working conditions, the PPM concentration of dissolved gas in transformer is low, but it will be very high in case of transformer failure. This will cause the distribution of dissolved gas concentration to be partial rather than normal, and these extreme points will affect the modeling method. In order to relieve this effect, we use log transform to preprocess DGA data, the processed data is shown in Fig. 3. In order to ensure the stability of the numerical operation, we preprocess the DGA data after the logarithmic transformation with the mean value and variance normalization, and then select 80% of the data as the training set and the remaining 20% as the test set.

4.2   Evaluation Criteria and Parameter Setting

The evaluation rules play important role in evaluating the fault identification ability of the model and guiding the establishment of the model. The commonly used evaluation rule is the prediction error of the model on the test set, which can be obtained by the number of confusion matrix elements of statistical classification, as shown in Table 1.

$$\text{Acc} = \frac{\text{TP} + \text{TN}}{\text{TP} + \text{TN} + \text{FP} + \text{FN}} \tag{7}$$

In the actual application scenario, the frequency of fault data is less, and the data often appears unbalanced distribution. Therefore, in addition to evaluating the prediction error of the model, it is necessary to combine the receiver operating characteristic curve (ROC) and precision-recall curve (PRC) to comprehensively evaluate the fault identification ability. In order to further discuss the effectiveness of the proposed method, this paper uses dynamic DPCA-SVM [19], CNN method and CNN-LSTM method to carry out online fault diagnosis on DGA data set. Gauss radial basis function is used as kernel function in dynamic PCA-SVM. The parameter is set to $\gamma = 1/d_f$, where d_f are the number of features extracted by dynamic PCA. The parameter settings of CNN and CNN-LSTM are shown in Table 2.

Table 1. Confusion matrix for classification

Category	Positive output from model	Negative output from model
Positive class	True positive, TP	False negative, FN
Negative class	False positive, FP	True negative, TN

Table 2. The parameters for CNN and CNN-LSTM

Methods	Convolution	Optimizer	Number of nodes	Activation	Dropout
CNN	2D	Adam	–	ReLU	0.25
CNN-LSTM	1D	Adam	100	ReLU	0.2

4.3 Comparative Analysis

As shown in Fig. 4, the loss and accuracy of the proposed CNN-LSTM diagnosis method are stable in the modeling stage, and the proposed method has faster convergence speed than CNN method. For testing the fault identification capability on DGA data test set, we first take ROC curve as the evaluation criterion, as shown in Fig. 4.

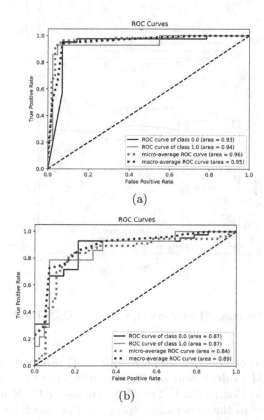

(a)

(b)

Fig. 5. ROC curves of DPCA-SVM, CNN and CNN-LSTM ($t = 10$).

It can be seen from Fig. 5(a) that in the case of time step, ROC curve of DPCA-SVM is mainly concentrated in the middle part of TP rate, while CNN model and CNN-LSTM model are basically in the upper left corner of coordinate axis, as shown in Fig. 5(b) and Fig. 5(c). This shows that CNN and CNN-LSTM can distinguish the fault signal from the normal signal accurately. The AUC values of DPCA-SVM, CNN and CNN-LSTM were 0.77, 0.98, 1.0 respectively. 0.77 < 0.98 <1.0, this shows that CNN-LSTM model is more sensitive to transformer faults. To further analyze the model's ability for identify unbalanced data, the model was evaluated by using the PRC curve shown in Fig. 6.

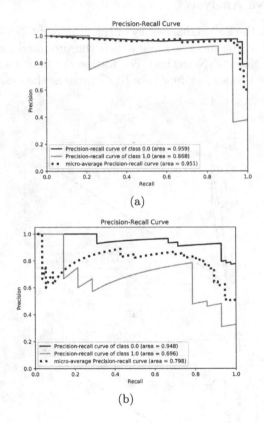

Fig. 6. PRC curves of DPCA-SVM, CNN and CNN-LSTM ($t = 10$).

From Fig. 6(a), it can be seen that the fault AUC value related to PRC curve fault of DPCA-SVM model is only 0.626, indicating that the fault state can not be identified well in this method. The PRC curves of CNN and CNN-LSTM models are basically in the upper right corner of the coordinate axis, as shown in Fig. 6(b) and (c), which show that the two methods can well deal with the imbalance of positive and negative sample proportions. From the AUC value of PRC curve, the fault AUC value related to CNN model is 0.966, which is less than the corresponding value 1 of CNN-LSTM model. This shows that CNN-LSTM model can well deal with data imbalance. Because the proposed method is an online modeling method, we need to analyze the sensitivity to time step. The 11 cases of time step $t = \{5, 8, 10, 13, 16, 18, 20, 23, 26, 28, 30\}$ are selected respectively. Through the comparison of the accuracy on the test set, it is found that the prediction accuracy of the proposed CNN-LSTM is basically maintained at about 92%, as shown in Fig. 7.

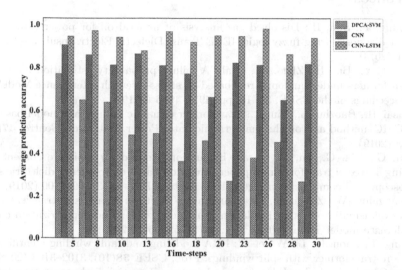

Fig. 7. Test accuracy of DPCA-SVM, CNN and CNN-LSTM over the different time steps.

5   Summary

In this paper, aiming to solving the problem of inaccurate feature vector extraction from DGA data by traditional methods, the convolution layer is used to extract spatial features, and give full play to the advantages of CNN model in spatial feature extraction. In order to further extract the time domain features from the data, LSTM model is used to deal with the timing and complex non-linear relationship of DGA data. Based on the feature extraction of space-time domain, combined with CNN and LSTM depth models, an online fault diagnosis method is proposed to effectively deal with the growing massive data in the transformer. The experimental results on DGA data of transformer show that the proposed CNN-LSTM model can identify faults well and the proposed method is not sensitive to the selection of time step.

Acknowledgements. This work was supported by National Natural Science Foundation of China (Grants No. 61703191), the Foundation of Liaoning Educational Committee (No. L2017LQN028) and the Scientific Research Foundation of Liaoning Shihua University (No. 2017XJJ-012).

References

1. Huang, Y., Sun, H.: Dissolved gas analysis of mineral oil for power transformer fault diagnosis using fuzzy logic. IEEE Trans. Dielectr. Electr. Insul. **20**(3), 974–981 (2013)
2. Cheng, Y., Bai, H., Zhao, Li., et al.: A failure probability estimation method of transformers under random stress based on stress-strength interference Model. In: Proceedings of the CSEE, 39, 16, pp. 4958–4966 (2019)
3. Saisai, H., Baozhu, L., Xin, A.: Transformer dynamic failure rate model based on MCMC method and oil chromatographic data. Power Syst. Prot. Control **47**(15), 1–8 (2019)
4. Wu, G., Xia, G., Su, M., et al.: Evaluation method for moisture content and aging degree of transformer oil-paper insulation based on frequency dielectric spectroscopy and compensation factor. High Voltage Eng. **45**(3), 691–700 (2019)
5. Guangning, W., Zongchao, D., Yuanguang, C., et al.: Analysis of transient moisture diffusion in oil-paper insulation using polarization and depolarization charge difference model. High Voltage Eng. **41**(2), 592–601 (2015)
6. Jiang, J., Zhou, L., Li, W., et al.: FRA modeling and fault winding identification of autotransformer with split windings. Proc. CSEE **38**(10), 3102–3108 (2018)
7. Xie, Q., Cheng, S., Lv, F.F.C., et al.: Location of partial discharge in transformer oil using circular array of ultrasonic sensors. IEEE Trans. Dielectr. Electr. Insul. **20**(5), 1683–1690 (2013)
8. IEEE Std C57.104-2019, IEEE Guide for the Interpretation of Gases Generated in Mineral Oil-Immersed Transformers (2019)
9. Wang, J.Y., Ling, M., Li, J.Z., et al.: DL/T 722–2014, Guide to the analysis and the diagnosis of gases dissolved in transformer oil. China Electric Power Press, Beijing (2015)
10. da Silva, I.N., Imamura, M.M., de Souza, A.N.: Application of neural networks to the analysis of dissolved gases in insulating oil used in transformers. In: Proceedings of IEEE International Conference on Systems, Man and Cybernetics, Nashville, October 8–11, pp. 2643–2648 (2000)
11. Su, Q., Mi, C., Lai, L.L., et al.: Fuzzy dissolved gas analysis method for the diagnosis of multiple incipient faults in a transformer. IEEE Trans. Power Syst. **15**(2), 593–598 (2000)
12. Chen, W.G., Pan, C., Yun, Y.X., et al.: Wavelet networks in power transformers diagnosis using dissolved gas analysis. IEEE Trans. Power Deliv. **24**(1), 187–194 (2009)
13. Wu, K., Kang, J.S., Chi, K.: Fault diagnosis method of power transformer based on improved multi-classification algorithm and correlation vector machine. High Voltage Eng. **42**(9), 3011–3017 (2016)
14. Zhang, Z.Y., Luo, S.H., Yue, H.T., et al.: Pattern recognition of acoustic signals of transformer core based on mel-spectrum and CNN. High Voltage Eng. **46**(2), 413–423 (2020)
15. Jia, J.L., Yu, T., Wu, Z.J., et al.: Transformer fault diagnosis method of convolutional neural network based on. Electric. Meas. Instr. **54**(13), 62–67 (2017)

16. Li, H., Zhang, Z.P., Zhang, Z.W.: Fault diagnosis of transformer based on convolutional neural network. J. Henan Polytechnic Univ. (Natural Sci.) **37**(6), 118–123 (2018)
17. Jiang, Y.Q., Huang, L., Wang, B., et al.: Transformer internal fault diagnosis based on DGA and deep belief network. Eng. J. Wuhan Univ **50**(6), 749–753 (2018)
18. Lecun, Y., Bottou, L., Bengio, Y., Haffner, P.: Gradient-based learning applied to document recognition. Proc. IEEE **86**(11), 2278–2324 (1998)
19. Dong, Y.N., Qin, S.J.: A novel dynamic PCA algorithm for dynamic data modeling and process monitoring. J. Process Control **67**, 1–11 (2018)

Robust Stability and Stabilization

Real-Time Path Planning with Application for Multi-UAVs to Track Moving Target

Xiaoyong Zhang[1] , Lili Li[1(✉)] , and Xin Ge[2]

[1] College of Marine Electrical Engineering, Dalian Maritime University,
Dalian 116026, China
lilili@dlmu.edu.cn
[2] College of Information Science and Technology, Dalian Maritime University,
Dalian 116026, China

Abstract. We propose a coordinated tracking strategy based on the Lyapunov Guidance Vector Field (LGVF) to solve the problem of moving target tracking in real-time path planning for multiple unmanned aerial vehicles (UAVs). First, a tracking strategy is proposed for high/low-altitude UAVs cooperative target tracking, the high altitude UAV monitors target, the low-altitude UAV detects and collects the status information of moving target. Second, to track a moving target and maintain wireless communication in a 3D environment, the LGVF method is improved by introducing flight height into the Lyapunov function, the limit cycle in the horizontal plane and the optimal height in the vertical plane are constantly being adjusted by attraction and repulsion function. Third, as the quality of the route is mainly influenced by the attraction parameters, repulsive parameters and the rate of convergence in the vertical direction, the real-time path can be planned by the rolling optimization according to the predicted motion. The experimental results validate the effectiveness of the tracking strategy.

Keywords: Real-time path planning · Lyapunov guidance vector field · Target tracking · Unmanned aerial vehicles

1 Introduction

In the last few years, UAVs have played an increasingly important role in military or civilian fields, e.g. moving target tracking, target searching and rescue operations, etc. [1, 2]. In complex tasks, path planning for UAVs is one of the critical UAV technologies to improve the work efficiency. Compared to the simple 2D path, a 3D route is more effective in improving UAV capabilities of low-altitude tracking [3]. The cooperative path planning problem of target tracking in a 3D environment is studied in this paper.

Recently, various studies have focused on target tracking using UAVs. In these works, many motion planning strategies have been proposed for multi-UAVs cooperative moving target tracking. To track the target in urban environments by UAVs, Wang et al. [4] propose a modified grey wolf optimizer with the Gaussian estimation of distribution strategy, which uses the Gauss probability model to select superior track

© Springer Nature Singapore Pte Ltd. 2020
J. Qian et al. (Eds.): ICRRI 2020, CCIS 1335, pp. 165–177, 2020.
https://doi.org/10.1007/978-981-33-4929-2_12

direction. In Ref [5], a new system framework and a new tracker are proposed to complete target tracking and real-time path planning by a UAV, with the advantages of superior tracking performance and computation costs reduction. A novel Hammerstein model-based predictive control method is proposed for target tracking of UAV to improve tracking performance and enhance robustness under external disturbances [6]. In Ref [7], the decision-making problem is modeled in the framework of distributed multi-agent partially observable Markov decision processes, which express a method of executing sequential actions to get the expected movement decision. Sun et al. [8] propose a fast convergent LGVF for multi-UAVs target tracking, the optimal parameter in guidance functions is estimated to get the best standoff distance. To maintain persistent track of target, [9] presents a target tracking framework based on particle filter prediction. LGVF and nonlinear model predictive control-based methods are applied to improve target position prediction accuracy. In Ref. [10], a guidance vector field method is proposed to make the UAV converge to a desired path. Besides, there are other algorithms used in this area, such as bionic algorithm, deep learning [11, 12]. But the above methods are only effective in 2D target tracking. What's more, the quality of detection information that a UAV acquires from a moving target is affected by the high altitude.

Therefore, a target tracking method based on the cooperation of UAVs at different altitudes is proposed. In particular, this method maintains communication distance and get the optimal performance of standoff tracking for UAVs by revising the optimal vertical height of low-altitude UAV. Simulation examples demonstrate the effectiveness of the presented strategy.

In Sect. 2, the UAV motion model and the problem model of path planning are constructed. In Sect. 3, the cooperative target tracking is described by an improved LGVF. In Sect. 4, the rolling optimization strategy is described. The simulation results are analyzed in Sect. 5. Finally, Sect. 6 draws the conclusion.

2   Modeling of Path Planning

2.1   Problem Description

In this paper, an effective target tracking strategy is put forward to increase the detection and tracking capabilities of multi-UAVs. Besides, flight safety between UAVs is considered for cooperative mechanisms with a stable communication network. To ensure the feasibility of the route, the planning path needs to obey the constraints of speed, turn rate and flight path angle. The discrete waypoints are generated quickly with the absence of the aforementioned constraints, and the final flight path is constituted by connecting these discrete waypoints.

2.2 Modeling of UAV

UAVs are simplified into a three-degree-of-freedom particle model. The motion model can be described as:

$$\begin{cases} \dot{x}_u = V \cos \theta \cos \varphi \\ \dot{y}_u = V \cos \theta \sin \varphi \\ \dot{z}_u = V \sin \theta \end{cases} \tag{1}$$

where $P_u = (x_u, y_u, z_u)$ is the position coordinates of UAV, $(\dot{x}_u, \dot{y}_u, \dot{z}_u)$ is the velocity component of UAV; V is flight speed; the sampling time is Δt, for any waypoint $P_k = (x_k, y_k, z_k), \forall k = 1, 2, \cdots K$, θ is the flight path angle, φ is the heading angle, $\dot{\varphi}$ is the turn rate. The any waypoint should satisfy the following constraints:

$$\begin{cases} |\varphi_{k+1} - \varphi_k| \leq \dot{\varphi} \cdot \Delta t \\ \theta_{\min} \leq \theta \leq \theta_{\max} \\ V_{\min} \leq V \leq V_{\max} \\ h_{\min} \leq h \leq h_{\max} \end{cases} \tag{2}$$

where $h_{\min}$ and $h_{\max}$ are the minimum and maximum altitudes. In addition, to avoid collision and maintain wireless communication, the distance d between two UAVs should be bigger than the desired minimum safety distance $d_{\min}$ and smaller than the desired maximum communication distance $d_{\max}$.

2.3 Modeling of Target

Denote the position coordinates and the velocity components of a target as $P_t = (x_t, y_t, z_t)$ and $(v_{tx}, v_{ty}, 0) = (\dot{x}_t, \dot{y}_t, 0)$ respectively. It is assumed that the target's speed is always less than the UAV's speed. According to [3], UAVs should converge to the desired limit cycle above the target which is composed of the horizontal convergence distance R and the optimal height H are usually influenced by many factors. In this paper, we assume that R and H of low-altitude UAVs are constantly changing.

3 Cooperative Moving Target Tracking by LGVF

In Ref. [8], the LGVF is used for target tracking in a 2D environment making UAVs phase uniformly distributed on the limit cycle via steering control and speed control based on the Lyapunov phase function. For tracking missions in 3D space, the LGVF method will be improved for UAVs by introducing the vertical component.

3.1 Target Tracking by High-Altitude UAV

It is supposed that the UAV tracks a moving target $P_t = (x_t, y_t, z_t)$ at the constant horizontal speed $(v_{\min} \leq v_0 \leq v_{\max})$, $(v_x, v_y, v_z) = (\dot{x}_u, \dot{y}_u, \dot{z}_u)$ and vertical velocity

$v_z = 0$, flight path angle $\theta = 0$ and the optimal height $H = z_u$. The Lyapunov distance function can be expressed as:

$$V(R) = \frac{1}{2}(r^2 - R^2)^2 \tag{3}$$

Based on Eq. (3), the guidance vector field can be inferred, where the desired velocity $\mathbf{v}$ is calculated as follows:

$$\mathbf{v} = \begin{bmatrix} v_x \\ v_y \end{bmatrix} = \frac{v_0}{r \cdot \sqrt{r^2 + R^2}} \cdot \begin{bmatrix} -x_{ut}(r^2 - R^2) - y_{ut}(2rR) \\ -y_{ut}(r^2 - R^2) + x_{ut}(2rR) \end{bmatrix} \tag{4}$$

$r = \sqrt{x_{ut}^2 + y_{ut}^2}$ is the distance between the high altitude UAV and the target, $x_{ut} = x_u - x_t$, $y_{ut} = y_u - y_t$. Based on Eq. (4), the horizontal speed is always equal to v_0. Then it yields that:

$$\frac{dV(R)}{dt} = \begin{bmatrix} \frac{\partial V(R)}{\partial x} & \frac{\partial V(R)}{\partial y} \end{bmatrix} \begin{bmatrix} v_x \\ v_y \end{bmatrix} = \frac{-4rv_0(r^2 - R^2)^2}{r^2 + R^2} \tag{5}$$

From Eq. (5), it is obvious that $dV(R)/dt \leq 0$. In particular, when $dV(R)/dt = 0$, UAV is on the limit cycle. On the basis of the Lasalle invariance principle, the velocity $\mathbf{v}$ will guide UAV to converge to this limit cycle. It is easy to get that $\varphi = \arctan(v_x/v_y)$ and $\dot{\varphi} = 4v_0 \cdot \frac{R \cdot r^2}{(r^2 + R^2)^2}$ which gives that $\dot{\varphi}_{max} = \frac{v_0}{R}$ when the horizontal convergence distance R equals to r. To guarantee the feasibility of the planning path, the constraint condition $\frac{v_0}{R} \leq \dot{\varphi}_{max}$ should be fulfilled.

There is a moving ground target with velocity $v_t^T = \begin{bmatrix} v_{tx} & v_{ty} & v_{tz} \end{bmatrix}$ and $v_{tz} = 0$, the relative Lyapunov guidance vector field can be redefined as:

$$v = a \begin{bmatrix} v_x \\ v_y \end{bmatrix} + \begin{bmatrix} v_{tx} \\ v_{ty} \end{bmatrix} \tag{6}$$

where a is the horizontal speed correction coefficient calculated by:

$$(v_x^2 + v_y^2) \cdot a^2 + 2(v_x v_{tx} + v_y v_{ty}) + v_{ty}^2 + v_{tx}^2 - v_0^2 = 0 \tag{7}$$

Since the speed of the target is always less than the speed of the UAV, the inequality $a > 0$ holds, so it could be inferred that:

$$\frac{dV(R)}{dt} = \frac{-4arv_0(r^2 - R^2)^2}{r^2 + R^2} \leq 0 \tag{8}$$

For the mission of tracking a moving target, the UAV can still converge to the limit cycle, and the velocity component of UAV is $[av_x + v_{tx} \quad av_y + v_{ty} \quad 0]^T$. Then one has $\varphi = \arctan\left(\frac{av_y + v_{ty}}{av_x + v_{tx}}\right)$, and $\dot{\varphi}$ is calculated to meet $\dot{\varphi} \leq 4(a^2 + a)v_0 \cdot \frac{r^2 R}{(r^2 + R^2)^2}$. To ensure that the path is feasible, the condition $R \geq \frac{v_0(a^2 + a)}{\dot{\varphi}_{max}}$ must be fulfilled.

3.2 Target Tracking by Low-Altitude UAV

The low-altitude UAV not only need to monitor moving targets, but also need to maintain communication with high-altitude UAV. To meet the urgent needs of the surveillance missions, the LGVF method is improved for low-altitude UAV by introducing a vertical component into the Lyapunov function, and the generated velocity component can guide the UAV to converge to the optimal height in the vertical plane. As the relative positions between the high-altitude UAV and the target is constantly changing, the optimal horizontal distance R and optimal vertical height H of low-altitude UAV need to be constantly adjusted.

Suppose a low-altitude UAV tracks a moving ground target $P_t = (x_t, y_t, 0)$ at a constant flight speed v_0. By introducing the vertical component, the Eq. (3) can be redefined as:

$$V(d) = \frac{1}{2}(r^2 - R^2)^2 + \frac{1}{2}(z_{ut}^2 - H^2)^2 \tag{9}$$

From Eq. (9), the velocity $\mathbf{v}$ of the low-altitude UAV can be deduced as follows:

$$\mathbf{v} = \begin{bmatrix} v_x \\ v_y \\ v_z \end{bmatrix} = \frac{v_0}{r \cdot \sqrt{(r^2 + R^2)^2 + \lambda^2(z_{ut}^2 - H^2)^2}} \cdot \begin{bmatrix} -x_{ut}(r^2 - R^2) - y_{ut}(2rR) \\ -y_{ut}(r^2 - R^2) + x_{ut}(2rR) \\ -\lambda r(z_{ut}^2 - H^2) \end{bmatrix} \tag{10}$$

where $z_{ut} = z_u - z_t$ is the vertical distance, λ is the convergence coefficient in the vertical direction. Then the velocity $\mathbf{v}$ is considered as the low-altitude UAV velocity, which yields that:

$$\frac{dV(R)}{dt} = \begin{bmatrix} \frac{\partial v_0}{\partial x} & \frac{\partial v_0}{\partial y} & \frac{\partial v_0}{\partial z} \end{bmatrix} \begin{bmatrix} v_x \\ v_y \\ v_z \end{bmatrix} = \frac{-2rv_0(r^2 - R^2)^2 - 2\lambda z_{ut}v_0(z_{ut}^2 - H^2)^2}{\sqrt{(r^2 + R^2)^2 + \lambda^2(z_{ut}^2 - H^2)^2}} \tag{11}$$

The UAV is always above the targets, so $\frac{dV(R)}{dt} \leq 0$ always holds. When $r = R$ and $z_{ut} = H$, $\frac{dV(R)}{dt} = 0$ holds.

To meet the monitoring target and maintain wireless communications, we define the continuous function to adjust the optimal horizontal distance R and optimal vertical height H. The continuous function is as follows:

$$\beta(d) = \begin{cases} 1 + \vartheta \frac{2}{\pi} \arctan(\frac{1}{d} - \frac{1}{d_{max}})^2, & d > d_{max} \\ 1, & d_{max} > d > d_{min} \\ 1 - \vartheta \frac{2}{\pi} \arctan(\frac{1}{d_{min}} - \frac{1}{d})^2, & d < d_{min} \end{cases} \quad (12)$$

where ϑ is the regulate parameters. The H and R at the time t_{k+1} can be redefined as:

$$H(k+1) = f(d) \cdot H(k)$$
$$R(k+1) = \sqrt{d_t^2 - H(k+1)^2} \quad (13)$$

where d_t is the optimal detection range. Next, we analyze the feasibility of the path, mainly considering φ and θ. For discrete waypoint, $\varphi = \arctan\left(\frac{v_x}{v_y}\right)$. Hence, we can deduce:

$$\dot{\varphi} = 4v_0 \frac{r^2 + R^2}{\sqrt{(r^2+R^2)^2 + \lambda^2 (z_{ut}^2 - H^2)^2}} \cdot \frac{Rr^2}{(r^2+R^2)^2} \quad (14)$$

When $z_{ut} = H$ and $r = R$, the low-altitude UAV is on the limit cycle, which yields $\dot{\varphi}_{max} = \frac{v_0}{R}$. To ensure the feasibility of the planned route, we need to fulfill the condition $R \geq \frac{v_0}{\dot{\varphi}_{max}}$. The flight-path angle is $\theta = \arcsin(\frac{v_z}{v_0})$. The minimum value is $\theta_s = $

$\arcsin(\frac{-\lambda(h_{max}^2 - H^2)}{\sqrt{R^4 + \lambda^2 (h_{max}^2 - H^2)^2}})$ when $z_{ut} = h_{max}$ and $r = R$. The maximum value is $\theta_b = $

$\arcsin(\frac{-\lambda(h_{min}^2 - H^2)}{\sqrt{R^4 + \lambda^2 (h_{min}^2 - H^2)^2}})$ when $z_{ut} = h_{min}$ and $r = R$. h_{max} and h_{min} are the maximum and minimum altitude. The flight-path angle θ satisfies the constraint $\theta_{min} \leq \theta_s \leq \theta \leq \theta_b \leq \theta_{max}$.

When tracking a moving ground target with velocity v_t and $v_{tz} = 0$, to adjust the height of the low-altitude UAV, we obtain the relative LGVF according to Eq. (6):

$$v = a \begin{bmatrix} v_x \\ v_y \\ v_z \end{bmatrix} + \begin{bmatrix} v_{tx} \\ v_{ty} \\ v_{tz} \end{bmatrix} \quad (15)$$

where a is the speed correction coefficient, the velocity should remain the constant v_0 by Eq. (15), so the following condition holds:

$$(v_x^2 + v_y^2 + v_z^2) \cdot a^2 + 2a(v_x v_{tx} + v_y v_{ty} + v_z v_{tz}) + (v_{tx}^2 + v_{ty}^2 + v_{tz}^2) - v_0^2 = 0 \quad (16)$$

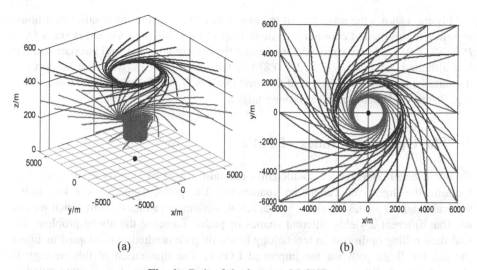

(a)   (b)

Fig. 1. Path of the improved LGVF

Figure 1 and Fig. 2 illustrate the UAVs from a series of start points ($x_u \in [-6, 6], y_u \in [-6, 6], z_u \in [0.1, 0.3]$) by improved and traditional LGVF functions respectively. Assuming that the target position is $P_t(0, 0, 0)$ km, the limit cycle radius of the high-altitude UAV is $R = 2$ km, the limit cycle radius of the low-altitude UAV is $R = 1$ km, and the optimal vertical height of low-altitude UAV is $H = 0.1$ km. When the UAV starts to move, the improved LGVF will always guide it gradually to converge to the limit cycle at the optimal vertical height. However, the traditional LGVF only guides the UAV gradually to converge to the limit cycle on the horizontal plane.

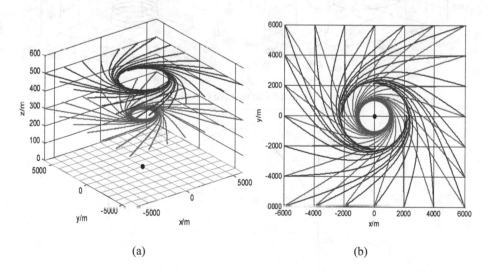

(a)   (b)

Fig. 2. Path of the traditional LGVF

Figure 3 shows the processes of tracking a moving target in the relation coordinate system and the ground coordinate system, respectively. Suppose the target starts from $P_t = (0, 2, 0)$ km in the sinusoidal motion on the horizontal plane, and the start point of low-altitude UAV is $P_u = (1.4, 2.8, 0.8)$ km, the surveillance range is $d_t = 0.8$ km and the initial optimal vertical height of the limit cycle is $H = 0.3$ km. As can be seen from Fig. 3, the UAV can track the moving target.

4   The Optimizing Index of Path

As the selected convergence coefficient λ and adjustment parameters ϑ are aimless or random, the flight path cannot be guaranteed. The paths, with $d_t = 0.8$ km, initial vertical height $H = 0.3$ km and $\lambda = 0.5, 1.5, 4$, are shown in Fig. 4, from which we can see that different λ yields different shapes of paths. To solve the above problem, the real-time rolling optimization technology based on path prediction is adopted to adjust the reliable flight path via the improved LGVF. The illustration of this strategy is shown in Fig. 5. When the UAV reaches S_k, a local path $\{S_k, \cdots S_{k+N}\}$ represented as a sequence of the forthcoming N discrete waypoints is generated by rolling optimization. Then the UAV flies from S_k to S_{k+1} and the above optimization process is executed again to get a new path $\{S_{k+1}, \cdots S_{k+N+1}\}$.

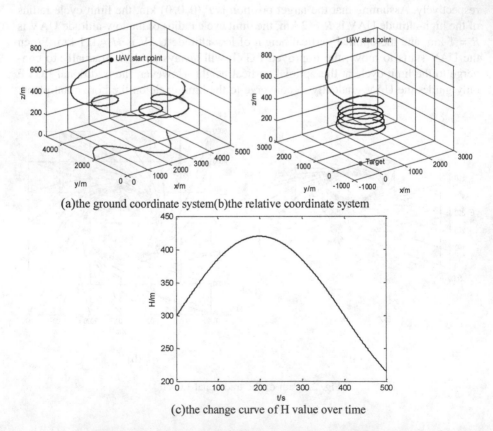

(a)the ground coordinate system(b)the relative coordinate system

(c)the change curve of H value over time

Fig. 3. Tracking of a moving ground target by low-altitude UAV

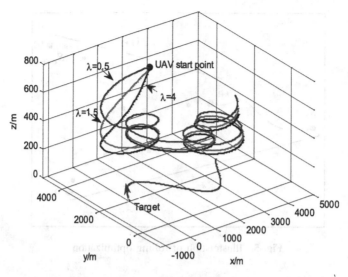

Fig. 4. Path of moving target tracking with $\lambda = 0.5, 1.5, 4$

We define the evaluation function $J(k)$ including the index of target tracking $M(k)$ and path smoothness $L(k)$:

$$J(k) = w_1 \cdot M(k) + w_2 \cdot L(k) \tag{17}$$

where w_1 and w_2 are the weighting factors of two indexes with $w_1 + w_2 = 1$.

The initial velocity guides the UAV to track the target through LGVF. Therefore, the deviation value of the adjustment velocity relative to the initial one can be regarded as the target tracking index.

$$M(k) = \sum_{i=k}^{k+N+1} \langle \mathbf{v}_{k+1}, \mathbf{v}_k \rangle \tag{18}$$

where $\langle \mathbf{v}_{k+1}, \mathbf{v}_k \rangle$ is the angle between $\mathbf{v}_{k+1}$ and $\mathbf{v}_k$. N is the rolling optimization step size. Besides, $L(k)$ should be introduced to evaluate the smoothness of the path:

$$L(k) = \frac{1}{N} \sum_{i=k}^{k+N+1} \left(u_1 \cdot \left| \varphi_{i+1} - \varphi_i \right| + u_2 \cdot \left| \theta_{i+1} - \theta_i \right| \right) \tag{19}$$

Where u_1 and u_2 are the corresponding weight factors.

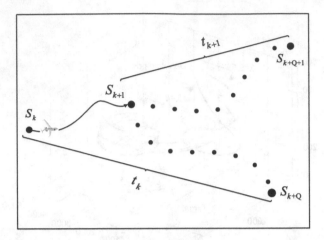

Fig. 5. Illustrastion of rolling optimization

5 Simulation

The proposed method is simulated in MATLAB R2014a on the computer. The necessary parameters of path planning are listed in Table 1. Suppose that the target moves in the horizontal plane with the motion model $x = 10 * t$, and $y = 2000 + 1000 * \sin(0.01781 * t)$, $z = 0 * t$, $t \in [0, 400]$. Two UAVs will track a target with the initial positions $(600, 1600, 800)$m and $(1600, 1600, 600)$m.

Figure 6(a) shows the paths with the high/low-altitude tracking strategy via the LGVF method. All the paths converge to the respective limit cycle and optimal vertical height. The vertical information of paths is shown in Fig. 6(b). Figure 6(c) illustrates the distance among UAVs. It is obvious that when the distance between two UAVs is greater than d_{max}, the value of H gradually increases, which means that the distance between two UAVs gradually decreases until it is less than the communication distance. It can be seen that the gravitational function and the repulsion function have good characteristics of adjusting the communication distance.

Figure 7 shows the paths under the one-step planning strategy with the LGVF which means the rolling optimization strategy is not used. As shown in Fig. 7, the path quality, especially the smoothness, is poor than Fig. 6(a). If the rolling optimization is adopted, the path will be much smoother.

The simulation results indicate that the improved LGVF method and high/low-altitude tracking strategy can resolve the 3D cooperative ground target tracking mission.

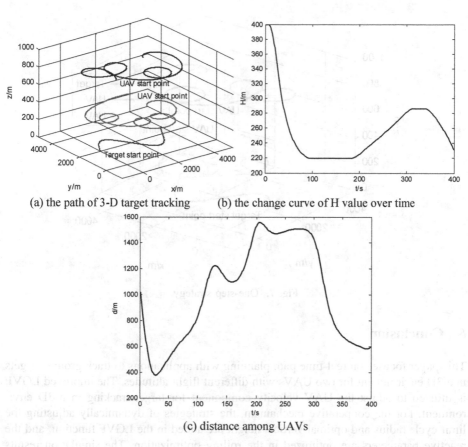

(a) the path of 3-D target tracking   (b) the change curve of H value over time

(c) distance among UAVs

Fig. 6. Cooperative path planning

Table 1. Parameters of path planning.

Parameter	Simple	Value
Sampling time	Δt(s)	1
Flight velocity	v_0(Km/s)	50
Minimum and maximum flight path angle	θ_{min}(deg) and θ_{max}(deg)	−35,35
Minimum and maximum flight altitude	h_{min} and h_{max}(m)	0,800
Minimum and maximum distance	d_{min} and d_{max}(m)	400,1400
Weight values of path cost	w_1 and w_2	0.7, 0.3
Weight values of smoothness	u_1 and u_2	0.5, 0.5
Range of planning steps	N	5
Maximum UAV turn rate	$\dot{\varphi}$(deg/s)	35

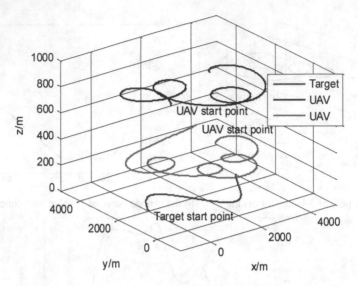

Fig. 7. One-step strategy

6 Conclusion

This paper focuses on real-time path planning with applications to track ground targets in a 3D environment for two UAVs with different flight altitudes. The improved LGVF is utilized to adjust the UAV velocity component for target tracking in a 3D environment. For the cooperative mechanism, the strategies of dynamically adjusting the limit cycle radius and optimal vertical height are added in the LGVF function, and the reactive parameters are optimized in the rolling optimization. The simulation results show that the planned path can guide the UAVs to track targets under certain optimization criteria. In the future, we will use this strategy on real UAVs.

References

1. Yin, C., Xiao, Z., Cao, X., et al.: Offline and online search: UAV multiobjective path planning under dynamic urban environment. IEEE Internet Things J. **5**(2), 546–558 (2018)
2. Zhang, Q., Wang, R., Yang J., Ding, K., et al.: Modified collective decision optimization algorithm with application in trajectory planning of UAV. Appl. Intell. **48**(8), 2328–2354 (2018)
3. Yao, P., Wang, H., Su, Z.: Real-time path planning of unmanned aerial vehicle for target tracking and obstacle avoidance in complex dynamic environment. Aerosp. Sci. Technol. **47**, 269–279 (2015)
4. Wang, X., Zhao, H., Han, T., et al.: A grey wolf optimizer using Gaussian estimation of distribution and its application in the multi-UAV multi-target urban tracking problem. Appl. Soft Comput. **78**, 240–260 (2019)

5. Liu, Y., Wang, Q., Hu, H., et al.: A novel real-time moving target tracking and path planning system for a quadrotor UAV in unknown unstructured outdoor scenes. IEEE Trans. Syst. Man Cybern. Syst. **49**(11), 2362–2372 (2019)
6. Altan, A., Hacıoğlu, R.: Model predictive control of three-axis gimbal system mounted on UAV for real-time target tracking under external disturbances. Mech. Syst. Sig. Process. **138**, 1–23 (2020)
7. Zhao, Y., Wang, X., Wang, C., et al: Systemic design of distributed multi-UAV cooperative decision-making for multi-target tracking. Autonom. Agents Multi-Agent Syst. **33**(1–2), 132–158 (2019)
8. Sun, S., Wang, H.P., Liu, J., et al.: Fast lyapunov vector field guidance for standoff target tracking based on offline search. IEEE Access **7**, 124797–124808 (2019)
9. Hyondong, O., Seungkeun, K.: Persistent standoff tracking guidance using constrained particle filter for multiple UAVs. Aerosp. Sci. Technol. **84**, 257–264 (2019)
10. Rezende, A.M.C., Gonçalves, V.M., Raffo, G.: Robust fixed-wing UAV guidance with circulating artificial vector fields. In: IEEE International Conference on Intelligent Robots and Systems, pp. 5892–5899. Institute of Electrical and Electronics Engineers Inc., Madrid, Spain (2018)
11. Tian, G., Zhang, X., Bai, X.: Real-time dynamic track planning of multi-UAV formation based on improved artificial bee colony algorithm. In: Proceedings of the 37th Chinese Control Conference, pp. 10055–10060. IEEE Computer Society, Wuhan, China (2018)
12. Li, B., Wu, Y.: Path planning for UAV ground target tracking via deep reinforcement learning. IEEE Access **8**, 29064–29074 (2020)

Research on Energy Management Strategy of Isolated Microgrid with Photovoltaic Power Generation

Wenrui Li[1]([✉]), Cunxu Wang[2], Dongxu Dai[3], Na Zhao[4],
Yuanyuan Su[1], and Kai Guo[4]

[1] Graduate Department, Shenyang Institute of Engineering,
Shenyang 110136, Liaoning, China
13835041569@163.com
[2] Institute of Automation, Shenyang Institute of Engineering,
Shenyang 110136, Liaoning, China
[3] State Grid Benxi Electric Power Supply Company, State Grid Liaoning
Electric Power Supply Co., Ltd., Benxi 117000, Liaoning, China
[4] State Grid Huludao Electric Power Supply Company, State Grid Liaoning
Electric Power Supply Co., Ltd., Huludao 125000, Liaoning, China

Abstract. The globalized stone energy crisis and the climate and environmental crisis have promoted the adjustment and transformation of the energy industry structure worldwide. As a key area of the energy development strategy, the power industry will develop towards clean and efficient renewable energy. With the development of new material technology and control technology, the construction of micro-grid is bound to become a trend, and the large-scale development of micro-grid will affect the operation of the power grid, so the use of advanced control strategies and energy management technologies to improve the performance of micro-grid imperative. This paper first establishes the mathematical model of the independent micro-grid operation mode, and establishes the objective function from the constructed mathematical model. Then, through the improved genetic algorithm to coordinate and optimize the multi-objective. Finally, a micro-grid structure is built on the MATLAB platform to simulate the control strategy of the independent operation mode of the micro-grid. The experimental results show that the strategy proposed in this paper is feasible and correct for energy optimization of the independent micro-grid containing photovoltaic power generation Grid energy coordination and optimization improves the utilization rate of energy, maximizes the consumption of renewable energy, reduces the operating costs of micro-grids, and improves the comprehensive economic benefits of micro-grids.

Keywords: Independent micro-grid · Energy management · Photovoltaic power generation · Goal optimization

J. Qian et al. (Eds.): ICRRI 2020, CCIS 1335, pp. 178–190, 2020.
https://doi.org/10.1007/978-981-33-4929-2_13

1 Introduction

With the rise of new energy technology, combined with energy storage technology, micro-grid as a new power grid technology has become a research focus. Part of the intermittent power supply is constrained by natural factors and cannot be stable for a long time. For example, wind power and photovoltaic power generation are severely constrained by meteorological factors. There is great randomness, volatility and uncertainty. The impact is small, but when it is developed to a certain scale, if it is not controlled, direct grid connection will have an impact on the power grid and affect the stable operation of the power grid. As an important technical component of the smart grid, micro-grid technology can be combined with various available resources in the area for full coordinated scheduling and utilization, which can greatly improve the power grid's ability to absorb intermittent power supplies; improve the efficiency of renewable energy utilization, To reduce power grid losses and optimize the operation mode of the power grid; to ensure the power supply of critical loads in the event of failure; can be used to solve the power consumption problems of users in remote areas.

At present, there have been related literature on the research of energy optimal dispatching of micro-grid. From the point of view of the operation mode of micro-grid, it can be divided into two types: the research on the energy optimal scheduling in the mode of isolated operation of micro-grid and the research on the energy optimal scheduling in the mode of grid-connected operation. Yin et al. aimed at the frequency stability problem during the process of grid-connected micro-grid transfer to islands, through two-stage, three-frequency energy management strategy to achieve stable control of micro-grid frequency [1]; Zhang et al. mainly operates in micro-grid islands An energy management control method was proposed for home application scenarios [2]; Zhang et al. considered the impact of large disturbances on the micro-grid during the operation of the isolated grid, and used energy storage devices to provide priority frequency control strategies for large disturbance events [8]. To maintain the stability of the system under the big disturbance of the micro-grid as the operating goal, rationally arrange energy storage and distributed power output, to achieve the micro-grid economic and environmentally friendly optimized operation in the isolated grid mode; Zhou et al. for wind and solar storage micro-grid To ensure the reliability of system power supply during operation, the distributed capacity allocation problem was studied, and the multi-objective optimization problem of micro-grid economy, environmental protection and energy utilization efficiency was established, and the model was solved using the electromagnetic simulation algorithm [9]; Liao. for energy optimization management of micro-grid, two links of day-to-day scheduling and day-to-day scheduling were established [11]. Based on the day-to-day load and photovoltaic forecast data, considering the factors such as the purchase and sale of electricity prices from the grid, the economic goal optimization function of the micro-grid is established. Intra-day scheduling is based on the day-to-day scheduling. The power fluctuations caused by the inaccuracy of the prediction using energy storage batteries Carry out stabilization and introduce additional costs to obtain a more accurate total operating cost of the micro-grid cycle. At present, most of the literature uses genetic algorithms [10], particle swarm optimization [12] and other evolutionary algorithms (Evolutionary

Algorithms, EAs). However, the above evolutionary algorithm cannot obtain reasonable values for multi-objective optimization of micro-grid energy with different structures and different operating modes, which makes micro-grid energy not optimally utilized.

In view of the above problems, this paper proposes an improved genetic algorithm to study the energy management of independent micro-grids containing photovoltaic power generation. First, the mathematical model of the independent micro-grid operation mode is established, and the optimization objective function is established accordingly. Then, through the improved genetic algorithm to coordinate and optimize the multi-objective. Finally, a micro-grid structure was built on the MATLAB platform to simulate the control strategy of the independent operation mode of the micro-grid. The result analysis shows that the proposed algorithm has important guidance for energy optimization of the independent micro-grid containing photovoltaic power generation. To increase the utilization of energy and maximize the consumption of renewable energy.

2   Research on Energy Management of Independent Micro-Grid Considering Randomness

2.1   Energy Management Strategy for Independent Operation of Micro-Grid

When the micro-grid is running off-grid, there is no external grid support. It is necessary to improve the stability of the micro-grid operation as much as possible through energy management strategies, control the frequency and voltage stability of the micro-grid, and ensure the uninterrupted power supply of important loads. During off-grid operation, the micro-grid can only obtain electrical energy through intermittent power supply, energy storage, and backup power because it cannot obtain electrical energy from the power grid. Considering that there are certain errors in the wind and light forecasting and load forecasting, the scheduling instructions may not be able to control the stable operation of the micro-grid, and the system may generate too much power and the energy storage cannot be charged or the emergency power supply is full and the energy storage cannot be discharged Extreme situation where load demand cannot be met. At this time, we can only consider maintaining the stability of the system by "abandoning the light and abandoning the wind" or cutting off the load. This situation needs to be avoided as much as possible. The basic control strategy for the independent operation of the micro-grid is as follows:

1) For renewable energy power generation, such as wind power generation, photovoltaic power generation, etc., it should be ensured that they operate in the maximum power tracking mode as much as possible. New energy power generation works near the maximum power point and can maximize the use of natural resources such as wind and light.

2) Energy storage keeps the charge level at a reasonable percentage as much as possible, and can meet the needs of charging or discharging at various points in time.

When renewable energy is abundant and the output is greater than the load demand, the energy storage is charged first, and the output of the diesel generator can be reduced. The energy storage must keep the charge level lower than the maximum allowable charge, and the output of the diesel generator is not lower than the minimum limit, otherwise, the excess wind and light output will be treated in a way of abandoning the wind and the light. When the diesel generator is full, and the power generated by the micro-grid cannot meet the load demand, the energy storage is discharged, but if the energy storage power is not higher than the minimum power, only some non-important loads can be cut to maintain the stability of the system.

3) The diesel generator is used as a backup power source and can be used as the main power source to maintain the micro-grid operation when the micro-grid is off-grid. The diesel engine must have a stable power output and continue to operate when the micro-grid is off-grid.

Figure 1 is a flow chart of the basic control strategy for off-grid operation of the micro-grid, where $f_{N,\min}$ represents the lower limit of the allowed system frequency fluctuations, and $f_{N,\max}$ represents the upper limit of the allowed system frequency fluctuations.

2.2   Dispatch Flexibility of Independent Operation of Micro-Grid

The power system should have the ability to deal with the randomness factors in the system, that is, the flexibility of the power system. At present, there is no uniform definition of power system flexibility. However, based on the current research at home and abroad, power system flexibility should include reliability, directionality, timeliness, supply and demand balance and other factors [3]. NERC believes that power system flexibility is the ability of the system to respond to random changes on both sides of supply and demand. NERC believes that the flexibility of the power system can also be expressed as the ability of the system to store energy; the efficient unit combination and the ability of the system to economically optimize [4]. Literature [5] believes that flexibility is used to reflect the system's ability to cope with volatility and randomness on both sides of supply and demand, and maintain the system's stable operation under certain economic costs. Literature [6] discusses the utility of flexibility in dispatch operation of power systems containing intermittent energy.

The evaluation of system flexibility mainly depends on the difference between the adjustment capability provided by the system and the adjustment capability required by the system. At present, most of the indicators of equipment flexibility evaluation mainly consider capacity constraints, minimum output constraints, and climbing constraints, start and stop time constraints. There are usually intermittent power sources ouoh as wind and solar in the micro-grid, which is limited by the current data prediction errors. Micro-grid operations often face greater randomness. Therefore, this paper draws on the evaluation of flexibility in the power system to measure the off-grid operation of the micro-grid. Ability to deal with randomness. Controllable micro-source such as diesel generators and energy storage in the micro-grid, as well as schedulable loads, can be used as sources of the micro-grid's ability to cope with

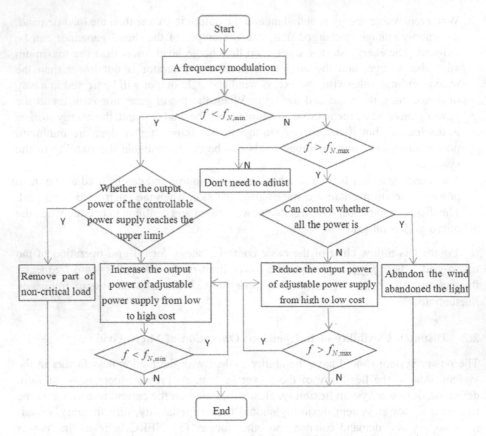

Fig. 1. Basic control strategy of micro-grid operation under island coed.

randomness. The randomness factors faced by the micro-grid are mainly caused by intermittent power sources such as wind and solar power generation. Errors also produce uncertainties. This paper introduces the concept of flexible margin insufficiency rate [7], which is applied to the optimization model of micro-grid island operation energy management, and evaluates the flexibility of micro-grid off-grid operation by calculating the adjustable margin of the controllable micro-source output in each period. There are two directions of flexibility and flexibility. For the generator model, the calculation formula [3] is as follows:

$$F_G^{up} = \min \begin{pmatrix} \left(P_G^{max} - P_G(t), \Delta t \cdot R_G^{up}, \Delta t \cdot R_G^{up} - \Delta P_G\right) \\ \left(P_G(t+1) - P_G(t) + \Delta t \cdot R_G^{up}\right) \end{pmatrix} \tag{1}$$

$$F_G^{down} = \min \begin{pmatrix} \left(P_G(t) - P_G^{min}, \Delta t \cdot R_G^{down}, \Delta t \cdot R_G^{down} + \Delta P_G\right) \\ \left(P_G(t) - P_G(t+1) + \Delta t \cdot R_G^{down}\right) \end{pmatrix} \tag{2}$$

where: $P^{down}(t)$ represents the downward fluctuation power that may occur in the Δt time period, $P^{up}(t)$ represents the upward fluctuation power that may occur in the Δt time period, f_{IP}^{up} represents the upward flexible margin insufficient rate, and f_{IP}^{down} represents the downward flexible margin insufficient rate.

3 Construction of Independent Micro-Grid Operation Model with Photovoltaic Power Generation System

In this paper, the micro-grid dispatching model takes 24 h as a dispatching cycle, and the optimization goals include economy and environment. The basic model is as follows:

$$
\begin{aligned}
\min F(X) = \min[f_1(X), f_2(X), \cdots, f_n(X)] \\
s.t. \ X \in \Omega
\end{aligned}
\tag{3}
$$

$$
G(X) = 0
\tag{4}
$$

$$
L(X) \leq 0
\tag{5}
$$

where: F includes optimization goals such as comprehensive economy and ability to deal with randomness; f_i represents the third sub-goal; i is the decision variable; X is the feasible solution space of the decision variable; Ω represents the equality constraint; L represents the inequality constraint.

3.1 Objective Function

(1) Economic goals

The economic objectives mainly include maintenance costs during the operation of the micro-grid, fuel costs for some micro-power sources, and initial construction costs of the micro-grid. Unless otherwise specified, the cost unit is yuan, the power unit is kW, and T is the optimization period.

The fuel cost calculation formula is as follows:

$$
C_{FUEL} = \sum_{t=1}^{T} \left(\sum_{i=1}^{N} e_{PRICE,i} \cdot P_i(t) \cdot \Delta t \right)
\tag{6}
$$

where: $e_{PRICE,i}$ is the market price of fuel i; $P_i(t) \cdot \Delta t$ refers to the output energy of the generator in period Δt, and t refers to the time.

The maintenance cost calculation formula is as follows:

$$
C_{MTEN} = \sum_{t=1}^{T} \left(\sum_{i=1}^{N} (K_{OM,i} \cdot P_i(t)) \cdot \Delta t \right)
\tag{7}
$$

where: $K_{OM,i}$ is the operation and maintenance coefficient of the equipment i; $P_i(t) \cdot \Delta t$ is the amount of equipment capacity to be maintained during the period of Δt.

The formula for calculating the overall economic cost C_{ECO} is as follows:

$$\min C_{ECO} = \min C_{FUEL} + \min C_{MTEN} \tag{8}$$

(2) Environmental cost

The energy production process of micro-grid operation may produce pollutant emissions. Considering the impact of different pollutants on the environment, this paper uses the following formula to calculate the environmental cost C_{EN}:

$$\min C_{EN} = \min \sum_{t=1}^{T} \left(\sum_{i=1}^{N} \left(\sum_{j=1}^{M} \left(K_{TRE,i,j} \cdot K_{EX,i,j} \cdot f_i(t) \right) \cdot \Delta t \right) \right) \tag{9}$$

where: $f_i(t) \cdot \Delta t$ is the pollution emitted by the equipment i in the Δt period; $K_{EX,i,j}$ is the emission coefficient of the corresponding pollutant j; $K_{TRE,i,j}$ is the treatment cost of the corresponding pollutant j.

3.2 Restrictions

Constraints of system power balance:

$$P_{GRID}(t) + P_{BAT}(t) + P_{DG}(t) = P_{LOAD}(t) - P_{PV}(t) \tag{10}$$

$$P_{LOAD}(t) = P_{TRAN,LOAD}(t) + P_{UNTR,LOAD}(t) \tag{11}$$

where: $P_{BAT}(t)$ represents the charge-discharge power value of the energy storage battery at t, $P_{BAT}(t) > 0$ represents the discharge, $P_{BAT}(t) < 0$ represents the charge; $P_{DG}(t)$ represents the power value of the diesel generator output at t; $P_{GRID}(t)$ represents the power value exchanged between the external grid and the micro-grid at t, $P_{GRID}(t) > 0$ indicates that the power flows from the large grid to the micro grid, $P_{GRID}(t) < 0$ indicates that the power flows from the micro grid to the large grid; $P_{PV}(t)$ indicates the magnitude of the photovoltaic power generation at t; $P_{LOAD}(t)$ indicates the magnitude of the load power at t. $P_{TRAN,LOAD}(t)$ is a translatable load and $P_{UNTR,LOAD}(t)$ is a non-translatable load.

In order to ensure the service life of the energy storage system, over-charging and over-discharging of the energy storage are generally avoided. The energy storage is generally restrained within a certain range, and the constraint conditions are as follows:

$$SOC^{\min} \leq SOC(t) \leq SOC^{\max} \tag{12}$$

$$P_{BAT}^{\min} \leq P_{BAT}(T) \leq P_{BAT}^{\max} \tag{13}$$

where: $SOC(t)$ is the state-of-charge of the battery, SOC^{min} is the lowest level of charge the battery should maintain, and SOC^{max} is the highest state of charge the battery should maintain. P_{BAT}^{min} is the maximum discharge power of the battery, and P_{BAT}^{max} is the maximum charging power of the battery.

The constraints for diesel generators are:

$$P_{DG}^{min} \leq P_{DG}(t) \leq P_{DG}^{max} \tag{14}$$

where: P_{DG}^{min} is the minimum power of the diesel generator; P_{DG}^{max} is the maximum power of the diesel generator.

4 Multi-objective Optimization Processing of Micro-Grid

4.1 Multi-objective NSGA-II Optimization Algorithm

NSGA-II is a classic multi-objective optimization algorithm proposed by Deb and Srinivas on the basis of NSGA in 2002. It is one of the most popular multi-objective optimization algorithms at present. NSGA-II adopts the fast non-dominated sorting method as the basis for individual ranking in the population, and the calculation complexity is greatly reduced compared with NSGA. NSGA-II uses crowded distance comparison method on the basis of fast non-dominated sorting to further sort the individuals at the same level after non-dominated sorting, which promotes the decentralized arrangement of individuals, and finally allows Pareto front-end to be distributed more ideally, ensuring the diversity of the population.

4.2 The NSGA-II Algorithm Process

The NSGA-II algorithm process

(1) First set the algorithm parameters, initialize the N individuals of the internal population, the internal population cross-mutation produces external N individuals, evaluate 2N individuals of the two populations, and calculate the fitness of all individuals.

(2) Perform non-dominated sorting on the 2N individuals of the two populations, and then select individuals according to the sorting level. When a non-dominated class makes the number of selected individuals initially reach or exceed N individuals, the non-dominated class of individuals is crowded Distance sorting, further sorting the non-dominated individuals according to the crowding distance sorting until the end of the selection of N individuals, the selected N individuals constitute the original population in the next iteration.

(3) The original population variation crosses to generate the external population and calculates the fitness of all individuals.

(4) Repeat process (2).

(5) Determine whether the iteration termination condition is reached. If it is, terminate the iteration to obtain a most solvable set; if not, return to step (3) to continue.

(6) The optimal solution is selected from the solution set.

5   Example Analysis

5.1   Basic Data of Micro-Grid

The micro-grid model constructed in the analysis of the examples in this paper includes photovoltaic power generation system (PV), diesel generator (DG), micro gas turbine (MG), and energy storage system (BAT). The micro-grid topology structure diagram is as follows (Fig. 2):

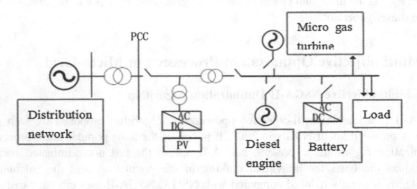

Fig. 2. Topology diagram of the micro-grid.

With a cycle of 24 h a day, the step size is 0.5 h. As shown in the above figure, the photovoltaic capacity is 100 kW; the rated power of the micro gas turbine is 65 kW, the minimum power is set to 9 kW, the climbing rate is set to 40 kW/h; the energy storage battery energy is 100 kW·h; the rated power of the diesel generator is 50 kW, The minimum power is set to 9 kW, the climbing rate is set to 40 kW/h; the micro-grid does not exchange power with the large power grid during independent operation; the initial SOC of the energy storage battery is set to 20%, and the SOC variation range of the energy storage battery is set to 20%–80%.

Take the micro-grid project data of an area as an example for analysis (Fig. 3) (Table 1):

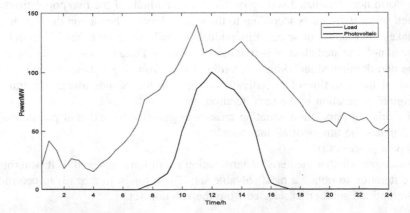

Fig. 3. Load and photovoltaic output curve.

Table 1. Datas of micro-grid.

Micro power supply	Power/kW		Life/year	Installation costs/yuan	Maintenance coefficient	Environmental costs/ [yuan · $(kW·h)^{-1}$]	The cost of fuel/ [yuan · $(kW·h)^{-1}$]
	The lower limit	Ceiling					
PV	0	125	25	65500	0.0092	0	0
BAT	0	65	15	36000	0.0012	0	0
DG	0	60	20	15000	0.0860	0.7610	1.2
MT	0	80	15	16650	0.0012	0.1632	0.82

6 Optimization Results

An optimization strategy that does not consider flexibility is called strategy 1, and an optimization strategy that considers flexibility but does not include storage flexibility is called strategy 2.

Comparison of optimization strategies considering flexibility and not considering flexibility.

Under the above information conditions, the micro-grids in the micro-grid are compared with the previous output curve, which includes two optimization strategies that consider the flexibility index and do not consider the flexibility index. The results are shown in the following (Fig. 4, 5, 6, 7 and Table 2):

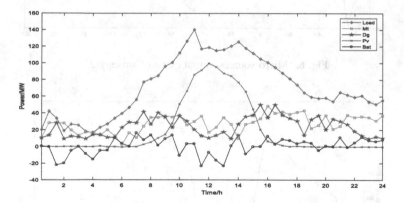

Fig. 4. Micro sources output curve of strategy 1.

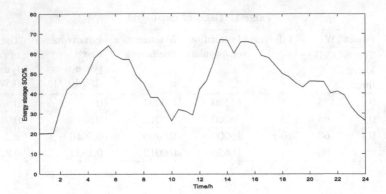

Fig. 5. Energy storage SOC curve of strategy 1.

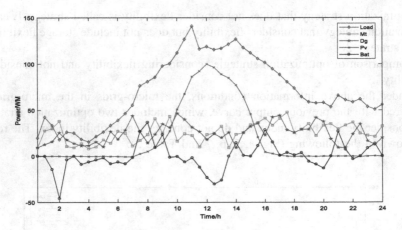

Fig. 6. Micro sources output curve of strategy 2.

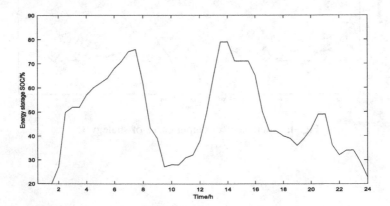

Fig. 7. Energy storage SOC curve of strategy 2.

Table 2. Optimization results of strategy 1 and strategy 2.

	Comprehensive economic cost/yuan		Flexible margin deficiency rate/%	
	The optimal solution	The average	The optimal solution	The average
Strategy 1	2223.7	2234.6	19.1	16.2
Strategy 2	2259.4	2290.1	3.2	1.2

As shown in the above table, it can be seen that the overall economic cost of optimization strategy 1 micro-grid scheduling optimization results is slightly better than optimization strategy 2, but strategy 1 optimization results are less flexible than strategy 2 optimization results.

It can be seen from the analysis of the example that the economics of the optimization strategy considering the micro-grid flexibility index are slightly less economical than the optimization strategy not considering the flexibility index, but in terms of the micro-grid's ability to cope with uncertainties when operating independently In other words, the optimization strategy considering the flexibility index is much higher than the optimization strategy not considering the flexibility index. From the above analysis, it can be concluded that to improve the ability of the micro-grid to cope with randomness during independent operation requires a certain economic cost, but in the actual operation of the micro-grid, higher flexibility can reduce wind and light or forced removal the probability of load.

7 Conclusion

In this paper, in the energy management of independent micro-grids containing photovoltaic power generation, an optimization strategy considering the micro-grid flexibility index is proposed, and the multi-objective NSGA-II optimization algorithm is used to analyze the proposed micro-grid optimization strategy. The results of the calculation example show that the stability of the optimization strategy considering the flexibility index of the micro-grid is higher than that of the optimization strategy not considering the flexibility index. In the actual operation process, the randomness of the photovoltaic output makes the operation of the micro-grid suffer. Certain volatility, therefore, considering the flexibility index of micro-grid operation can increase the reliability of micro-grid power supply, reduce the removal of clean energy, and improve energy utilization.

References

1. Yin, L.L., Li, Z.: Optimized management of independent operation energy of wind and solar household micro-grid based on load classification. Electric. Appl. Energy Efficiency Manage. Technol. (03), 84–90 (2017)
2. Zhang, Y., Li, X.S.: Multi-objective optimized self-healing control of micro-grid based on frequency adjustment strategy. Power Syst. Technol. **41**(03), 831–839 (2017)
3. Yang, L.J., Li, H.Q., Yu, X.Y., Zhao, J.S., Liu, W.Y.: Multi-objective optimal scheduling method for island micro-grids considering flexibility. Power Syst. Technol. 1–8 (2018)
4. Shi, T., Zhu, L.Z., Yu, R.Y.: Overview of research on power system flexibility evaluation. Power Syst. Protect. Control **44**(5), 146–154 (2016)
5. Ma, J., Silva, V., Belhomme, R., et al.: Evaluating and planning flexibility in sustainable power system. IEEE Trans. Sustain. Energy **4**(1), 200–209 (2012)
6. Lannoye, E., Flynn, D., Malley, M.: Transmission, variable generation, and power system flexibility. IEEE Trans. Power Syst. **30**(1), 57–66 (2014)
7. Lannoye, E., Flynn, D., Malley, M.: Evaluation of power system flexibility. IEEE Trans. Power Syst. **27**(2), 922–931 (2012)
8. Zhang, H.T., Qin, W.P., Han, X.Q., Wang, P., Guo, X.L.: Multi-time scale micro-grid energy management optimization scheduling scheme. Power Syst. Technol. **41**(05), 1533–1542 (2017)
9. Zhou, Z.C., Wang, C.S., Jiao, B.Q., Guo, L., Xu, W.: Optimal control of the independent micro-grid system of Fengchai storage biomass. Proc. CSEE **35**(14), 3605–3615 (2015)
10. Tan, Y., Lv, Z.L., Li, J.: Multi-objective capacity optimization configuration of distributed micro-grid distributed power based on improved ELM for wind/photo/diesel/storage micro-grid. Power Syst. Protect. Control **44**(08), 63–70 (2016)
11. Liao, M.Y.: Research on optimization of economic operation of micro-grid with CCHP. South China University of Technology (2014)
12. Kang, J., Jin, B., Duan, X.J., Shang, X.H., Li, W.: Optimal operation of micro-grid based on Bayesian-particle swarm optimization. Power Syst. Protect. Control **46**(12), 32–41 (2018)
13. Luo, Z., Wu, Z., Li, Z.Y., Cai, H.Y., Li, B.J., Gu, W.: A two-stage optimization and control for CCHP microgrid energy management. Appl. Thermal Eng. **125**, 513–522 (2017)
14. Han, Y., Zhang, G.R., Li, Q., You, Z.Y., Chen, W.R., Liu, H.: Hierarchical energy management for PV/hydrogen/battery island DC microgrid. Int. J. Hydrogen Energy **44**, 5507–5516 (2018)
15. Julio, P., Javier, B., Pablo, S., Luis, M.: Energy management strategy for a renewable-based residential micro-grid with generation and demand forecasting. Appl. Energy **158**, 12–25 (2015)

Quantitative Analysis of Oil Drilling Fluid Based on Partial Least Square Method

Weihang Han and Guoliang Wang[✉]

School of Information and Control Engineering, Liaoning Shihua University,
Fushun 113001, Liaoning, China
glwang@lnpu.edu.cn

Abstract. The types and contents of chemical bonds in the mixture are analyzed more and more widely by means of Raman spectroscopy. In view of the particularity of the mixture of crude oil and drilling fluid, the qualitative and quantitative analysis of the mixture by Raman spectroscopy is a more accurate method. In this paper, partial least square regression is used to analyze the data of Raman spectrum, the data is modeled on the basis of data preprocessing, and the content of unknown concentration data is analyzed. In the simulation part, the real data of the test is processed by grouping and the corresponding analysis results are given. Through the data modeling and verification results, we finally determined that partial least square regression algorithm is suitable for qualitative and quantitative analysis of oil drilling fluid mixture, and conducted a variety of data preprocessing to ensure the accuracy of data results.

Keywords: Partial least squares regression algorithm · Raman spectrum · Mixture of oil and drilling fluid · Quantitative analysis

1 Introduction

Raman spectrum originates from the change of molecular polarizability. The spectrum range is 1–4000 cm^{-1}, which reflects the relationship between the frequency of Raman scattering line and the molecular structure [1]. For the same material, Raman displacement has nothing to do with incident light frequency and represents the characteristic physical quantities of molecular vibration and rotational energy level, especially the determination of molecular skeleton [2]. One of the most attractive features of Raman spectroscopy is that it is simple to sample, almost no sample preparation is required, and the samples can be tested directly through glass packaging without interference from water [3]. Raman spectroscopy is a beneficial supplement to infrared spectroscopy. Samples with weak infrared absorption can usually obtain strong Raman spectroscopy. Therefore, Raman spectroscopy has been widely used in biological systems, inorganic and organic analysis and polymer analysis. In terms of quantitative analysis, surface-enhanced Raman spectroscopy has a very low detection limit, and there

© Springer Nature Singapore Pte Ltd. 2020
J. Qian et al. (Eds.): ICRRI 2020, CCIS 1335, pp. 191–207, 2020.
https://doi.org/10.1007/978-981-33-4929-2_14

are many reports on the detection of organic molecules [4]. Based on the above advantages of Raman spectroscopy, aiming at the qualitative and quantitative analysis of the mixture of petroleum crude oil and drilling fluid, we used Raman spectroscopy of samples with different concentrations to extract data, analyze the types of chemical bonds corresponding to substances in the mixture, and further determine the content [5].

We choose different concentration data of the Raman spectra of the mixture are extracted, and the full wave data extraction, concentration of each 60 group experiment preparation, we extract the data, after finishing to simply determine the dimensions of the data, can be divided into high-dimensional data, which need to be the dimension reduction of data preprocessing operations.

For any data analysis system, the data cannot be guaranteed to be clean, such as incomplete data set, overlap, redundancy, "bad value" and the existence of noise, which seriously affect the quality of data analysis results [6]. Redundant data will lead to chaotic data analysis, resulting in unreliable output, and a large number of redundant data may affect the performance of the data analysis system, reducing the reliability and stability of the system [7]. Data preprocessing can improve the quality of data sets and improve the accuracy and performance of data analysis systems. As high-quality decision is inevitably dependent on high-quality data, data preprocessing as early as possible will be highly rewarded in the decision system [8].

The methods of data preprocessing include data cleaning, data integration and transformation, and data specification [9]. Data cleaning can remove noise in the data and correct inconsistencies. Data integration combines data from multiple sources into consistent data stores, such as data warehouses and data silos. Data transformation (such as normalization) transforms data into a form suitable for data mining. For example, normalization can improve the accuracy and effectiveness of mining algorithms involving distance measurement [10]. Data conventions can be compressed by means of clustering, deleting redundant features or clustering, such as data party clustering, dimensional conventions, data compression, numerical conventions and discretization, which can be used to obtain the data's canonical representation, so as to minimize the loss of information content [11].

It is generally believed that the data of 10, 20 or more dimensions should be included in the category of high-dimensional data, and the dimension of feature space should not be too high [12]. For our test data, the dimension of the data reaches 1961 dimension, so we must divide it into the range of high-dimensional data for preprocessing. Due to the huge difference between the performance of data in high dimensional space and that in low dimensional space, most traditional classification methods cannot obtain good classification results [13]. Therefore, the classification and analysis of high dimensional data has become a challenging problem in the field of data mining. In the research center of classification and discrimination of high-dimensional data, there are mainly two aspects: one is the dimensionality reduction of these high-dimensional data; the other is the establishment of mathematical model of high-dimensional data classification [14]. PCA (principal component analysis) is one of the most commonly used high dimensional data dimension reduction method, its main idea is to pass dimension reduction, extract the convenience of a few principal components, the variable has the largest divergence in low dimensional space and the correlation between the minimum, and that a few principal components almost contains all useful information about the high-dimensional data [15]. PLS (partial least square method) is based on PCA and USES the principle of PCA to calculate the principal component (latent variable) by linear combination of the original variables, and also refers to the principal component as PLS factor. Then the PLS factor is carried out multiple linear regression to obtain the obtained regression coefficient matrix [16].

This paper focuses on the Raman spectra of crude oil and oil drilling fluid mixture, has carried on the corresponding data extraction, and then to the operation of the data preprocessing, and combined with the characteristics of data after pretreatment the establishment of mathematical model, based on the mathematical model of unknown concentration data percentage content of material in the simple prediction, we will to continue down.

2 Establishment of Mathematical Model – Partial Least Squares Algorithm PLS

The purpose of partial least squares is to find some variables in the matrix space of explanatory variables, and combine these variables linearly to form latent variables, so that these latent variables can contain most useful information of the original data, and then extract the correlation between explanatory variables in a more concise and efficient form [15]. Assuming that all explanatory variables are related to reaction variables, the basic idea of PLS can be shown in the following figure (Fig. 1).

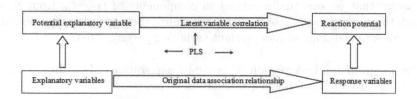

Fig. 1. Working principle of Single-chip Microcomputer.

Above reveals the basic idea of PLS, namely, interpretation and response variables under the PLS algorithm processing, respectively explain the latent variables and latent variables reaction, among them, explain the number of latent variables is far less than the number of variables, when the number of response variable is greater than 1, reaction the number of latent variables also should be less than the number of variables, namely explain latent variable matrix space dimensions should be far less than the dimension of variable matrix space, latent variable matrix space dimension is less than the reaction variable matrix dimensions of the space, so, to establish the link between the two lower dimensional space matrix, then it can reflect the relation between the original data, namely the relation between the two high-dimensional space matrices, and build the mathematical model between the high-dimensional data through the mapping of this relation, so as to realize the applications of PLS regression, classification and data prediction [17].

The implementation steps of partial least squares algorithm are as follows:

First, we define the variables in the modeling process:

Dependent variables q: $\{y_1 \cdots y_q\}$, independent variables p: $\{x_1 \cdots x_q\}$;

Data table of independent variables and dependent variables composed of n sample points: $X = [x_1, \cdots , x_p]_{n \times p}$, $Y = [y_1, \cdots , y_q]_{n \times q}$;

We extract components t_1, u_1 from X and Y respectively by partial least squares regression algorithm. Of course, when extracting these two components, we hope to meet the following requirements:

(1) t_1, u_1 carry "variation information" in their respective data tables as far as possible;
(2) the degree of correlation between t_1, u_1 can be reached to the maximum extent [18].

The requirements we put forward indicate that we hope that t_1, u_1 can better express the information covered by the data tables X and Y. Meanwhile, the component t_1 of the independent variable we extracted has the best explanatory ability relative to the component u_1 of the dependent variable we extracted [19].

In extraction of partial least-squares regression algorithm is after the first pair of component t_1, u_1, separately for t_1 return of the X and Y in the regression of u_1, until we get the regression equation of the extracted meet precision. Otherwise, we will use the residual information after X is interpreted by t_1 and Y is interpreted by u_1 for the second round of principal component extraction [20]. We assume that we eventually extract m components of $t_1 \cdots t_m$ from X, and through the regression of y_k to $t_1 \cdots t_m$, we realize the regression equation of y_k with respect to the original independent variable $x_1 \cdots x_p$, where $k = 1, 2, \cdots , q$ [21].

The program implementation of partial least squares algorithm is mainly carried out through the following calculation steps:

(1) the pre-processed data is standardized, and the standardized data table is denoted as follows $E_0 = (E_{01}, \cdots , E_{0p})_{n \times p}$, $F_0 = (F_{01}, \cdots , F_{0q})_{n \times q}$;
(2) t_1 as the first component of E_0, $t_1 = E_0 w_1$ and w_1 as the first principal axis of E_0, is the unit vector. Similarly, we think that u_1 as the first component of F_0, $u_1 = F_0 c_1$, and c_1 as the first principal axis of F_0, are also unit vectors; and w_1 is the largest eigenvector of $E_0' F_0 F_0' E_0$, and c_1 is the largest eigenvector of $F_0' E_0 E_0' F_0$;
(3) Solve the three necessary regression equations of and to component:

$$E_0 = t_1 p_1' + E_1 \tag{1}$$

$$F_0 = u_1 q_1' + F_1^* \tag{2}$$

$$F_0 = t_1 r_1' + F_1 \tag{3}$$

The corresponding regression coefficient vector is solved as follows:

$$p_1 = \frac{E_0' t_1}{\|t_1\|^2} \tag{4}$$

$$q_1 = \frac{F_0' u_1}{\|u_1\|^2} \tag{5}$$

$$r_1 = \frac{F_0' t_1}{\|t_1\|^2} \tag{6}$$

where, E_1, F_1^*, F_1 are respectively the residual matrix of the above three regression equations;

(4) The aforementioned residual matrices E_1, F_1 are used to replace the aforementioned original matrices E_0, F_0. The same method is used to solve the second principal axis w_2 and c_2, and the corresponding second principal components t_2 and u_2, the regression coefficients p_2, r_2 are also calculated, and the following regression equation can be obtained:

$$E_1 = t_2 p_2' + E_2 \tag{7}$$

$$F_1 = t_2 r_2' + E_2 \tag{8}$$

(5) We repeat the above operations to extract the components of the residual matrix, assuming that A principal components are extracted, the following equation can be obtained:

$$E_0 = t_1 p_1' + \cdots + t_A p_A' \tag{9}$$

$$F_0 = t_1 r_1' + \cdots + t_A r_A' + F_A \tag{10}$$

and then we can revert to the regression form of $y_k^* = F_{0k}$ with respect to $x_j^* = E_{0j}$, as follows:

$$y_k^* = a_{k1} x_1^* + \cdots + a_{kp} x_p^* + F_{Ak}, k = 1, \cdots, q \tag{11}$$

and F_{Ak} is the KTH column of the residual matrix F_A [22].

Partial least square method is a method of latent variables. Latent variables are obtained by linear combination of original variables, and these latent variables are queued. Among the latent variables acquired in sequence, the latent variables at the top of the list contain the most information [23]. In terms of importance, it is obvious that the first principal component and the first PLS factor are the most important, followed by the second principal component and the second PLS factor, and so on. Therefore, it is not necessary to consider which latent variables to choose for modeling, just to determine the appropriate number of latent variables, starting from the first principal component and the first PLS factor [24].

Partial least square method directly targets multivariate correction on prediction, so the principle of determining PLS factor data is to minimize prediction error. The prediction error is generally calculated by using the residual adjustment method of prediction, PRESS, and its formula is

$$PRESS = \sum_{i=1}^{p} (\hat{y}_i - y_i)^2 \qquad (12)$$

where, p is the number of samples used for prediction, y_i is the real situation of the ith sample, and $\hat{y}_i$ is the prediction result of the ith sample [25].

The number of potential variables in the partial least squares process is determined by cross validation. The cross validation method is to divide all the sample data into two parts according to a certain proportion, one part is called the correction set, the other part is the test set. The PLS prediction model was established by using the correction set, the PLS mathematical model was used to predict the test set, the prediction results of each test sample were obtained, and then the prediction results were compared with the actual results, and the prediction error was more PRESS [26]. For the PLS prediction model with different number of latent variables, the PRESS value will be different. When the number of potential variables selected is small, the PRESS value is high, the prediction ability of the model is poor, and the correlation is low. When the number of potential variables selected is too large, the over-fitting linearity exists and the PRESS value is also high, although the model correlation is high. Only when the appropriate number of latent variables is selected can the minimum PRESS value be obtained. In this case, the model has the moderate correlation and the strongest prediction ability, and the number of latent variables is the optimal number. According to the different selection methods of correction set and prediction set, the cross validation methods can be divided into three forms: de-one cross validation, fixed prediction sample cross validation and block cross validation [27].

As a full-spectrum analysis method, PLS can make full use of the useful information in multiple wavelengths without the need to deliberately choose the wavelength, and can filter the noise of the original data, improve the signal-to-noise ratio, and solve the nonlinear problem of interaction, mainly solving the problem of multi-factor regression [28].

3   Establishment and Verification of Mathematical Model

The data used in this paper are all from the laboratory, and we give a brief description and introduction of the test conditions.

Test equipment: gloves; Glue head dropper; Glass bottle; Label paper; Drilling fluid; Crude oil; Water bath pot; Raman instrument

Test steps:

(1) Take different contents of water-based drilling fluid and put them into 6 medium-sized and large glass bottles;
(2) Take the crude oil and heat it in a water bath to 80 to reduce the viscosity;
(3) Take a certain amount of sandstone cuttings and grind them into small particles;
(4) Respectively add the crude oil, cuttings and drilling fluid with the content shown in Fig. 2 into each glass plane, and shock it fully. After mixing, let it stand for a period of time, and shock it fully again when the temperature of the crude oil in the solution drops from 80 to normal temperature;
(5) Put the six glass bottles into the water bath box, and the water bath heating temperature is 60;
(6) When waiting for the heating temperature of water bath to rise, laser Raman irradiation shall be carried out on pure crude oil, pure drilling fluid and pure cuttings to determine the power, irradiation time and irradiation frequency used by Raman, so as to prevent the situation of exceeding its melting point;
(7) After the water bath heating temperature remains unchanged and the oil drilling fluid is continuously added for 30 min, it is fully shaken by taking out the oil drilling fluid and taking it into the darkroom for Raman spectral irradiation. For a sample, different points are taken for 50 times of irradiation, and it is fully shaken by each time of irradiation;
(8) Repeat step 5. (Note: considering that the temperature of the drilling fluid containing oil and cuttings will decrease with the passage of time after taking out the water bath pot, move the water bath pot into the darkroom. After 25 times of irradiation, put the water bath pot into the water bath pot for 30 min and then take out the water bath pot for another 25 times of irradiation)

Data at this time as follows, we with the single material and of crude oil and oil drilling fluid in the determination of the Raman spectra of the material, the mixture concentration distribution is as follows: 5%, 25%, 45%, 65%, 85%, 95%, there are 6 different concentrations of the mixture, combined with simple material, drilling fluid, crude oil, 100%, 0%, the concentration data for a total of eight groups, each group concentration data at this time we chose 60 times of test results, the concentration of 8 kinds of 60 sets of data with the first data preprocessing, and then after preprocessing the data and the division of modeling data and test data.

The general distribution of the data form extracted from Raman spectrum is shown in the following table. The spectral length we selected is $300\text{--}4300\,\text{cm}^{-1}$, with $300\text{--}4300\,\text{cm}^{-1}$ total of 1961 wavelengths, which corresponds to the 1961 key values. We extracted the following data from the spectrum. The corresponding abscissa of the table is different test times, and the ordinate is the key value intensity value obtained under the same test. In the same test, with the increase of wavelength, the bond strength of our data increased and decreased. The corresponding value of the same wavelength varies in different amplitude with the number of tests, but has a certain trend. For example, in the wavelength range corresponding to a certain type of chemical bond, if the number of tests changes, the bond value intensity fluctuation can be within a certain range, beyond the fixed intensity range, it can be considered as invalid data, and the test results of that time can be eliminated.

Label	Drilling fluid content/ml	Content of crude oil/ml	The quality of crude oil/g	Cuttings content/ml	The proportion of crude oil /%	The temperature/℃
1	4.75ml	0.25ml	0.9g	37.5g (75%)	5	60
2	4.5ml	1.5ml	4.5g	35g (70%)	25	60
3	3.85ml	3.15ml	8.1g	32.5g (65%)	45	60
4	2.7ml	5.2ml	11.7g	30g (60%)	65	60
5	1.35ml	7.65ml	15.3g	27.5g (55%)	85	60
6	0.5ml	9.5ml	17.1	25g (50%)	95	60

Fig. 2. Oil - containing cuttings drilling fluid component content table. The crude oil density of Liao he is about $0.9\,\mathrm{g/cm^3}$. Drilling fluid density is $1.8\,\mathrm{g\ /cm^3}$.

Table 1. Data distribution table.

Name	Number
Raman wave Numbers	300 1500 4000
Test once	0 368.96 27.15
Test the secondary	0 422.42 262.67
Test three times	0 423.81 216.44

We have the following processing methods for the above data, and the corresponding results and evaluations are given after each processing method. We selected the whole band spectrum of Raman spectrum for data extraction. The data used was the strength value of the chemical bond corresponding to each wavelength. We used the relationship between the bond strength and the corresponding concentration percentage for quantitative analysis.

Processing method one our data procession of the concentration of whole directly PLS arithmetic as well as the concentration of each separate PLS operation, found that deal directly with the means to get the model of the corresponding key strength density deviation is larger, and the real value of the original density could not be better repetition, and the concentration of each single model calculation results show that the deviation between the model corresponds to the density and the real value is small, can be ignored, but for those we choose the concentration of the data, because of the difference of the concentration of the corresponding data, the computing results show that there are differences, and will lead to the calculation results of instability, and this method is not applicable.

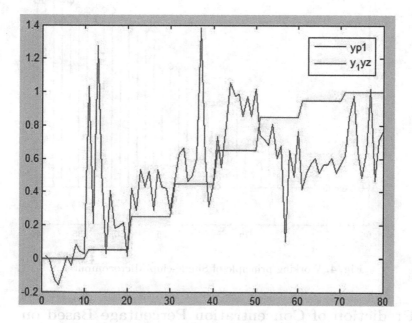

Fig. 3. Working principle of Single-chip Microcomputer. (Color figure online)

The solid red line in Fig. 3 represents the known true concentration, and the solid blue line represents concentration calculated by the model. The following images in this paper are also presented in the same way.

Processing method two: PLS we select all band data involved in operations, each time to participate in the operation of data dimension is 1961 dimensions, far higher than that of low dimensional space, the maximum allowable dimensions and due to get the data in the process of the test due to operation, equipment, etc. Under the same concentration of different experiments, the output spectrum of the result, and then we set of test results under the same concentration of 60, according to their characteristics, the data show that the data classification, and the classification of data preprocessing.

Through the classification of the data, we get the classification of several groups of different data, we has carried on the processing for each type of data, because in belongs to a class of data, most of the wavelength of the corresponding key values under different test times the size of the area can be divided into a certain range, the data after PLS operation by the mathematical model between calculated value and actual value of gap is small, we think the data on the numerical classification is a effective treatment method (Fig. 4).

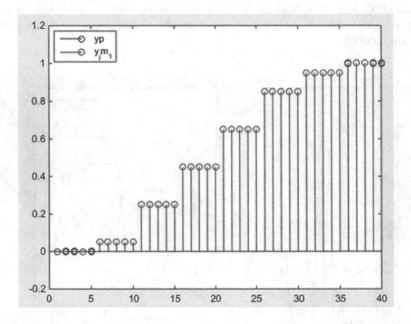

Fig. 4. Working principle of Single-chip Microcomputer.

4   Prediction of Concentration Percentage Based on Mathematical Model

We divided the data into building module set and verification data set. Through the pretreatment of modeling data and the establishment of mathematical model, the "unknown" concentration percentage content of crude oil and drilling fluid mixture was quantitatively analyzed. Modeling method based on the above, we choose different categories of data, it is divided into the modeling data set and validation data sets, using PLS modeling, data validation data set for the concentration percentage content of calculation, and we know the true value of contrast, for each type of data processing, the following validation results are obtained, as shown (Fig. 5).

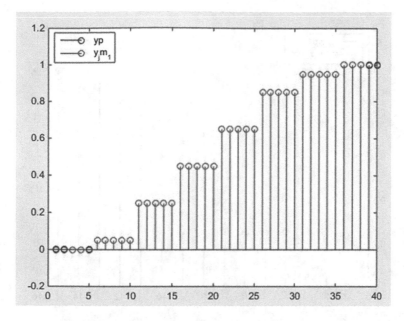

Fig. 5. Working principle of Single-chip Microcomputer.

We figure based on the results of the above analysis, the effect of modeling of the display model of the deviation between calculated value and actual value is small, but the validation data set of data concentration percentage content of calculation, the calculation values under different test data, the deviation between the calculation results and the true shows that the concentration of the different. And the prediction accuracy of the data decreases with the increase of the concentration. We believe that due to various factors, we deal with the interference factor that the data contained in the still more, even if the data preprocessing, we now PLS modeling results still can't comprehensively covers all categories of data information, leading to the we compare the validation data set of data to calculate the large deviation, based on this result, we decided to other data processing, in order to improve our calculation accuracy of the data validation data set.

We each type of data for the first time the PLS model is established in this paper, according to the calculated value of the first and the known real value deviation between calculated, to eliminate the deviation is significant, the thought out of data for the interference, to eliminate the data after the internal group, divided into the modeling data set and validation data sets, the data of the modeling data set PLS operation, get the corresponding mathematical model, based on the mathematical model of our validation data set was calculated, and the following results are obtained, as shown in Fig. 7 and Fig. 8.

We analyzed the verification results. At this point, it can be seen that the accuracy of the calculated value for the low-concentration data is higher than

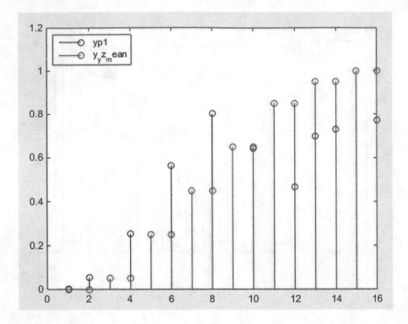

Fig. 6. Working principle of Single-chip Microcomputer.

that for the high-concentration data, but the deviation of the calculated result is still not negligible compared with the effect of the modeling data. At this point, we considered the influence of the dimension. Due to high dimensional data under different test results of the data deviation, and the difference degree is bigger, at this point does not guarantee that each dimension corresponds to the classification of the pretreatment of data belonging to the same category, but because most of the dimension specified in the same latitude, and in the process of PLS model show the good performance, but because of the influence of a small number of large deviation data, cause PLS mathematical modeling of greater influence on the results of the validation data sets of data, and deviation between the calculated results and the known density is more obvious. In view of the above results and processing process, we consider a new data processing.

Due to the influence of the high dimension of the data and other factors of the data obtained by Raman spectrum, we performed a new pretreatment operation on logarithm at this time. At this time, the data were divided into "high concentration" data set and "low concentration" data set. Considering the reality, we mainly processed and verified the data in the low concentration data set. At this time, the low-concentration data set was divided into 0%, 5% and 25% data. On the basis of data classification, "normalization" and PCA dimensionality reduction were performed on the data. Then, we established the PLS mathematical model for the dimensionality reduction data. In addition, two PLS operations were used to compare the calculated value with the real value

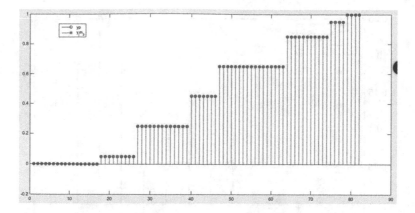

Fig. 7. Working principle of Single-chip Microcomputer.

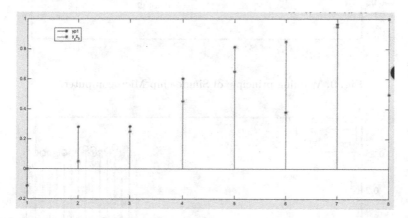

Fig. 8. Working principle of Single-chip Microcomputer.

of the modeling data and the verification data, and the results were shown as follows (Fig. 9 and Fig. 10).

We analyzed Fig. 9 and Fig. 10. At this time, the first 15 groups of each concentration were selected for the modeling data, while the last 5 groups were selected for the verification data. We performed PLS operation on the modeling data and obtained 6 principal components. It can be seen from the verification results that the deviation between the calculated value of 0% concentration and the real value can be ignored. The deviation between the calculated value and the real value of 2 groups of data of 5% concentration is relatively ideal, and the deviation of the remaining 3 groups of data is within 0.05. In the 25% concentration data, the deviation between the calculated value and the real value of the two groups of data can be ignored. The deviation of the two groups of data is within 0.05, while the deviation of the one group is relatively large. Our verification results can be limited to 5%.

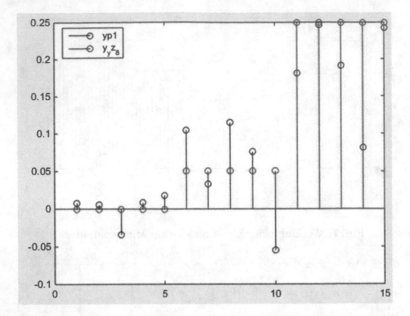

Fig. 9. Working principle of Single-chip Microcomputer.

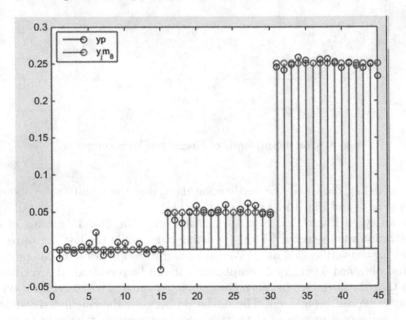

Fig. 10. Working principle of Single-chip Microcomputer.

We divided based on the above, the concentration of eight groups into a data set "high" and "low density" data sets, for "high" data at this point, we through mathematical modeling and calculating the corresponding concentration of "unknown" data, modeling effect is not good at this time of reaction concentration of "unknown" realities of the data set, the calculation results and the reality gap is more obvious, we compared the same test under the condition of the Raman spectra of multiple sets of data, found that "high" data set in the Raman peaks of crude oil and the Raman peaks of the drilling fluid, high contact ratio, that is to say, with the increase of oil content in the mixture, The Raman spectra of the two substances have a high coincidence degree, and at this time the Raman spectra can no longer reflect the real existence of the two substances, so our mathematical model is no longer applicable. In response to this problem, we will make corresponding changes in the future, for the "high concentration" data set, to be processed again.

Comparing the original PLS modeling effects and the corresponding verification effect, we have to deal directly with the data, the calculated value of the validation data set and the deviation between the "true value" is bigger, the maximum deviation at around 0.6, much less we above 0.05 the deviation of the results, further certification we finally choose PLS arithmetic of data preprocessing and concentration groups of data modeling data set information extraction, and then used to calculate the concentration of the corresponding percentage of each data validation data set precision, compared with the original method of calculation, we at this time of relatively high accuracy.

5 Conclusion

The Raman spectra of oil-drilling fluid mixtures are presented in the form of high-dimensional data. Based on the 1961 dimension data of 60 sets of experiments, we established mathematical models for mixtures with different percentage contents, and predicted the percentage of oil contained in mixtures with "unknown" percentage contents. By means of data pretreatment, normalization, dimensionality reduction of PCA and data classification, the Raman spectral data of the mixture of crude oil and drilling fluid were processed, and the results showed that the deviation between the calculated value and the real value of the modeling data and the verified data in the "low concentration" data set could be limited to 5%. We believe that the data processing method and mathematical model are effective.

In the future, we will improve the matching method of drilling fluid reagent used in the test. This is because we in repeated tests, compared the "high" the Raman spectra of the sample for many times, found that high concentrations of samples at this time has the problem of drilling fluid color too deep, it will lead to most of the wavelength of the data and drilling data of crude oil by Raman spectrum reflecting the basic information of overlap, making data to feedback the actual circumstances of the sample solution, which makes the "high" data set modeling effect of "unknown" data cannot be directly used for validation. We consider the replacement of drilling fluid formula method, such as using saturated salt water drilling fluid, it can eliminate the problem of the oil color too deep, can better identify the peak value of crude oil, make up for the original system of polysulfide drilling fluid in terms of detection, in addition, due to the complex system of polysulfide drilling fluid composition material impact on containing oil detection in drilling fluid is larger, and the salt water drilling fluid is relatively small, the material impact on the crude oil can be ignored, to ensure the accuracy and authenticity of the experimental data. In the future, we will conduct more in-depth research and data processing for the replacement of test samples.

References

1. Cheung, W., Shadi, I.T., Xu, Y., Goodacre, R.: Quantitative analysis of the banned food dye Sudan-1 using enhanced Raman scattering with multivariate chemometrics. J. Phys. Chem. C. **114**(16), 7285–7290 (2010)
2. Zhang, L., Li, Q., Tao, W., Yu, B., Du, Y.: Quantitative analysis of the thymine with surface-enhanced Raman spectroscopy and partical least squares(PLS) regression. Anal. BioanalChem. **398**(4), 1827–1832 (2010)
3. Gelder, J.D., Gussem, K.D., Vandenabeele, P., Moens, L.: J. Raman Spectrosc. **38**(9), 1133 (2007)
4. Efremov, E.V., Ariese, F., Gooijer, C.: Achievements in resonance Raman spectroscopy: review of a technique with a distinct analytical chemistry potential. J. Anal. Chimica Acta **606**(2), 119–134 (2008)
5. Bahler, D., Navarro, L.: Method for combining heterogeneous sets of classifiers. In: 17th National Conference On Artificial Intelligence (AAAI), Workshop on New Research Problems for Machine Learning, vol. 24, pp. 35–37 (2000)
6. Nafie, L.A.: Recent advances in linear and nonlinear Raman spectroscopy. J. Raman Spectrosc. **43**(12), 1845–1863 (2012)
7. Pogieter-Vermaak, S., Maledi, N., Wagner, N., et al.: Raman spectroscopy for the analysis of coal: a review. J. Raman Spectrosc. **42**, 123–129 (2011)
8. Jain, A.K., Duin, R.P.W., Mao, J.: Statistical pattern recognition: a review. IEEE Trans. Pattern Anal. Mach. Intell. **22**(1), 4–37 (2000)
9. Webb, A.R.: Statistical Pattern Recognition, pp. 34–40. Oxford University Press, New York (1999)
10. Bishop, C.M.: Pattern Recognition and Machine Learning (6–10) (2006)
11. Kearns, M.J.: Learning Boolean formulate for finite automata is as hard as factoring. Technical report TR-14-88. Harvard University Aiken Computation Laboratory, Cambridge (1998)
12. Krogh, A., Vedelsby, J.: Neural network ensembles, cross validation, and active learning. J. Adv. Neural Inf. Process. Syst. **7**, 231–238 (1995)
13. Breiman, L.: Bagging predictors. Mach. Learn. J. **24**(2), 123–140 (1996)
14. Breiman, L.: Arcing classifiers. Ann. Statist. J. **26**(3), 801–849 (1998)
15. Hansen, L.K., Salamon, P.: Neural network ensembles. IEEE Trans. Pattern Anal. Mach. Intell. J. **12**(10), 993–1001 (1990)
16. Duda, R.O., Hart, P.E., Stork, D.G.: Pattern Classification. Wiley, Hoboken (2001)
17. Xu, L., Krzyzak, A., Suen, C.Y.: Methods of combining multiple classifiers and their applications to handwriting recognition. J. IEEE Trans. Systems Man Cybern. **6**, 54–59 (1992)
18. Bahler, D., Navarro, L.: Methods for combining hetergeneous sets of classifiers. In: 17th National Conference On Artificial Intelligence(AAAI), Workshop on New Research Problems for Machine Learning, vol. 24, pp. 35–37 (2000)

19. Guyon, I., Elisseeff, A.: An introduction to variable and feature selection. J. Mach. Learn. Res. **3**, 1157–1182 (2003)
20. Vladimir, N.V., Zhang, X.: The Nature of Statistical Learning Theory, vol. 9. Tsinghua University Press, Beijing (2000)
21. Bian, Z., Zhang, X.: Pattern Recognition, vol. 9. Tsinghua University Press, Beijing (2000)
22. Geladi, P., Herman, W.: The father of PLS. Chemometr. Intell. Lab. Syst. **15**(1), 7–8 (1992)
23. Wold, H.: Path Least Squares. In: Samule, K., Johson, N.L. (eds.) Encyclopedia of Statistical Sciences, vol. 6, pp. 581–591. Wiley, New York (1985)
24. Wold, S., Erikson, L.: PLS-regression: a basic tool of chemometrics. Chemometr. Intell. Lab. Syst. **58**, 109–130 (2001)
25. Hoskuldsson, A.: PLS regression methods. J. Chemometr. **2**, 211–228 (1988)
26. Frank, I.E.: Intermediate least squares regression method. J. Chemometr. **1**, 233–242 (1987)
27. Bi, Y., Xie, Q., Peng, S., Lu, W.: Slice transform-based weight updating strategy for PLS. J. Chemometr. **12**, 10–12 (2012)
28. Boulesteix, A.L., Strimmer, K.: Partical least squares: a versatile tool for analysis of high-dimensional genomic data. J. Brief Bioninform. **8**(1), 32–44 (2007)

Blurring Detection Based on Selective Features for Iris Recognition

Jianping Wang, Qi Wang$^{(\boxtimes)}$, Yuan Zhou, Yuna Chu,
and Xiangde Zhang

College of Sciences, Northeastern University, Shenyang 110819, China
{wangjianping, wangqimath, zhangxdneu}@mail.neu.edu.cn,
664251755@qq.com, chuyunamath@163.com

Abstract. Iris recognition is one of the most accurate biometric technologies. Iris images are easily blurred because of motions or defocusing. Blurring iris image easily leads to mismatch in recognition. So, these blurred iris images should be excluded before recognition to improve the recognition performance. This paper presents a blurring detection method for iris image based on local feature. The proposed method firstly employs RST (Radial Symmetry Transform) to localize pupil boundary. Then 32 indicators are calculated on several ROIs (Region of Interest) for each iris. After that, LASSO (Least Absolute Shrinkage and Selectionator Operator) is used to select features. Finally, SVM (Support Vector Machine) algorithm is used to separate the blurred iris images from ideal ones based on the selected 14 indicators. The developed method is experimented on a self-built database. And the correct classification rate reaches 97.87%.

Keywords: Iris recognition · Blurring detection · LASSO · SVM

1 Introduction

Iris recognition is one of the most accurate biometric technologies. It has broad applications in different fields. It attracts quite a lot of interests both in industrial and academic research.

A typical iris recognition process generally includes several steps, which are image acquisition, image quality evaluation, segmentation, feature extraction and recognition.

Because high quality iris image is the base of excellent iris recognition system, iris image acquisition is an important fundamental step in iris recognition. In real applications, the captured images may include different kinds of poor-quality iris images, such as blurring, large spot, low contrast, occlusion, and deformation.

Blurred irises are easy to cause mis-matching in iris recognition [2]. So it is necessary to exclude these blurred iris images. But to detect blurred iris images is still a quite challenging job.

Classical blurring detection methods are mainly based on frequency and space analysis. Frequency information could reflect some of blurring characters for iris images. Daugman [6] proposes to extracts high frequency energy information by filter to estimate the clarity of iris image. Ma etc. [8] divide the two-dimensional Fourier

© Springer Nature Singapore Pte Ltd. 2020
J. Qian et al. (Eds.): ICRRI 2020, CCIS 1335, pp. 208–223, 2020.
https://doi.org/10.1007/978-981-33-4929-2_15

Transform spectrum of two horizontal iris sub-regions into high, medium and low frequency bands for blurring detection. Chen etc. [3] use high frequency energy band of wavelet packet decomposition to measure the quality of iris image. Chen etc. [5] propose an assessment method by dividing the entire iris region into eight concentric ring bands and measure the frequency content using a Mexican Hat Wavelet. Saad etc. [10] classified images by extracting feature of block by DCT (Discrete Cosine Transform).

Besides frequency methods, researchers also develop evaluation methods based on some special information. Zhang etc. [15] calculate the average value of pupil edge gradient S and the difference of average gray value between iris and pupil H, then they compute the focusing factor F = S/H to evaluate focusing degree. Anish etc. [1] use a GGD (Generalized Gaussian Distribution) to capture the tail behavior of MSCN (Mean Subtracted Contrast Normalized coefficients).

On the other hand, the ROI (Region of Interest) regions, where the blurring features are extracted, are also key points. Saad etc. [10] and Anish etc. [1] extract features from whole iris image for blurring detection. However, features extracted from whole iris are easily affected by glass, eyelashes, spot, etc. And the texture information from local area [3, 5, 6, 8] also has some drawbacks: the texture feature varies with iris images. If iris images are estimated just depending on the texture of some iris regions. Some kinds of irises are easily rejected, such as the iris with sparse or light texture.

Blurring detection for iris recognition is still an open problem. This paper aims to detect blurring just depending on single frame iris image. This is a non-reference image quality assessment problem, which is quite challenging.

According to our analysis, the clarity of large gradient edge could be used to evaluate blurring images. Thus, this paper extracts high frequency features via DCT on some selected pupil edge regions. Besides this, we use multiple feature indexes on some additional iris regions. To lower down the computation cost, we adopt LASSO to select the most effective feature combination from the feature set. Then the selected features are classified by SVM.

The whole paper is organized as follows: Sects. 2, 3 and 4 introduce the details of the proposed method including feature extraction, feature selection and feature classification. Section 5 illustrates the experimental result and analysis. The conclusion of proposed method is presented in Sect. 6.

Among these low-quality images, blurring is a common problem. Blurring is generated from defocus and motion generally. Defocusing blur arises from improper distance between eye and lens. While motion blur is caused by the relative movement between camera and eye. When it comes to mobile, long distance and non-cooperative iris recognition, blurring occurs more easily and seriously in general.

2 Technical Details

In order to separate blurring iris images from ideal ones, we firstly select several appropriate image regions. Then feature set is constructed by spatial and frequency domain features.

2.1 Pupil Localization

Firstly, pupil boundary is roughly estimated by MRST [13]. Figure 1 shows a typical localizing result. According to [13], some of low-quality iris images can be excluded in this step, such as close eye and severe out-of-focus images.

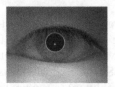

Fig. 1. Pupil location result.

2.2 Statistical Features on Space Domain

ROI Selection. We evaluate the quality of iris image on local iris image's information in this paper. The texture of noise pixel such as eyelash, spot is deeper than that of iris texture, so the key point is how to select regions avoided these noise pixels. We primarily choose two sides of horizontal downward pupil region sized of 66×55 as candidate regions, and apply formula (1) to determine whether they contain noise:

$$h_1^{(k)} = \begin{cases} 1 & \text{if } \forall I_{un}(i,j) \in [T_{min}, T_{max}], i \in [1, M], j \in [1, N] \\ 0 & otherwise \end{cases} \tag{1}$$

Where, $I_{un}(i,j)$ is the input candidate region, $i \in \{1, 2, \cdots, M\}$, $j \in \{1, 2, \cdots, N\}$, M and N are the height and width of candidate region. T_{min} is the threshold of eyelash noise, T_{max} is the threshold of light spot noise, $k = 1\ or\ 2$ stands for the left or right candidate region, respectively. The processing is illustrated in Fig. 2.

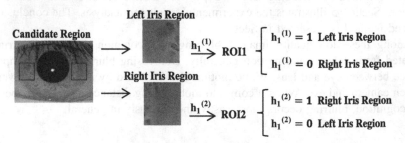

Fig. 2. Selecting ROI of iris.

Mean Subtracted Contrast Normalized (MSCN). Ruderman [9] employed a non-linear pretreatment operation to minus the local average of intensity image, and normalize the local variance, which can decrease the correlation between background and texture. The non-operation can be defined as:

$$MSCN(i,j) = \frac{I(i,j) - \mu(i,j)}{\sigma(i,j) - C} \tag{2}$$

Where $I(i,j)$ is the given iris region intensity image, $i \in \{1, 2, \cdots, M_1\}$, $j \in \{1, 2, \cdots, N_l\}$, M_1, N_1 are the height and width iris region. C is a positive constant which limit to zero, that is used to avoid instabilities from occurring when $\sigma(i,j)$ tends to zero. $\mu(i,j)$ is the local average value, $\sigma(i,j)$ is the local standard variance of $I(i,j)$.

$$\mu(i,j) = \sum_{a=-A}^{A} \sum_{b=-B}^{B} \omega_{a,b} I_{a,b}(i,j) \tag{3}$$

$$\sigma(i,j) = \sqrt{\sum_{a=-A}^{A} \sum_{b=-B}^{B} \omega_{a,b} \left(I_{a,b}(i,j) - \mu(i,j)\right)^2} \tag{4}$$

Where, $\omega_{a,b}$ notes a weighted coefficient generated by 2D normalized Gaussian function, $A = B = 2$ in the experiments. This operation represents Mean Subtracted Contrast Normalized coefficients (MSCN) intuitively. Acquired MSCN is provided with specific statistical characteristic which is prone to change by the presence of distortion and that quantifying these changes could predict iris region sharpness. As shown in Fig. 3, the normalized frequency of MSCN changes differently to ideal iris image and blurred iris image.

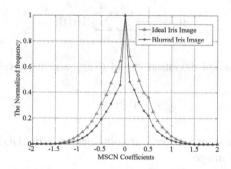

Fig. 3. Histograms of MSCN coefficients for an ideal iris image and a blurred iris image.

GGD Shape Parameter Feature of MSCN (MSCN). Considering MSCN operation includes a step of minus average value, we select GGD with zero mean ($\mu = 0$) to fit the MSCN, then extract feature γ.

GGD[16] is a wide distribution, which could reflect the tail variation of the histogram.

The GGD is defined as:

$$f(x; \gamma, \sigma, \rho) = \frac{\gamma}{2\beta\Gamma\left(\frac{1}{\gamma}\right)} e^{-\left(-\left(\frac{|x|}{\beta}\right)^{\gamma}\right)}. \tag{5}$$

where

$$\beta = \sigma \sqrt{\frac{\Gamma\left(\frac{1}{\gamma}\right)}{\Gamma\left(\frac{3}{\gamma}\right)}} \tag{6}$$

Γ is the gamma function:

$$\Gamma(s) = \int_0^{+\infty} x^{s-1} dx \tag{7}$$

σ is the standard deviation of $\hat{I}(i,j)$:

$$\sigma = \sqrt{\frac{1}{\hat{M}\hat{N}} \sum_{i=1}^{\hat{M}} \sum_{j=1}^{\hat{N}} \left(\hat{I}(i,j) - \mu\right)^2} \tag{8}$$

Where, $\hat{I}(i,j)$ is the input pre-processing model. $\hat{M}$, $\hat{N}$ are the height and width of $\hat{I}(i,j)$, μ is the average value.

$$\mu = \frac{1}{\hat{M}\hat{N}} \sum_{i=1}^{\hat{M}} \sum_{j=1}^{\hat{N}} |\hat{I}(i,j)| \tag{9}$$

ρ is the ratio of standard deviation σ and average value μ, respectively:

$$\rho = \frac{\sigma}{\mu} \tag{10}$$

As illustrated in Fig. 4, parameter γ are obtained by using GGD to fit the MSCN of left or right iris region image.

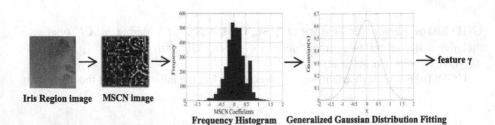

Fig. 4. Flowchart of MSCN feature extraction

Improved Differencing Signal. In addition to analyzing MSCN frequency of image, gradient variety of image could also contribute to assessment for image quality. Based on this idea, we extract feature with one-order derivative of iris region image which is called differencing signal. Wang [11] evaluated quality of JPEG image captured in visible light via by calculating the differencing signal along horizontal and vertical direction. We find that it is also sensitive to evaluate the quality of infrared image, and modify the method to make sure it fit the iris texture. The differencing signal along horizontal and vertical direction can be defined as:

$$d_k(i,j) = \begin{cases} I(i,j+1) - I(i,j) & k = 1 \\ I(i+1,j) - I(i,j) & k = 2 \end{cases} \tag{11}$$

Initially, the intensity difference is calculated every 3 pixels in both two directions.

Activity Level of Overall Block. Activity level of overall block B_k is selected to reflect the differencing signal intensity of global region. It is defined as the average difference across block boundaries, and is calculated as formula 12.

$$B_k = \begin{cases} \dfrac{1}{M([N/3]-1)} \sum_{i=1}^{M} \sum_{j=1}^{[N/3]-1} |d_k(i,3j)| & k = 1 \\ \dfrac{1}{N([M/3]-1)} \sum_{i=1}^{[M/3]-1} \sum_{j=1}^{N} |d_k(3i,j)| & k = 2 \end{cases} \tag{12}$$

Activity Level of Local Block. Activity level of local block Ak is more applicable to reflect the differencing signal intensity of iris texture. It is defined as the absolute difference between in-block image, and is calculated as formula 13.

$$A_k = \begin{cases} \dfrac{1}{2} \left| \dfrac{3}{M(N-1)} \sum_{i=1}^{M} \sum_{j=1}^{N-1} |d_k(i,j)| - B_k \right| & k = 1 \\ \dfrac{1}{2} \left| \dfrac{3}{N(M-1)} \sum_{i=1}^{M-1} \sum_{j=1}^{N} |d_k(i,j)| - B_k \right| & k = 2 \end{cases} \tag{13}$$

Number of Low Signal Intensity. In addition, feature Z_k is calculated by the zero-crossing of differencing signal in traditional algorithm, which manifests as the differencing signal elements go from positive to negative or reverse.

$$Z_k = \dfrac{1}{M(N-2)} \sum_{i=1}^{M} \sum_{j=1}^{N-2} z_k(m,n) \tag{14}$$

Where

$$z_k(m,n) = \begin{cases} 1 & \text{zero-crossing at } d_k(i,j) \\ 0 & \text{otherwise} \end{cases} \tag{15}$$

However, for uncompressed image, image blurring represents are not only less at the transitions in the background region but also flatter at the transitions in the edge region. It means that the differencing signal elements of the iris region are less than 2 except several special ones. Consider Z_k^* as the number of low signal intensity with threshold adjustment strategy to the iris region image, which has performance superior to Z_k. We modify formula Z_k^* to be the number of horizontal or vertical direction differencing signal elements which is less than 2.

$$Z_k^* = \frac{1}{M(N)} \sum_{i=1}^{M} \sum_{j=1}^{N} z_k^*(m,n) \qquad (16)$$

Where

$$z_k^*(m,n) = \begin{cases} 1 & d_k(i,j) < 2 \\ 0 & otherwise \end{cases} \qquad (17)$$

Differencing signal feature vector (A, B, Z^*) is obtained through calculating average value of feature vector in two directions.

$$B = \frac{B_1 + B_2}{2}, A = \frac{A_1 + A_2}{2}, Z^* = \frac{Z_1^* + Z_2^*}{2} \qquad (18)$$

Finally, the features on spatial domain of iris regions are shown as follows: $Spat_{ROI1} = (\gamma^{RI}, A^{RI}, B^{RI}, Z^{*RI})$ is the feature vector of left iris region; $Spat_{ROI2} = (\gamma^{R2}, A^{R2}, B^{R2}, Z^{*R2})$ is the feature vector of right iris region.

2.3   Statistical Feature on Frequency Domain

ROI selection. In this paper, frequency features are extracted on pupil edge region. The same as iris ROIs, to avoid noise from eyelash, we select two horizontal sides of pupil edge sized 20×40 as candidate regions. Since light spot noise distributes one side for the most part, formula 19 is applied to judge whether the left region is polluted or not:

$$h_2 = \begin{cases} 1 & if \; \forall I_{un}(i,j) \in [0, T_{max}], i \in [1,M], j \in [1,N] \\ 0 & otherwise \end{cases} \qquad (19)$$

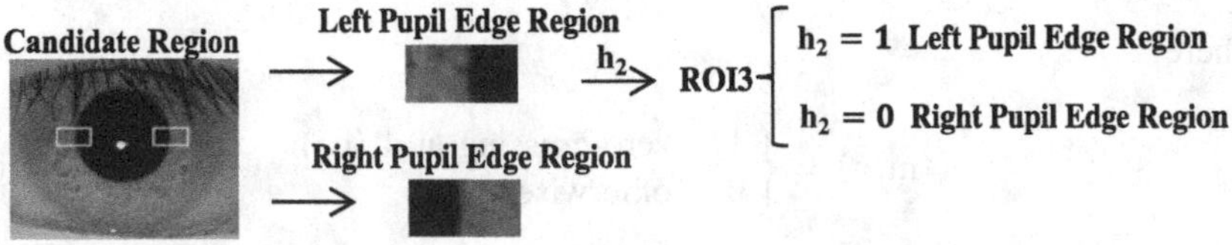

Fig. 5. Illustration of selecting pupil edge ROI.

If $h_2 = 1$, it is regarded as the left ROI, otherwise, another side is chosen as shown in Fig. 5.

Extracting Statistical Features on Frequency Domain. Discrete cosine transform (DCT) is real signal. Compared with DFT, DCT is time-saving. For its fine energy compression performance, DCT is widely used as the main method of image compression and it can also capture blurred factor. For higher efficiency, we adopt DCT to extract pupil edge ROI features in this paper. Just as presented in Fig. 6, we firstly extract features to the pupil edge region by 9×9. Then make a multiple scale by 5×5 low-pass filtering and down-sampling the image, and extract features to the sampled map by 5×5. Finally make the multiple scale again to the sampled map and extract features to the re-sampled map by 3×3.

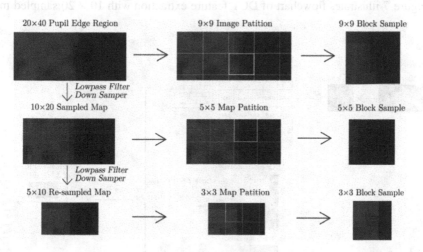

Fig. 6. Illustration of down-sampling and partition

Referred to Saad's [10] method, feature extraction is provided the following 3 forms:

1. For each sub-block, we apply formula (5) to fit a GGD, then obtain features $(\gamma_{i,j}, \rho_{i,j})$, where, $i = [1, [M_2/m]], j = [1, [N_2/m]]$. M_2, N_2 are the height and width of pupil edge or sub-sampled map, m is the length of block side. Next, count all feature of all sub-image blocks, extract first frequency feature vector $(\overline{\gamma_1}, \overline{\gamma_{1\uparrow}}, \overline{\rho_1}, \overline{\rho_{1\downarrow}})$. $\overline{\gamma_1}, \overline{\rho_1}$ note the average of all feature blocks $\gamma_{i,j}, \rho_{i,j}$, $\overline{\gamma_{1\uparrow}}, \overline{\rho_{1\downarrow}}$ are the average value of ranked top 10% of sorted $\overline{\gamma_1}, \overline{\rho_1}$ (percentile pooling).

2. For each sub-block, divide three regions through main diagonal direction, and fit GGD function respectively, then we'll get $\rho_{i,j,1}, \rho_{i,j,2}, \rho_{i,j,3}$. Compute variance $\sigma_{i,j}^2$ of $\rho_{i,j,1}, \rho_{i,j,2}, \rho_{i,j,3}$, count average value of all the $\sigma_{i,j}^2$ and top 10% rank of sorted $\sigma_{i,j}^2$. These construct the second feature vector $(\overline{\sigma_2}^2, \overline{\sigma_{2\downarrow}}^2)$.

3. For each sub-block, utilize formula (18) to extract three frequency features of high frequency, mid frequency, low frequency $E_{i,j,1}, E_{i,j,2}, E_{i,j,3}$ by anti-diagonal direction of 30, 60, 90 degrees regions. Then get the energy differences $\gamma_{i,j,1}, \gamma_{i,j,2}$. Compute average of them to $\bar{\gamma}_{i,j}$, and for all $\bar{\gamma}_{i,j}$, save average and average value of rank 10% sorted sequences, we finally acquire the third feature vector $(\bar{\gamma}_3, \overline{\gamma_{3\uparrow}})$.

$$E_{i,j,n} = \sigma_{i,j,n}^2 \tag{20}$$

$$r_{i,j,n} = \frac{\left|E_{i,j,n} - \frac{1}{n-1}\sum_{k<n} E_{i,j,k}\right|}{E_{i,j,n} + \frac{1}{n-1}\sum_{k<n} E_{i,j,k}} \tag{21}$$

Figure 7 illustrates flowchart of DCT feature extraction with 10×20 sampled map.

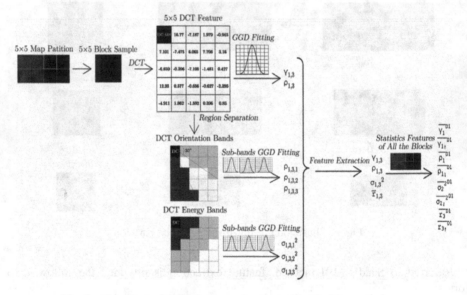

Fig. 7. Flowchart of DCT feature extraction with 10×20 sampled map

Finally, the features on frequency domain of pupil edge region are shown as follows:

$Freq_{ROI3} = \left(\bar{\gamma}_1^{R3}, \overline{\gamma_{1\uparrow}}^{R3}, \overline{\rho_1}^{R3}, \overline{\rho_{1\downarrow}}^{R3}, \overline{\sigma_2}^{2R3}, \overline{\sigma_{2\downarrow}}^{2R3}, \overline{\gamma_3}^{R3}, \overline{\gamma_{3\uparrow}}^{R3}\right)$ is the feature vector of pupil edge region;

$Freq_{down1} = \left(\bar{\gamma}_1^{D1}, \overline{\gamma_{1\uparrow}}^{D1}, \overline{\rho_1}^{D1}, \overline{\rho_{1\downarrow}}^{D1}, \overline{\sigma_2}^{2D1}, \overline{\sigma_{2\downarrow}}^{2D1}, \overline{\gamma_3}^{D1}, \overline{\gamma_{3\uparrow}}^{D1}\right)$ is the feature vector of sampled map;

$Freq_{down2} = \left(\bar{\gamma}_1^{D2}, \overline{\gamma_{1\uparrow}}^{D2}, \overline{\rho_1}^{D2}, \overline{\rho_{1\downarrow}}^{D2}, \overline{\sigma_2}^{2D2}, \overline{\sigma_{2\downarrow}}^{2D2}, \overline{\gamma_3}^{D2}, \overline{\gamma_{3\uparrow}}^{D2}\right)$ is the feature vector of resampled map.

2.4 Classify Feature via SVM

3 Feature Selection via LASSO

To reduce the mutual correlation and computational complexity, we employ LASSO
[12] to select the features. The multivariate linear model is defined as:

$$y = X\beta + \varepsilon \tag{22}$$

Where $y = (y_1, y_2, \cdots, y_n)^T$, $X = (x_1, x_2, \cdots, x_n)$, $x_j = (x_{1j}, x_{2j}, \cdots, x_{nj})^T$, $j = 1, 2, \ldots,$
d. β is the parameter and $dim(\beta) = d$. ε is the error term. The mathematical model is:

$$\hat{\beta} = arg\min_{\beta} \|y - X\beta_i\|^2, \text{ subject to } \sum_{i=1}^{d} |\beta_i| \leq t \tag{23}$$

In this paper, LASSO is used to select features and the result is shown in Table 1. To
obtain the best balance between time efficiency and accuracy, we finally reserve the
following features:

1. $Spat'_{ROI1} = (\gamma^{R1}, A^{R1}, B^{R1})$ in the feature vector of left iris region;
2. $Spat'_{ROI2} = (\gamma^{R2}, Z^{*R2})$ in the feature vector of right iris region;
3. $Freq'_{ROI3} = (\overline{\gamma_1}^{R3}, \overline{\sigma_2}^{R3}, \overline{\gamma_3}^{R3}, \overline{\gamma_{3\uparrow}}^{R3})$ in the feature vector of pupil edge region;
4. $Freq'_{down1} = (\overline{\gamma_1}^{D1}, \overline{\rho_1}^{D1}, \overline{\rho_{1\downarrow}}^{D1}, \overline{\gamma_{3\uparrow}}^{D1})$ in the feature vector of sampled map;
5. $Freq'_{down2} = (\overline{\gamma_1}^{D2})$ in the feature vector of resampled map.

The selected 14D feature vector $LFSF = (Spat'_{ROI1}, Spat'_{ROI2}, Freq'_{ROI3}, Freq'_{down1}, Freq'_{down2})$.

Table 1. Regression coefficients of the 32D features

ROI1	Regression coefficient	ROI3	Regression coefficient of pupil edge image	Regression coefficient of sampled map	Regression coefficient of resampled map
γ	−0.139	$\overline{\gamma_1}$	0	−0.827	−0.231
A	−0.446	$\overline{\gamma_{1\uparrow}}$	−0.005	0	0
B	0.145	$\overline{\rho_1}$	0	−0.013	0
Z^*	0	$\overline{\rho_{1\downarrow}}$	0	0.480	0
ROI2	Regression coefficient	$\overline{\sigma_2}^2$	0.071	0	0
γ	−1.512	$\overline{\sigma_{2\downarrow}}^2$	0	0	0
A	0	$\overline{\gamma_3}$	−0.288	0	0
B	0	$\overline{\gamma_{3\uparrow}}$	−0.274	−0.281	0
Z^*	−0.652				

4   Feature Classification via SVM

SVM is a classifier based on structure risk minimization (SRM) inductive principle, and has good characteristics of simple structure, global optimum, and strong generalization ability. It constructs an optimal hyper plane in a high dimensional feature space utilizing a small set of vectors near boundary [12 13].

For given train set $\{(x_1, y_1), (x_2, y_2)\ldots(x_1, y_i)\ldots(x_n, y_n)\}, x_i \in R^d, y_i \in \{-1, +1\}$ is the sample label, when sample could be classified linearly, positive samples and negative samples would be distributed on both sides of hyper plane $\omega \cdot x + b = 0$. SVM could be considered as the solution of problem (24):

$$\begin{cases} \min \frac{1}{2}\|\omega\|^2 \\ y_i[(\omega \cdot x_i) + b] - 1 \geq 0 \quad i = 1, \ldots, n \end{cases} \tag{24}$$

Here the problem is solved via Lagrange algorithm:

$$\min(\omega, b, a) = \frac{1}{2}\|\omega\|^2 - \sum_{i=1}^{n} \alpha_i \{y_i[(\omega \cdot x_i) + b] - 1\} \tag{25}$$

Where, α_i is Lagrange coefficients and $\alpha_i > 0$. Assign $\frac{\partial}{\partial \omega}$ and $\frac{\partial}{\partial b}$ to zero, solution is:

$$\omega = \sum_{i=1}^{n} \alpha_i y_i x_i$$

For linearly separable problem, the decision function is:

$$f(x) = sgn\left\{\sum_{i=1}^{n} \alpha_i y_i (x_i \cdot x) + b\right\} \tag{27}$$

Linearly inseparable problem can change into a separable one through selecting the proper kernel function. The decision function is:

$$f(x) = sgn\left\{\sum_{i=1}^{n} \alpha_i y_i K(x_i \cdot x) + b\right\} \tag{28}$$

Liner kernel function C-SVC is adopted in our experiments. Kennel function is $K(x_i \cdot x) = (x_i \cdot x) = x_i^T x$. We apply SVM to classify 14D feature set *LFSF* and obtain the result.

5   Performance Evaluation

We use monocular iris reader TCI301 to establish a database include 8 subjects, 16 eyes. Iris image was captured from long distance to close gradually, and permit object wear glasses or contact lenses. The image resolution is 640×480. We test proposed method on selected 1169 iris images which the number of blurred images is 682 and the number of ideal images is 487 (Fig. 8).

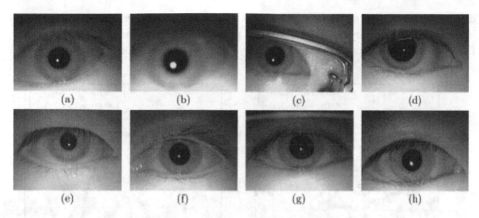

Fig. 8. Blurred and ideal image examples: (a) Light blurred image (b) Severe blurred image (c) Blurred image with glasses (d) Sparse iris texture; (e)(f)(g) Ideal image (h) Ideal image with contact lenses.

We apply SVM to classify the sample with features we extracted. As demonstrated in Fig. 9, compared the scatterplot of feature γ on randomized 300 global iris images and their ROI images, it's obvious that the feature outliers of ROI images are less than the global. And feature γ on ROI images has more robust performance.

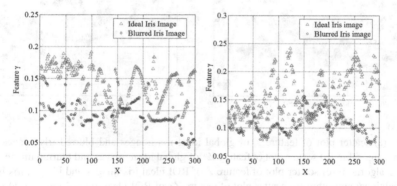

Fig. 9. (1) scatter plot of feature γ of global ideal iris images and blurred iris images (2) scatter plot of feature γ of ROI ideal images and blurred images

As observed from the Fig. 10, decidability increases after we modified differencing signal feature z_K^* that supports improvement to this feature.

Besides, Fig. 11 plots the compare of global differencing signal features and local features, local features have a better separation between ideal iris images and blurred iris images.

Figure 12 shows the relation between the number of selection features with Lasso and detection accuracy. It can be inferred that the detection accuracy increased first and then decreased with the increasing of the number of selection features. This result proves that different combination of features can effectively raise detection performance. However, information redundancy is obvious defected as the input dimension increase. The use of LASSO has played a critical role in accuracy and time consumption.

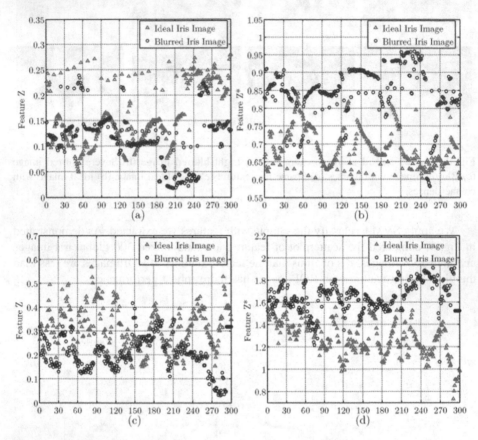

Fig. 10. (a) scatter plot of feature Z of global ideal iris images and blurred iris images using traditional algorithm (b) scatter plot of feature Z∗ of global ideal images and blurred images using improved algorithm (c) scatter plot of feature Z of ROI ideal iris images and blurred iris images using traditional algorithm (d) scatter plot of feature Z∗ of ROI ideal images and blurred images using improved algorithm.

Table 2. Comparison of different kernel functions

Methods	The proposed method (without LASSO)	The proposed method (LASSO)
Radial Basis Function(RBF)	96.80%	97.87%
Linear	**96.59%**	**97.87%**
2st-order polynomial	92.75%	96.38%
3st-order polynomial	91.47%	95.74%
4st-order polynomial	89.98%	91.26%
5st-order polynomial	88.91%	89.98%

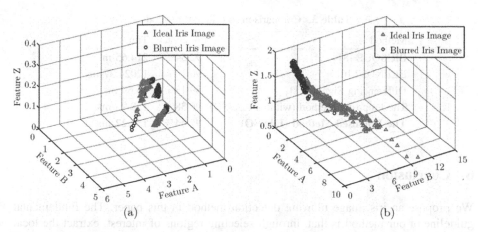

(a)   (b)

Fig. 11. (1) 3-D scatter plot of feature (A,B,Z) of global ideal iris images and blurred iris images using traditional algorithm (2) 3-D scatter plot of feature (A,B,Z*) of ROI ideal images and blurred images using improved algorithm.

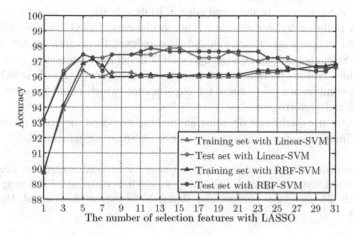

Fig. 12. The relation between the number of selection features with Lasso and detection accuracy

In the experiment, we also trained SVM with Radial Basis Function kernel, Linear-kernel, and higher order polynomial-kernel. Preferable detection performance is obtained with radial basis function-kernel and Linear-kernel, and illustrates the feature data is linearly separable, as indicated in Table 2. We ultimately choose Linear-kernel SVM in order to reduce the time complexity.

A comparative test among our proposed method and other four blurring detection algorithms is done with these 1169 images. The result in Table 3 shows that the proposed method performs well on accuracy and efficiency.

Table 3. Comparison of different methods.

Method	Accuracy	Time
BRISQUE [8]	92.54%	214.65 ms
BLIINDS2 [6]	90.19%	12024.37 ms
Differencing Signal [11]	89.13%	46.69 ms
The Proposed Method (without LASSO)	96.38%	178.95 ms
The Proposed Method (LASSO)	**97.44%**	**153.22 ms**

6   Conclusion

We propose an iris image blurring detection method in this paper. The fundamental guideline of our method is that, through selecting regions of interest, extract the local feature both on space domain and frequency domain, then construct feature set based on these local indexes and LASSO feature selection. We choose SVM as the classifier to train our features, which performs well in distinguishing ideal iris image and blurred iris image. The innovative points of this paper include: (1) Propose effective strategies to eliminate the noise on iris and pupil edge in ROIs selecting. (2) Select the iris region that is less polluted by eyelash and other occlusions. Based on this, MSCN and improved differencing signal are calculated as indexes for evaluating the iris image quality. (3) Select pupil edge as ROI, apply DCT to get high frequency information on this ROI which realizes more accurate on blurring detection. (4) Use LASSO to obtain the optimal combination of features for classification which could improve the algorithm speed and performance.

The results of experiments demonstrate that the proposed method can classify low quality iris image satisfying the requirement of real-time judgement.

Acknowledgement. This research is partly supported by National Natural Science Funds of China, No. 61703088, the Doctoral Scientific Research Foundation of Liaoning Province, No. 20170520326 and "the Fundamental Research Funds for the Central Universities", N160503003.

References

1. Anish, M., Anush Krishna, M., Alan Conrad, B.: No-reference image quality assessment in the spatial domain. IEEE Trans. Image Process. **21**(12), 4695–4708 (2012)
2. Belcher, C., Du, Y.: Information distance-based selective feature clarity measure for iris recognition. In: Electronic Imaging (2007). International Society for Optics and Photonics (2007)
3. Chen, J., Guangshu, H.U., Jin, X.U.: Iris image quality evaluation method based on wavelet packet decomposition. J. Tsinghua Univ. **43**(3), 377–380 (2003)
4. Chen, X., Li, Y., Harrison, R., Zhang, Y.Q.: Type-2 fuzzy logic-based classifier fusion for support vector machines. Appl. Soft Comput. **8**(3), 1222–1231 (2008)

5. Chen, Y., Dass, S.C., Jain, A.K.: Localized iris image quality using 2-d wavelets. In: Zhang, D., Jain, A.K. (eds.) ICB 2006. LNCS, vol. 3832, pp. 373–381. Springer, Heidelberg (2005). https://doi.org/10.1007/11608288_50
6. Daugman, J.: How iris recognition works. Handb. Image Video Process. 1(1), 1251–1262 (2005)
7. Fatemi, M.H., Gharaghani, S.: Prediction of selectivity coefficients of univalent anions for anion-selective electrode using support vector machine. Electrochim. Acta 53(12), 4276–4282 (2007)
8. Ma, L., Tan, T., Wang, Y., et al.: Personal identification based on iris texture analysis. IEEE Trans. Pattern Anal. Mach. Intell. 25(12), 1519–1533 (2003)
9. Ruderman, D.L., Bialek, W.: Statistics of natural images: Scaling in the woods. J. Policy Anal. Manage. 21(1), 121–125 (2002)
10. Saad, M.A., Bovik, A.C., Charrier, C.: Blind image quality assessment: a natural scene statistics approach in the dct domain. IEEE Trans. Image Process. 21(8), 3339–3352 (2012)
11. Sharififi, K., Leon-Garcia, A.: Estimation of shape parameter for generalized gaussian distributions in subband decompositions of video. IEEE Trans. Circuits Syst. Video Technol. 5(1), 52–56 (1995)
12. Tibshirani, R.: Regression shrinkage and selection via the lasso. J. Royal Stat. Soc. 58(1), 267–288 (1996)
13. Wang, Q., Wang, H., Zhang, T.: A fast iris image assessment procedure. In: Industrial Engineering, Machine Design and Automation (IEMDA 2014) and Computer Science and Application (CCSA 2014), vol. 12, no (14), pp. 386 – 392 (2014)
14. Wang, Z., Sheikh, H.R., Bovik, A.C.: No-reference perceptual quality assessment of JPEG compressed images. In: Proceedings of IEEE 2002 International Conference on Image Processing, pp. 477–480 (2002)
15. Zhang, G.H., Sensar, I.: Method of selecting the best enroll image for personal identification. Maple Shadenj Us (1999)

Minimize State Jumps and Guaranteed Cost Control for a Class of Switched Descriptor Systems

Shuhui Shi[✉] and Jie Hao

Shenyang Institute of Engineering, Shenyang 110136, People's Republic of China
shishuhuishuxue@163.com

Abstract. This paper deals with the problem of state jumps and guaranteed cost control for a class of switched descriptor systems. State feedback guaranteed cost controller is adopted to make the resulting closed-loop system admissible and have an minimum state jump and upper bound of cost function. An explicit formula of state jumps for switched descriptor systems is given. Based on single Lyapunov function and convex combination technique, a switching law is designed and a sufficient condition of the existence of such controller is presented. By means of variables substitution and matrix inequality transformation, the condition can be solved from an optimization problem. The advantage of method presented in this paper is illustrated by an example.

Keywords: Switched descriptor systems · Guaranteed cost control · State jump · Inconsistent initial condition

1 Introduction

Switched descriptor systems is composed of a finite family of singular subsystems and a switching signal that governs the switching between them. The kind of system arises from, for instance, electrical networks, complex networks and economic systems [1–3]. Recently, there has been increasing interest in analysis and synthesis for switched descriptor systems [4–10].

However, the existing results did not consider the problem of state jump owing to inconsistent initial condition from a descriptor subsystem to another descriptor subsystem at switching instants. For descriptor systems, it is well known that inconsistent initial condition can cause finite instantaneous jumps. If the jumps reach a certain extent, the system can be destroyed [11].

For the switched descriptor system, state jumps occur not only at initial starting point, but also at switching points with inconsistent initial conditions. Such instantaneous jumps should be avoided since the systems will be destroyed when the jumps strength is too large. Therefore, it is significant to eliminate such jumps at the switching instants. Unfortunately, the methods of dealing with state jump at the initial point for single descriptor system cannot be applied to

© Springer Nature Singapore Pte Ltd. 2020
J. Qian et al. (Eds.): ICRRI 2020, CCIS 1335, pp. 224–235, 2020.
https://doi.org/10.1007/978-981-33-4929-2_16

switched descriptor systems directly. This is because the switching time and the activated subsystem at switching points will be arbitrary when the switching sequence is arbitrary. Usually, it cannot be guaranteed that the states at switching points satisfy the consistent initial condition of the next activated subsystem. This problem has attracted much attention [12–16]. [15]investigated the initial instantaneous jumps and gave a sufficient stability condition for the switched descriptor system with both stable and unstable subsystems. At arbitrary switching instant, inconsistent state jump for switched descriptor systems can be compressed in [16]. However, to our best knowledge, minimizing jumps and upper of cost function of switched descriptor systems at the switching instants is still an open problem.

In addition to the simple stabilization, there have been various efforts in assigning certain performance criteria when designing a delay-dependent guaranteed cost controller [17]. One approach to this problem is the so-called guaranteed cost control first proposed in [18]. Its essential idea is to construct a controller that not only stabilizes the systems but also guarantees a desire level of performance represented by a quadratic cost function. Many constructive results on guaranteed cost control have been reported in [19–25]. By using a quadratic performance function with derivative of state, [19] studied the design of guaranteed cost control law for a class of descriptor system with parameter uncertainties. A sufficient condition of guaranteeing the performance described by a quadratic cost function for discrete-time Takagi-Sugeno (T-S) fuzzy systems was derived in [20].

In this paper, single Lyapunov function combining with convex combination technique are adopted for a class of switched descriptor systems, an condition is presented to ensure the admissible of the resulting closed-loop systems while guaranteeing a minimum state jumps and upper bound of cost function. A state feedback guaranteed controller and a switching law are designed. To show the effectiveness of state feedback guaranteed cost controller and switching law, an example is given in the paper.

2 Problem Statement and Preliminaries

Consider the following switched descriptor system

$$E\dot{x}(t) = A_{\sigma(t)}x(t) + B_{\sigma(t)}u(t), \qquad x(0) = x_0 \qquad (1)$$

where $\sigma(t) : [0, +\infty) \rightarrow \chi = \{1, 2, \cdots, m\}$ is the switching signal. $x \in R^n$ is the state, $u_{\sigma(t)} \in R^m$ is the control input. $E \in R^{n \times n}$ and $rank(E) = r \leq n$, $A_{\sigma(t)}, B_{\sigma(t)}$ are known constant matrices. x_0 is the initial state.

Some definitions and assumptions on switched descriptor systems are given as follows.

Definition 1. *Consider the unforced switched descriptor system*

$$E\dot{x}(t) = A_i x(t) \qquad (2)$$

1) *For arbitrary $i \in \chi$, the pair (E, A_i) is said to be regular if there exists a constant scalar $s \in C$ such that $det(sE - A_i) \neq 0$.*
The system (2) is said to be regular if every pair (E, A_i) is regular, $i \in \chi$.
2) *For arbitrary $i \in \chi$, the pair (E, A_i) is said to be impulse free if $deg(det(sE - A_i)) = rank(E)$ holds.*

The system (2) is said to be impulse free if every pair (E, A_i) is impulse free, $i \in \chi$.

The unforced switched descriptor system (2) is said to be admissible if system (2) is regular, impulse free and stable.

Assumption 1. *Each descriptor system (E, A_i) is regular, impulse-free and stabilizable.*

In the paper, $\{t_k\}$ denotes a switching sequence, where $k = 1, 2, \cdots, 0 < t_1 < t_2 < \cdots < t_k < \cdots (k \to \infty, t_k \to \infty)$. we always assume that the jth subsystem is activated in $t \in (t_{k-1}, t_k]$ and the ith subsystem is activated in $t \in (t_k, t_{k+1}]$, that is, the system is switched from the jth subsystem to the ith subsystem at $t = t_k$. Denote $x(t_k) = x(t_k^-) = \lim_{h \to 0^+} x(t_k - h)$, $x(t_k^+) = \lim_{h \to 0^+} x(t_k + h)$.

It is well known that even if there is no impulse, the switched descriptor system can be still have finite instantaneous state jump due to inconsistent initial conditions. Also, it cannot be guaranteed that the states are always consistent with the next activated subsystem at arbitrary switching instant. If the state is not the consistent initial state of the next singular subsystem, instantaneous jumps will occur at switching instant. In this paper, the switching sequence is arbitrary, which implies that the state at the switching instant is unknown. Furthermore, which the descriptor subsystem will be activated at each switching instant can be arbitrary, even stochastic. This makes the analysis and investigation problem of instantaneous state jumps become more complex and difficult than that of normal descriptor systems.

Associated with the system (1) is the cost function

$$J = \int_0^{+\infty} [x^T(t)Qx(t) + u^T(t)Ru(t)]dt \tag{3}$$

where Q and R are given positive-definite matrices.

The objective of this paper is to construct state feedback controller of the following form

$$u(t) = K_{\sigma(t)}x(t) \tag{4}$$

where $K_{\sigma(t)}$ is constant matrices to be determined, and the corresponding switching strategy such that the resulting closed-loop system

$$E\dot{x}(t) = (A_{\sigma(t)} + B_{\sigma(t)}K_{\sigma(t)})x(t) \tag{5}$$

is sdmissible and the cost function (3) satisfies $J \leq J^*$, where J^* is some specified constant.

Definition 2. *Consider the switched descriptor system (1) and the cost function (3), if there exist a state feedback controller of the form (4) and a positive scalar J^* such that the closed-loop system (5) is admissible and the cost function (3) satisfies $J \leq J^*$, then J^* is said to be a guaranteed cost and (4) is said to be a guaranteed cost controller of the system (1).*

3   Admissible Control and Guaranteed Cost Control

In this section, first of all, a sufficient condition of the existence of guaranteed cost controller (4) for the switched descriptor systems (1) is presented.

Theorem 1. *For the switched descriptor system (1) and cost function (3), if there exist scalars $\eta_i < 0$, $\sum\limits_{i=1}^{m} \eta_i = 1$ and matrix P such that the following equality and inequality*

$$E^T P = P^T E \geq 0 \tag{6}$$

$$\sum_{i=1}^{m} \eta_i S_i < 0 \tag{7}$$

hold for a given $\alpha_i > 0$ and $\forall i \in \chi$, where

$$S_i = A_i^T P + P^T A_i + Q - \alpha_i P^T B_i (I - \frac{1}{4}\alpha_i R) B_i^T P \tag{8}$$

Then, the closed system (5) exists a guaranteed cost controller

$$u(t) = K_i x(t) = -\frac{1}{2}\alpha_i B_i^T P x(t) \tag{9}$$

The corresponding switching signal is

$$\sigma(t) = \arg\min\{x^T [A_i^T P + P^T A_i + Q - \alpha_i P^T B_i (I - \frac{1}{4}\alpha_i R) B_i^T P]x, i \in \chi\} \tag{10}$$

A upper bound of cost function is

$$J^* = x_0^T E^T P x_0 \tag{11}$$

Proof. Choose controller as

$$K_i = -\frac{1}{2}\alpha_i B_i^T P \tag{12}$$

the corresponding closed system is

$$E\dot{x}(t) = (A_i - \frac{1}{2}\alpha_i B_i B_i^T P)x(t) \tag{13}$$

Choose the switching signal (10) and single Lyapunov function

$$V(x) = x^T(t)E^T Px(t) \tag{14}$$

where P satisfying (6).

Then one has

$$\begin{aligned}
\dot{V}(x) &= \dot{x}^T E^T Px + x^T E^T P\dot{x} \\
&= (E\dot{x})^T Px + x^T P^T E\dot{x} \\
&= x^T[(A_i - \frac{1}{2}\alpha_i B_i B_i^T P)^T P + P^T(A_i - \frac{1}{2}\alpha_i B_i B_i^T P)]x
\end{aligned} \tag{15}$$

and

$$\begin{aligned}
J &= \int_0^\infty [x^T(t)Qx(t) + u^T(t)Ru(t)]dt \\
&= \sum_{i=1}^m \sum_{j=1}^\infty \int_{t_{i_j}}^{t_{i_j}+1} [x^T(t)Qx(t) + u^T(t)Ru(t) + \dot{V}(x(t))]dt \\
&\quad - \sum_{k=0}^\infty \int_{t_k}^{t_{k+1}} \dot{V}(x(t))dt
\end{aligned} \tag{16}$$

According to the design of switching time sequence, the construction of switching signal (10) and the condition (7), for $\forall t \geq 0$, there must exist a i satisfying

$$x^T S_i x < 0 \tag{17}$$

Thus, when $\sigma(t) = i$, one gets

$$x^T(t)Qx(t) + u^T(t)Ru(t) + \dot{V}(x(t)) < 0 \tag{18}$$

This implies that $\dot{V}(x(t))$ decreases along the trajectory of system (5). Thus,

$$(A_i - \frac{1}{2}\alpha_i B_i B_i^T P)^T P + P^T(A_i - \frac{1}{2}\alpha_i B_i B_i^T P) < 0$$

holds. So, the closed-loop system is admissible and it follows that

$$\begin{aligned}
J &< -\sum_{k=0}^\infty \int_{t_k}^{t_{k+1}} \dot{V}(x(t))dt \\
&= V(x(t_0)) \\
&= x_0^T E^T Px_0
\end{aligned} \tag{19}$$

Thus, (9) is a state feedback guaranteed cost controller and an upper bound of cost function is $x_0^T E^T Px_0$. □

Remark 1. The condition (6) is a non-strict inequality and the condition (7) is a non-linear inequality, so it is difficult to verify whether the conditions are satisfied or not. To solve this problem, by means of variables substitution and Schur complement, the inequalities (6) and (7) can be turned into an linear matrix inequality, which is given in the following Theorem 2.

Theorem 2. *If there exist matrices* $Z > 0 \in R^{n \times n}, W \in R^{(n-r) \times n}, U \in R^{n \times (n-r)}, rank(U) = n - r, EU = 0$ *and* $\eta_i < 0$ *for a given* $\alpha_i > 0, \delta_i > 0, \gamma > 0, \varepsilon_i$ *and* $\forall i \in \chi$ *such that the following linear matrix inequality*

$$\begin{bmatrix} \Upsilon & (ZE + WU)^T \\ (ZE + WU) & \mu^{-1}I \end{bmatrix} < 0 \qquad (20)$$

hold, where $\Upsilon = \mathcal{A}^T(ZE + WU) + (ZE + WU)^T \mathcal{A} + Q - \lambda(ZE + WU) - \lambda(ZE + WU)^T + \mathcal{D}$, $\mathcal{A} = \sum\limits_{i=1}^{m} \eta_i A_i$, $\lambda = \sum\limits_{i=1}^{m} \eta_i \varepsilon_i \sqrt{\alpha_i}$, $\mathcal{D} = \sum\limits_{i=1}^{m} \eta_i \varepsilon_i^2 [B_i(I - \frac{1}{4}\alpha_i R)B_i^T + \delta_i I]^{-T}$, $\mu = \sum\limits_{i=1}^{m} \eta_i \alpha_i \delta_i$.

Then, the closed system (5) exists a guaranteed cost controller

$$u(t) = K_i x(t) = -\frac{1}{2}\alpha_i B_i^T(ZE + WU)x(t) \qquad (21)$$

The corresponding switching signal is

$$\sigma(t) = \arg\min\{x^T[A_i^T(ZE + WU) + (ZE + WU)^T A_i + Q$$
$$-\alpha_i(ZE + WU)^T B_i(I - \frac{1}{4}\alpha_i R)B_i^T(ZE + WU)]x, i \in \chi\} \qquad (22)$$

A upper bound of cost function is

$$J^* = x_0^T(ZE + WU)x_0 \qquad (23)$$

Proof. In order to deal with nonlinear term $B_i(I - \frac{1}{4}\alpha_i R)B_i^T$ in (8) of Theorem 1, we introduce a $\delta_i > 0$ to satisfy $B_i(I - \frac{1}{4}\alpha_i R)B_i^T + \delta_i I > 0$. Then (8) is equivalent to

$$S_i = A_i^T P + P^T A_i + Q - \alpha_i P^T[B_i(I - \frac{1}{4}\alpha_i R)B_i^T + \delta_i I]P + \alpha_i \delta_i P^T P \qquad (24)$$

Thus $\forall \varepsilon_i$,

$$\{\varepsilon_i^{-1}\sqrt{\alpha_i}P - [B_i(I - \frac{1}{4}\alpha_i R)B_i^T + \delta_i I]^{-1}\}^T[B_i(I - \frac{1}{4}\alpha_i R)B_i^T + \delta_i I]$$
$$\{\varepsilon_i^{-1}\sqrt{\alpha_i}P - [B_i(I - \frac{1}{4}\alpha_i R)B_i^T + \delta_i I]^{-1}\} \geq 0 \qquad (25)$$

holds. Then one has

$$\varepsilon_i^{-2}\alpha_i P^T[B_i(I - \frac{1}{4}\alpha_i R)B_i^T + \delta_i I]P - \varepsilon_i^{-1}\sqrt{\alpha_i}P - \varepsilon_i^{-1}\sqrt{\alpha_i}P^T$$

$$+[B_i(I - \frac{1}{4}\alpha_i R)B_i^T + \delta_i I]^{-T} \geq 0$$

$$-\varepsilon_i^{-2}\alpha_i P^T[B_i(I - \frac{1}{4}\alpha_i R)B_i^T + \delta_i I]P \leq -\varepsilon_i^{-1}\sqrt{\alpha_i}P - \varepsilon_i^{-1}\sqrt{\alpha_i}P^T$$

$$+[B_i(I - \frac{1}{4}\alpha_i R)B_i^T + \delta_i I]^{-T}$$

$$-\alpha_i P^T[B_i(I - \frac{1}{4}\alpha_i R)B_i^T + \delta_i I]P \leq -\varepsilon_i\sqrt{\alpha_i}P - \varepsilon_i\sqrt{\alpha_i}P^T$$

$$+\varepsilon_i^2[B_i(I - \frac{1}{4}\alpha_i R)B_i^T + \delta_i I]^{-T} \qquad (26)$$

So, (24) must be less than or equal to

$$Y_i = A_i^T P + P^T A_i + Q - \varepsilon_i \sqrt{\alpha_i} P - \varepsilon_i \sqrt{\alpha_i} P^T$$
$$+ \varepsilon_i^2 [B_i (I - \frac{1}{4}\alpha_i R) B_i^T + \delta_i I]^{-T} + \alpha_i \delta_i P^T P \qquad (27)$$

Thus, one deduces that (7) holds if the following inequality

$$\sum_{i=1}^{m} \eta_i Y_i < 0 \qquad (28)$$

holds.

By the Schur complement, (28) is equivalent to

$$\begin{bmatrix} \Theta & P^T \\ P & \mu^{-1} I \end{bmatrix} < 0 \qquad (29)$$

where, $\Theta = \mathcal{A}^T P + P^T \mathcal{A} + Q - \lambda P - \lambda P^T + \mathcal{D}$, $\mathcal{A} = \sum_{i=1}^{m} \eta_i A_i$, $\lambda = \sum_{i=1}^{m} \eta_i \varepsilon_i \sqrt{\alpha_i}$, $\mathcal{D} = \sum_{i=1}^{m} \eta_i \varepsilon_i^2 [B_i (I - \frac{1}{4}\alpha_i R) B_i^T + \delta_i I]^{-T}$, $\mu = \sum_{i=1}^{m} \eta_i \alpha_i \delta_i$.

By making use of matrix decomposition of Lyapunov matrix, (6) and (7) are equvilent to (20) in Theorem 2. The proof is completed. □

4   Minimize State Jumps and Guaranteed Cost Control

In this section, we will show state jumps at switching instant for the switched descriptor system firstly.

There exist nonsingular matrices M, N such that

$$MEN = \begin{bmatrix} I_r & 0 \\ 0 & 0 \end{bmatrix}, MG_i = \begin{bmatrix} G_{i1} \\ G_{i2} \end{bmatrix},$$

$$M(A_i + B_i K_i C_{y_i})N = \begin{bmatrix} \bar{A}_{i11} & \bar{A}_{i12} \\ \bar{A}_{i21} & \bar{A}_{i22} \end{bmatrix},$$

holds and instantaneous state jump vector $e(t_k)$ can be denoted by

$$e(t_k) = \Gamma_i x(t_k), \qquad (30)$$

where $\Gamma_i = -N \begin{bmatrix} 0 & 0 \\ \bar{A}_{i22}^{-1} \bar{A}_{i21} & I \end{bmatrix} N^{-1}$.

In the following, we try to reduce the jump strength as small as possible, which can be carried out by minimizing the Frobenius norm of Γ_i since $x(t_k)$ is unknown at switching point t_k. The Frobenius norm of Γ_i denoted by $\|\Gamma_i\|$, is defined as $\|\Gamma_i\| = (trace(\Gamma_i^T \Gamma_i))^{\frac{1}{2}}$. The minimization problem of $\|\Gamma_i\|^2$ will be transformed into the following optimization problem. Γ_i in (30) can be rewritten as

$$\Gamma_i = N \begin{bmatrix} 0 & 0 \\ 0 & -\bar{A}_{i22}^{-1} \end{bmatrix} \begin{bmatrix} 0 & 0 \\ \bar{A}_{i21} & \bar{A}_{i22} \end{bmatrix} N^{-1} \tag{31}$$

Then one gets

$$\|\Gamma_i\| = \|N \begin{bmatrix} 0 & 0 \\ 0 & -\bar{A}_{i22}^{-1} \end{bmatrix} \begin{bmatrix} 0 & 0 \\ \bar{A}_{i21} & \bar{A}_{i22} \end{bmatrix} N^{-1}\|$$

$$\leq \|N\| \cdot \| \begin{bmatrix} 0 & 0 \\ 0 & -\bar{A}_{i22}^{-1} \end{bmatrix} \| \cdot \| \begin{bmatrix} 0 & 0 \\ \bar{A}_{i21} & \bar{A}_{i22} \end{bmatrix} \| \cdot \|N^{-1}\|$$

$$= \|N\| \cdot \| - \bar{A}_{i22}^{-1}\| \cdot \| \begin{bmatrix} 0 & 0 \\ \bar{A}_{i21} & \bar{A}_{i22} \end{bmatrix} \| \cdot \|N^{-1}\| \tag{32}$$

In order to deal with the nonlinear term $\| - \bar{A}_{i22}^{-1}\| \cdot \| \begin{bmatrix} 0 & 0 \\ \bar{A}_{i21} & \bar{A}_{i22} \end{bmatrix} \|$, $-\bar{A}_{i22}^{-1}$ is rewritten as

$$- \bar{A}_{i22}^{-1} = (I - (I + \bar{A}_{i22}))^{-1} \tag{33}$$

Let

$$(I + \bar{A}_{i22})^T (I + \bar{A}_{i22}) < \zeta_i I, \quad 0 < \zeta_i < 1/(n - r) \tag{34}$$

Then,

$$\|I + \bar{A}_{i22}\| = (trace[(I + \bar{A}_{i22})^T (I + \bar{A}_{i22})])^{1/2}$$
$$= (trace(I + \bar{A}_{i22})^T (I + \bar{A}_{i22}))^{1/2}$$
$$< (trace(\zeta_i I))^{1/2}$$
$$= (\zeta_i(n - r))^{1/2}$$
$$< 1 \tag{35}$$

holds. $[I - (I + \bar{A}_{i22})]^{-1}$ is transformed into

$$[I - (I + \bar{A}_{i22})]^{-1} = I + (I + \bar{A}_{i22})[I - (I + \bar{A}_{i22})]^{-1} \tag{36}$$

According to norm consistency theory and triangle inequality, one has

$$\|[I - (I + \bar{A}_{i22})]^{-1}\| \leq \|I\| + \|(I + \bar{A}_{i22})[I - (I + \bar{A}_{i22})]^{-1}\|$$
$$\leq (n - r)^{1/2} + \|I + \bar{A}_{i22}\| \cdot \|[I - (I + \bar{A}_{i22})]^{-1}\| \tag{37}$$

By (34), one gets

$$\|[I - (I + \bar{A}_{i22})]^{-1}\| \leq \frac{(n - r)^{1/2}}{1 - \|I + \bar{A}_{i22}\|} \leq \frac{(n - r)^{1/2}}{1 - (n - r)^{1/2} \zeta_i^{1/2}} \tag{38}$$

Thus,

$$\| - \bar{A}_{i22}^{-1}\| \leq \frac{(n-r)^{1/2}}{1-(n-r)^{1/2}\zeta_i^{1/2}} \tag{39}$$

holds. So, the minimization of ζ_i is equivalent to the minimization of $\| - \bar{A}_{i22}\|$.

Therefore, the minimization problem of the upper bound of guaranteed cost function and inconsistent state jumps at switching point can be converted into an optimization problem below.

Theorem 3. *If the following optimization problem with respect to* $Z^*, U^*, \zeta_i^*, \psi_i^*$ *has solutions, for arbitrary* $i, j \in \chi$,

$$\min_{Z^*,U^*,\zeta_i^*,\psi_i^*} \zeta_i + trace(\phi_i) + trace(ZE + WU)$$

$$s.t. \ (i) \ \begin{bmatrix} -\zeta_i I & (I+\bar{A}_{i22})^T \\ * & -I \end{bmatrix} < 0$$

$$(ii) \ 0 < \zeta_i < 1/(n-r)$$

$$(iii) \ \begin{bmatrix} -\phi_i & \psi_i^T \\ * & -I \end{bmatrix} < 0$$

$$(vi) \ (20) \tag{40}$$

where, $\psi_i = \begin{bmatrix} 0 & 0 \\ \bar{A}_{i21} & \bar{A}_{i22} \end{bmatrix}$, *the system (1) is admissible and has smaller state jump strength*

$$\|N_i\| \cdot \| - \bar{A}_{i22}^{-1}\| \cdot \| \begin{bmatrix} 0 & 0 \\ \bar{A}_{i21} & \bar{A}_{i22} \end{bmatrix} \| \cdot \|N^{-1}\| \tag{41}$$

via state feedback guaranteed cost controller

$$u(t) = K_i^* x(t) = -\frac{1}{2}\alpha_i B_i^T (Z^*E + WU^*)x(t) \tag{42}$$

The corresponding switching signal is

$$\sigma(t) = \arg\min\{x^T[A_i^T(Z^*E+WU^*)+(Z^*E+WU^*)^TA_i+Q6$$

$$-\alpha_i(Z^*E+WU^*)^TB_i(I-\frac{1}{4}\alpha_iR)B_i^T(Z^*E+WU^*)]x, i \in \chi\} \tag{43}$$

A upper bound of cost function is

$$J^* = x_0^T(Z^*E+WU^*)x_0. \tag{44}$$

Proof: Actually, the condition (i) is equivalent to $\psi_i^T\psi_i < \phi_i$. Thus, minimizing $trace(\phi_i)$ can ensure the minimization of $trace(\psi_i^T\psi_i)$. And according to the above analysis and Theorem 1 and Theorem 2, the proof is easily completed and omitted.

5   Example

Consider the switched descriptor systems composed of two subsystems

$$E\dot{x}(t) = A_i x(t) + B_i u(t)$$

where

$$E = \begin{bmatrix} 2 & 0 \\ 0 & 0 \end{bmatrix}, A_1 = \begin{bmatrix} 4 & 1.6 \\ 1 & -2 \end{bmatrix}, A_2 = \begin{bmatrix} -1.5 & -3 \\ 0 & -6 \end{bmatrix}, B_1 = \begin{bmatrix} 0 \\ 1 \end{bmatrix}, B_2 = \begin{bmatrix} 1 \\ 0 \end{bmatrix},$$

$$Q = \begin{bmatrix} 1 & 0 \\ 0 & 1 \end{bmatrix}, R = 1, x_0 = \begin{bmatrix} 2 & 1 \end{bmatrix}$$

Choose constants $\eta_1 = 0.5, \eta_2 = 0.5$, by the optimization problem (40), a upper bound of cost function can be solved

$$J^* = 37.8922$$

and the following matrices can be solve

$$Z^* = \begin{bmatrix} 1.400 & -0.0000 \\ -0.0000 & 265.36 \end{bmatrix}, U^* = \begin{bmatrix} 1.0583 & 18.8946 \end{bmatrix}$$

By simply calculation, we get Lyapunov matrix

$$P = \begin{bmatrix} 3.8951 & 0 \\ 1.2351 & 23.9056 \end{bmatrix}$$

and guaranteed cost feedback control gain can be solved

$$K_1 = \begin{bmatrix} 23.5079 & 4.5672 \end{bmatrix}, K_2 = \begin{bmatrix} -36.7781 & 3.4560 \end{bmatrix}$$

The corresponding switching signal is chosen as

$$\sigma(t) = \arg\min\{x^T(t)\Xi x(t), i \in \chi\}$$
$$= \arg\min\{x^T(t)[A_i^T(Z^*E + WU^*) + (Z^*E + WU^*)^T A_i + Q$$
$$- \alpha_i(Z^*E + WU^*)^T B_i(I - \frac{1}{4}\alpha_i R)B_i^T(Z^*E + WU^*)]x(t), i \in \chi\}$$

where $\Xi_1 = \begin{bmatrix} 1336.1 & 467.4 \\ 467.4 & 170.3 \end{bmatrix}, \Xi_2 = \begin{bmatrix} 4259.2 & 113.5 \\ 113.5 & 20.6 \end{bmatrix}$

6   Conclusions

In the paper, we investigates the admissible control and the minimization of state jumps and upper bound of guaranteed cost function for switched descriptor systems by designing static guaranteed cost controllers. These can be solved through a convex optimization problems by means of matrix decomposition and matrix inequality transformation. Some sufficient conditions are obtained under which the closed-loop system can be admissible and has minimum upper bound of cost function and state jumps at switching instants by designed switching signal. The advantage of the designed controllers is illustrated through a numeric example. Minimizing jumps of switched descriptor systems with exogenous disturbance at the switching instants is still an open problem. This is the direction of future research.

Acknowledgement. This work is supported by the Shenyang Young and Middle-aged Innovative Scienc6e and Technology Talents Project under Grant No. RC180312 and Guidance plan of Liaoning Natural Science Foundation under Grant No. 2019-ZD-0528.

References

1. Meng, B., Zhang, J.F.: Output feedback based admissible control of switched linear singular systems. ACTA Automatica Sinica **32**, 179–185 (2006). https://doi.org/10.16383/j.aas.2006.02.003
2. Xiong, W.J., Ho, D.W.C., Cao, J.D.: Synchronization analysis of singular hybrid coupled networks. Phys. Lett. A **372**, 6633–6637 (2008). https://doi.org/10.1016/j.physleta.2008.09.030
3. Cantó, B., Coll, C., Sánchez, E.: Positive N-periodic descriptor control systems. Syst. Control Lett. **53**, 407–414 (2004). https://doi.org/10.1016/j.sysconle.2004.05.017
4. Xi, J.X., Liu, H., Yao, Z.C., Yang, X.G., Liu, G.B.: Distributed admissible consensus control for singular swarm systems with switching topologies. Int. J. Robust Nonlinear Control **25**, 2816–2828 (2015). https://doi.org/10.1002/rnc.3234
5. Zhang, D., Yu, L., Wang, Q.G., Ong, C.J., Wu, Z.G.: Exponential H∞ filtering for discrete-time switched singular systems with time-varying delays. J. Franklin Inst. **349**, 2323–2342 (2012). https://doi.org/10.1016/j.jfranklin.2012.04.006
6. Ding, X.Y., Liu, X., Zhong, S.M.: Delay-independent criteria for exponential admissibility of switched descriptor delayed systems. Appl. Math. Comput. **228**, 432–445 (2014). https://doi.org/10.1016/j.amc.2013.11.107
7. Zhao, J.M., Zhang, L.J., Qi, X.: A necessary and sufficient condition for stabilization of switched descriptor time-delay systems under arbitrary switching. Asian J. Control **18**, 266–272 (2016). https://doi.org/10.1002/asjc.1018
8. Wei, J., Zhi, H., Mu, X.: New stability conditions of linear switched singular systems by using multiple discontinuous lyapunov function approach. Int. J. Control Autom. Syst. **17**(12), 1–9 (2019). https://doi.org/10.1007/s12555-018-0480-4
9. Liu, X., Zhong, S.M., Ding, X.Y.: A Razumikhin approach to exponential admissibility of switched descriptor delayed systems. Appl. Math. Modell. **38**, 1647–1659 (2014). https://doi.org/10.1016/j.apm.2013.09.007
10. Charqi, M., Chaibi, N., Tissir, E.H.: H∞ filtering of discrete-time switched singular systems with time-varying delays. Int. J. Adapt. Control Signal Process. **34**(4), 444–468 (2020)
11. Liu, W.Q., Yan, W.Y., Teo, K.L.: On initial instantaneous jumps of singular systems. IEEE Trans. Automat. Contr. **40**, 1650–1655 (1995). https://doi.org/10.1109/9.412639

12. Meng, B., Zhang, J.F.: Reachability conditions for switched singular systems. IEEE Trans. Automat. Contr. **51**, 482–488 (2006). https://doi.org/10.1109/TAC.2005.864196
13. Yin, Y.J., Zhao, J., Liu, Y.Z.: H-infinity control for switched and impulsive singular systems. J. Control Theory Appl. **6**, 86–92 (2008). https://doi.org/10.1007/s11768-008-6140-0
14. Liberzon, D., Trenn, S.: On stability of linear switched differential algebraic equations. In: Proceedings of the 48th IEEE Conference on Decision and Control, pp. 2156–2161 (2009). https://doi.org/10.1109/CDC.2009.5400076
15. Zhou, L., Ho, D.W.C., Zhai, G.S.: Stability analysis of switched linear singular systems. Automatica **49**, 1481–1487 (2013). https://doi.org/10.1016/j.automatica.2013.02.002
16. Shi, S.H., Zhang, Q.L., Yuan, Z.H., Liu, W.Q.: Hybrid impulsive control for switched singular systems. IET Control Theory Appl. **5**, 103–11 (2011). https://doi.org/10.1049/iet-cta.2009.0444
17. Zhang, H.G., Wang, Y.C., Liu, D.R.: Delay-dependent guaranteed cost control for uncertain stochastic fuzzy systems with multiple time delays. IEEE Trans. Syst. Man Cybern. Part B: Cybern. **38**(1), 126–140 (2008)
18. Chang, S.S.L., Peng, T.K.C.: Adaptive guaranteed cost control of systems with uncertain parameters. IEEE Trans. Autom. Control **17**(4), 474–483 (1972)
19. Liu, D., Zhang, G.P., Xie, Y.H.: Guaranteed cost control for a class of descriptor systems with uncertainties. Int. J. Inf. Syst. Sci. **5**(3–4), 430–435 (2009)
20. Feng, Y.F., Chang, X.H.: Guaranteed cost control design of T-S fuzzy systems. Int. J. Inf. Syst. Sci. **5**(3–4), 516–521 (2009)
21. Yang, D., Liu, Y.Y., Zhao, J.: Guaranteed cost control for switched LPV systems via parameter and state-dependent switching with dwell time and its application. Optimal Control Appl. Method **38**(4), 601–617 (2017)
22. Hien, L.V., Trinh, H.: Switching design for suboptimal guaranteed cost control of 2-D nonlinear switched systems in the Roesser model. Nonlinear Anal. Hybrid Syst. **24**, 45–57 (2017)
23. Mohammadi, L., Alfi, A., Xu, B.: Robust bilateral control for state convergence in uncertain teleoperation systems with time-varying delay: a guaranteed cost control design. Nonlinear Dyn. **88**(2), 1413–1426 (2017). https://doi.org/10.1007/s11071-016-3319-7
24. Yang, X., Liu, D.R., Wei, Q.L., Wang, D.: Guaranteed cost neural tracking control for a class of uncertain nonlinear systems using adaptive dynamic programming. Neurocomput. **198**, 80–90 (2016)
25. Wang, Z.P., Wu, H.N.: Finite dimensional guaranteed cost sampled-data fuzzy control for a class of nonlinear distributed parameter systems. Inf. Sci. **327**, 21–39 (2016)

Intelligent Method Application

Entropy Based Ranking Method for Nodes on Weighted and Directed Networks

Chinenye Ezeh, Ren Tao[✉], Li Zhe, Zheng Wen Wu, and Yi Qi

College of Information Science and Engineering, Northeastern University, No. 195, Chuangxin Road, Hunnan District, 110169 Shenyang, China
chinarentao@163.com

Abstract. Numerous research works have been conducted to find suitable models to measure node centralities with a view to determine the most important and influential nodes on a complex network. Out of the numerous existing centrality metrics, only few deal with centrality at the subgraph level. We modified the subgraph centrality model based on Entropy centrality by considering weights on directed networks. The modified subgraph centrality model now captures the intensity of interactions between a pivot node and the direct influence it exerts on its neigbhours as well as its indirect influence on its 2 - hop neighbours. A performance evaluation was done on weighted and directed real world network datasets. We equally compared the modified subgraph centrality with Out-Degree Centrality, Betweeness Centrality, Out-Closeness Centrality, Eigenvector Centrality and Pagerank Centrality. The results show that the modified subgraph centrality can rank nodes distinctly on directed and weighted networks.

Keywords: Sub - graph entropy · Directed networks · Node ranking · Spreading efficiency

1 Introduction

The universe is made up of network of systems and the interacting components that characterize them. These networks are quite complex in nature and can be seen in such systems as biological, biochemical, communication, social and technological networks. These networks include but are not limited to the Internet, the World Wide Web, social networks, scientific collaboration networks, lexicon or semantic networks, neural networks, food webs, metabolic networks and protein-protein interaction networks [1–3].

Numerous research works have been conducted and more are currently ongoing to find suitable models to measure node centralities with a view to determine the most important and influential nodes in complex networks [2,4–7]. Among the numerous centrality metrics, only a significant few came up with measures to deal with centrality at the subgraph level as almost all the popular centrality metrics either offer information about nodes at the local or global level of a

© Springer Nature Singapore Pte Ltd. 2020
J. Qian et al. (Eds.): ICRRI 2020, CCIS 1335, pp. 239–252, 2020.
https://doi.org/10.1007/978-981-33-4929-2_17

network. Additionally, these metrics cannot efficiently measure influence based proximity between two nodes [7].

Estrada and Rodriguez-Velazquez observed that Closeness and Betweeness centralities are considered to be non suitable measures of a subgraph centrality at the local level of a network. They further note that subgraphs are as significant as triangles in real networks. These subgraphs are network motifs that capture specific patterns of interconnection characterizing the networks at the local level [2]. Palla *et al.* indicate that subgraphs and modules are very helpful in breaking down complex networks into its constituent components. They further point out that if carefully applied, these modules and subgraphs can aid service providers (banks, telecommunication companies, web specialists etc.) find out important groups of customers/users or help biomedical researchers in facilitating their search for individual target molecules and novel protein complex targets [8]. Qiao *et al.* did a work along this line of research and proposed an algorithm based on disintegration of graphs into subgraphs [5].

The rest of this work is organized as follows: in Sect. 2, we do a literature review of some related works. In Sect. 3, we briefly review the previous Sub-Degree centrality based on Entropy. We also give a description of the modified model and draw up an algorithm to quantify the Direct Influence of a node on its neighbours and the Indirect Influence on its 2-hop neighbours. Section 4 presents node ranking distinction and the experimental results and discussions are presented in Sect. 5. Finally, we offer recommendations for future research and conclude in Sect. 6.

2 Related Works

Subgraphs are very significant in real networks and this has been demonstrated by Estrada and Rodriguez-Velazque in their work [2]. They proposed a centrality metric where every node takes part in a subgraph of a network. Lu *et al.* proposed a new node influence metric in relation to Degree, H-index and Coreness centralities [9]. Chen *et al.* introduced a semi-local centrality metric to determine influential nodes. They leveraged on the high computational complexities of Betweeness and Closeness centralities and their inapplicability to large scale networks as well as Degree centrality's low importance to propose the Semi Local centrality [10]. D-B Chen *et al.* proposed a ClusterRank centrality which takes into account the number of neigbhours, their influence and their level of connectedness or interactions with respect to clustering coefficient [11]. Zhao *et al.* proposed a method to determine effective multiple spreaders in a complex network by relating colour problems from graph theory to complex networks [12]. Min *et al.* proposed a new means of identifying influential spreaders by examining human behaviour in social circles and how they connect with people of diverse groups [13].

Qiao *et al.* proposed an Entropy centrality model based on disintegration of graphs into subgraphs. This enabled them to calculate the entropy of neigbhour nodes inspired by Shannon's information theory on Entropy. Using entropy theory, they quantified the local influence of a node on its neigbhours and an indirect influence on its 2-hop neigbhours [5]. Nonetheless, the work has some limitations such that the subgraph centrality is calculated with the assumption that influence propagations along the paths between nodes and their 2-hop neighbors have equal weights which in this case is unity. Also, this model is specifically designed for undirected, unweighted and static networks.

Spurred by these limitations inherent in this work, we modified the subgraph centrality metric by considering weights on directed networks. The proposed centrality metric captures the intensity of interactions between a pivot node and the direct influence it exerts on its neigbhours as well as its indirect influence on its 2-hop neighbours. A performance evaluation was done on four real world weighted and directed datasets namely Moreno High School friendship network [14], US Airports network in 2010 [15], C. Elegans neural network [16] and Facebook-like social network weighted by number of messages [17]. We equally compared the proposed centrality (PC) with Out-Degree Centrality (OutDC), Betweeness Centrality (BC), Out-Closeness Centrality (OutCC), Eigenvector Centrality (EC) and Pagerank Centrality (PrC). Results show that the proposed method is very efficient in node distinction and ranking.

3   Methodology

A directed graph G is given by a finite set of vertices, links and weights $G(V, E, W)$ where, $V = v_i, i = 1, 2, 3, \ldots, N$ which represents the set of nodes and E represents the set of links connecting them and W is the weight set of E. A link e_{ij} is given by v_{ij} where $v_{ij} \in V$ with a weight $w_{ij} \in W$. A link $e_{ij} = 1$ if v_i and v_j are connected else $e_{ij} = 0$. A self connection $v_{ii} = 0$ and $v_{ij} \neq v_{ji}$. One hop neigbhour of v_i is its degree k_i and $\Gamma_2(v_i)$ is the set of node v_i's 2-hop neighbours.

3.1   Highlights of Information Entropy

Shannon's information entropy of a given random variable A is given by:

$$H(A) = - \sum_{i=1}^{n} P_i \log_2 P_i \tag{1}$$

For more details on Shannon's information entropy, refer to [5]. Matthias Dehmer derived a probability distribution of a graph off the information functional of such graph. He defined and interpreted the quantified structural information of a graph as the resultant graph entropy. i.e $P = P_1, P_2, P_3 \ldots P_n$ [18]. Qiao et al. using this probability distribution defined a relationship between a node v_i and its neigbhours at the subgraph level [5].

$$P_i = \frac{\lambda_i}{\sum_{j=1}^{n} \lambda_j} \tag{2}$$

$i = 1, 2, 3 \ldots n$ and λ_i representing the i_{th} non-negative integer. Then, Entropy $H(A)$ becomes:

$$H(A) = -\sum_{i=1}^{n} P_i \log_b P_i = \log_b(\sum_{j=1}^{n} \lambda_i) - \sum_{j=1}^{n} \frac{\lambda_i}{\sum_{j=1}^{n}} \log_b \lambda_i \qquad (3)$$

3.2   Direct Influence Computation

Inspired by the works of [18–22] in the original work, Qiao *et al.* defines Sub-Degree Centrality SDC as:

$$SDC_i = \sum_{j}^{M} b_{ij} \qquad (4)$$

with M being the set of one hop neighbours of node v_i. $b_{ij} = 1$ if e_{ij} exists but $b_{ij} = 0$ otherwise. They also define the set P_i as vertex probabilities and further define the set of tuples $(\lambda_1, \lambda_2, \lambda_3 \ldots \lambda_{M+1})$ in relation to the SDC [5]. Hence,

$$\lambda_i = SDC_i \qquad (5)$$

Qiao *et al.* further interpreted λ_i as a measure of a node's direct influence on its neighbours which they quantified as the entropy of neighbour nodes at the subgraph level [5]. Therefore, the Direct Influence is given by:

$$DI_i = \log_b \left(\sum_{i=1}^{M+1} SDC_i \right) - \sum_{i=1}^{M+1} \frac{SDC_i}{\sum_{i=1}^{M+1} SDC_i} \log_b SDC_i \qquad (6)$$

To accommodate edge weights, Eq. 5 is modified to Weighted Sub-Degree Centrality SDC^w. The need for application of weights is to gain meaningful information about the strength of interaction among nodes and their surrounding neighbours [23]. Hence,

$$\lambda_i = SDC_i^w \qquad (7)$$

For undirected networks,

$$SDC_i^w = \left(\frac{k_i'}{|v^i|} \right)^{\alpha} \times \sum_{j=1}^{n} w_j' \qquad (8)$$

where k_i' is the sub-degree of v_i, $|v'|$ is the total number of nodes on the subgraph, $\sum_{j=1}^n w_j'$ is the strength of node v_i and α is a tuning parameter. For directed networks,

$$SDC_i^w = \left(\frac{k_{j(in)}'}{|v'|}\right)^\alpha \times \sum_{j=1}^n w_{j(in)} \tag{9}$$

$k_{j(in)}'$ the in-degree of node v_i's neighbours at the subgraph level. Therefore,

$$DI_i^w = \log_b \left(\sum_{i=1}^{M+1}\right) - \sum_{i=1}^{M+1} \frac{SDC_i^w}{\sum_i^{M+1} SDC_i^w} \log_b SDC_i^w \tag{10}$$

3.3 Indirect Influence Computation

At this point, the Indirect Influence of node v_i on its 2-hop neighbours has to be established. An example network is shown in Fig. 1 for this purpose. A variable N_{ac} represents common 1-hop neigbhours between node v_a and node v_c. For simplicity, we call N_{ac} total number of paths leading from node v_a to node v_c. If $N_{ac} = 1$ then:

$$II_{ac}^w = I_{ab} \times I_{bc} = DI_a \times DI_b \tag{11}$$

But if $N_{ac} \geq 2$, then:

$$II_{ac}^w = \frac{(I_{ab} \times I_{bc}) + (I_{ad} \times I_{dc})}{2} = \frac{(DI_a \times DI_b) + (DI_a \times DI_d)}{2} \tag{12}$$

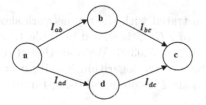

Fig. 1. Indirect Influence on two paths.

Therefore,

$$II_{ac}^w = \sum_{k=1}^{N_{ac}} \frac{DI_a \times DI_b}{N_{ac}} \tag{13}$$

The Indirect Influence of node v_a is given by:

$$II_a^w = \frac{\sum_{k=1}^{M_a} II_{ac}^w}{M_a} \tag{14}$$

In Eq. 14, M_a represents total number of node v_a's 2-hop neigbhour nodes, II_{ac}^w is node v_a's Indirect Influence on its 2-hop neighbour nodes. Wang and Street indicate that a set of two nodes influences each other as well as their immediate neighbours indirectly [7]. The path length is restricted to two degrees of freedom only for best performance [5, 24]. The Total Influence of node v_i is given by,

$$TI_i^w = (\beta LI_i^w) + (1 - \beta)II_i^w \tag{15}$$

where β is a tuning parameter.

The algorithmic procedure to calculate the total influence of a node is shown in Algorithm 1.

Algorithm 1. Computing algorithm for Total Influence of a node.

Input: $G(V, E, W)$.
Output: Total Influence (TI_i^w) of node v_i

1: **for** i $= 1 : |V|$ **do**
2:      G_i' ▷ Extract the subgraph of node v_i and its 2-hop neighbours
3:      **calculate** SDC_i^w ▷ use equation 9
4:      **calculate** DI_i^w ▷ use equation 10
5:      **calculate** II_i^w ▷ use equation 14
6:      **calculate** TI_i^w ▷ use equation 15
7: **end for**

An example is demonstrated with the toy network shown in Fig. 2.

The computed values of SDC_2^w is shown in Table 1.

From Table 1, $\sum SDC_i^w = 7.0366$. We note that if $\alpha \to 0$, then $SDC_i^w \to SDC_i$. We choose the base of the logarithm $b = 10$, $\alpha = 0.9$ and $\beta = 0.6$. The weighted entropy I_2^n of node 2 is computed using Eq. 10:

$$DI_2^w = I_2^n = \log_{10}\left(\sum_{i=1}^{3} SDC_2^w\right) - \sum_{i=1}^{3} \frac{SDC_i^w}{\sum_i^3 SDC_i^3} \log_{10} SDC_i^w = 0.1953$$

The Indirect Influence II_2^w of node 2 is computed using Eq. 14:

$$II_i^w = \frac{(I_2 \times I_3) + (I_2 \times I_4)}{2}$$

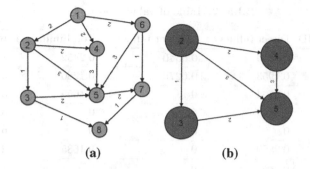

(a)   **(b)**

Fig. 2. (a) Example of a synthetic, directed and weighted network (b) Subgraph of node 2. Figure adapted from [5] ©Tong Qiao *et al.*, 2017 by Entropy

Table 1. Computation of weighted sub-degree

Node ID	$\left(\frac{k'_{j(in)}}{|v'|}\right)^{\alpha} \times \sum_{j=1}^{n} w'_{j(in)}$	SDC_i^w
3	$(\frac{1}{4})^{0.9} \times 1$	0.2872
4	$(\frac{1}{4})^{0.9} \times 2$	0.5743
5	$(\frac{3}{4})^{0.9} \times 8$	6.1751

Total Influence of node 2 is given by:

$$TI_2^w = (\beta DI_2^w) + ((1-\beta)II_2^w) = 0.1280$$

A value table of the Direct Influence, Indirect Influence, Total Influence and node rankings based on the Total Influence of the nodes are shown in Table 2.

4   Node Ranking Distinction

As a way to ascertain the validity of the proposed model, we conducted several experimental evaluations using weighted and directed network datasets of the following networks: Moreno High School friendship network [14], US Airports network in 2010 [15], C. Elegans neural network [16] and Facebook-like social network weighted by number of messages [17]. The details of the datasets are listed in Table 3. n is number of nodes, m is number of edges, $<k>$ is average degree, k_{max} is maximum degree, C is Clustering Coefficient, $<d>$ is average path length and R is assortativity.

Table 2. Table of values and rank

Node ID	Direct influence	Indirect influence	Total influence	Rank
1	0.3859	0.0340	0.2452	1
2	0.1953	0.0270	0.1280	4
3	0.2764	0	0.1659	3
4	0	0	0	5
5	0	0	0	6
6	0.2809	0	0.1685	2
7	0	0	0	7
8	0	0	0	8

Table 3. Topological features of the datasets

Network	n	m	$<k>$	k_{max}	C	$<d>$	R
Moreno HS	70	548	15.6571	38	0.404	5.8965	0.0830
US Airport	1574	28236	35.878	596	0.3525	3.14	−0.1229
C. Elegans	306	2345	15.3268	173	0.0432	5.9097	−0.1547
Facebook-like	1899	20296	21.3755	374	0.0297	3.9784	−0.1628

A careful examination of the table results in Moreno High School network in Table 4 shows that the top 10 ranked nodes of the proposed centrality (PC) appear randomly ranked among the first top 10 ranked nodes of out degree centrality (OutDC), betweeness centrality (BC), out closeness centrality (OutCC), eigenvector centrality (EC) and pagerank centrality (PrC). In US Airport network as shown in Table 5, among the top 10 ranked nodes, only one node ranked 5th in PC appeared in OutDC. In C. Elegans network as shown in Table 6, the top 4 ranked nodes of PC appear randomly among the other centrality metrics. Nodes ranked 5th, 6th, 8th, 9th and 10th in PC does not appear in any of the other centrality metrics in C. Elegans network. In Facebook-like network shown in Table 7, nodes ranked 3rd, 4th, 5th, 9th and 10th of PC does not appear in any of the other centrality metrics. To further ascertain the suitability of PC in ranking nodes, another experiment is carried out to demonstrate distinction of nodes based on the frequency of ranking by the different centrality metrics.

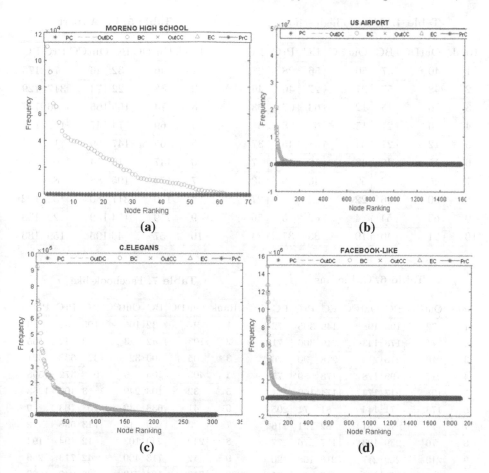

Fig. 3. (a) Moreno HS (b) US Airport (c) C. Elegans (d) Facebook-like.

In Fig. 3a, Betweeness Centrality does not rank nodes distinctly in Moreno High School network while PC does well in node distinction. In Fig. 3b, Betweeness Centrality does not distinguish node ranks properly in US Airport network. PC still does well in node distinction in US Airport network. In Fig. 3c, Betweeness centrality could not rank nodes distinctly in C. Eelegans network but PC could distinctly rank nodes on this network. In Fig. 3d, Betweeness centrality could not rank nodes distinctly in Facebook-like network. It can be noted from Fig. 3 that the proposed centrality (PC) does well in node distinction.

Table 4. Moreno high school

Rank	OutDC	BC	OutCC	EC	PrC	PC
1	40	37	50	56	28	28
2	48	57	21	27	46	46
3	56	15	12	63	67	63
4	6	21	15	6	68	6
5	12	27	54	58	45	37
6	15	46	48	59	43	7
7	46	22	59	40	66	40
8	50	16	22	64	7	27
9	63	54	4	62	5	56
10	21	40	19	33	37	4

Table 5. US Airport

Rank	OutDC	BC	OutCC	PrC	PC
1	46	32	46	74	175
2	88	22	174	317	129
3	74	165	165	46	177
4	69	74	147	165	188
5	165	147	69	10	174
6	147	10	74	69	92
7	174	195	88	88	72
8	150	317	80	32	102
9	159	174	159	22	138
10	57	46	195	159	183

Table 6. C. Elegans

Rank	OutDC	BC	OutCC	EC	PrC	PC
1	72	195	198	149	305	72
2	71	178	143	219	306	71
3	216	216	216	218	90	76
4	217	196	178	178	89	75
5	198	217	174	174	169	111
6	178	71	144	81	71	205
7	75	72	81	82	121	216
8	76	239	193	157	276	77
9	219	269	217	216	168	120
10	218	167	149	217	72	78

Table 7. Facebook-like

Rank	OutDC	BC	OutCC	EC	PrC	PC
1	9	32	105	105	32	3
2	103	42	3	249	42	105
3	105	400	32	32	638	321
4	400	105	9	9	372	1598
5	32	103	249	3	400	1283
6	41	638	42	103	103	9
7	3	9	12	713	598	12
8	249	249	103	12	194	194
9	42	713	400	42	713	266
10	713	194	713	638	249	1899

4.1   Epidemic Model

The Susceptible-Infectious SI epidemic model was used to carry out the spreading efficiency experiment of the proposed centrality method in comparison to the rest of the centrality metrics. This was done in order to ascertain the spreading efficiency of the seed nodes that are top ranked. This approach distinguishes which of the nodes are more important than the others in a network [25]. The SI model has two major compartments viz Susceptible and Infectious compartments. An infected node can infect a susceptible node with a transmission probability $\beta = \left(\frac{w_{ij}}{W_{max}}\right)^{\alpha}$ for weighted networks [25]. w_{ij} is the weight of the link between node v_i and node v_j and W_{max} the maximum weight on the network. The total number of infected nodes is represented by I_t after each t step.

5   Experimental Results

Using SI epidemic model, the results of the disease spreading experiments are shown in Figs. 4, 5, 6, and 7.

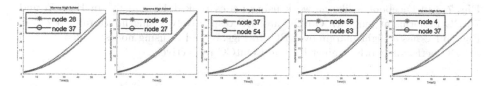

Fig. 4. In Moreno High School network using nodes ranked 1st and 2nd PC performs a little well over BC and almost at par with EC in Figs. 4a and 4b respectively. Using nodes ranked 5th, PC outperforms OutCC in Fig. 4c and performs a little below OutDC in Fig. 4d. In Fig. 4e using nodes ranked 10th PC performs poorly against PrC.

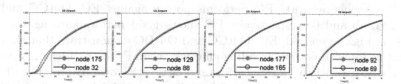

Fig. 5. In US Airport network using nodes ranked 1st, 2nd, 3rd and 6th in Figs. 5a, 5b, 5c and 5d respectively, PC performs at par with BC, OutDC, OutCC and PrC with the infection spread rising above and levelling up with BC eventually.

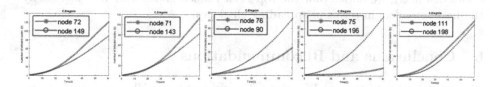

Fig. 6. In C.Elegans network using nodes ranked 1st, 2nd and 5th in Figs. 6a, 6b and 6e respectively PC performs poorly against EC, OutCC and DC respectively. Using nodes ranked 3rd and 4th in Figs. 6c and 6d, PC outperforms PrC and BC respectively.

In Moreno High School network, PC outperforms OutCC with a reasonable margin in Fig. 4c and performs just a little well over BC in Fig. 4a and almost at par with EC in Fig. 4b respectively. In Fig. 4e and Fig. 4d PC performs poorly against PrC and OutDC respectively. The results indicate that PC's performance in pinpointing nodes with high spreading efficiency is as good as that of BC, EC and better than OutCC.

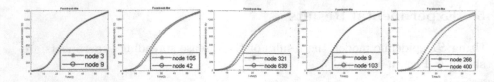

Fig. 7. In Facebook - like network using nodes ranked 1st and 6th in Figs. 7a and 7d PC performs at par with OutDC and EC respectively. Using nodes ranked 2nd in Fig. 7b PC has a slight margin over BC. In Figs. 7c and 7e using nodes ranked 3rd and 9th PC performs poorly against PrC and OutCC respectively.

In US Airport network, PC performs at par with BC, OutDC, OutCC and PrC in Figs. 5a, 5b, 5c and 5d respectively. The curves follow the same pattern in all the Figures except in Fig. 5a where PC rises just a little above BC after the initial infection spread before the curves merge around the 18th day. The results indicate that PC's performance on US Airport network is as good as those of BC, OutDC, OutCC and PrC.

In C. Elegans network, PC performs poorly against EC, OutCC and DC in Figs. 6a, 6b and 6e respectively. PC shows good performance against PrC and BC in Figs. 6c and 6d. EC, OutCC and OutDC demonstrate that they can pinpoint nodes with better spreading efficiency than PC on C. Elegans network. In Facebook-like network, PC performs a little well over BC in Fig. 7b and performs at par with OutDC and EC in Figs. 7a and 7d respectively. PC is outperformed by PrC and OutCC in Figs. 7c and 7e respectively.

The results obtained are quite instructive to demonstrate that the behaviour of nodes on directed networks can be significantly different compared to undirected networks. Even though that the proposed centrality method could rank nodes distinctly, we notice that the top most ranked nodes were not necessarily the best spreaders in terms of spreading efficiency.

6   Conclusions and Recommendations

This paper reports the modification of a subgraph model based on entropy used in unweighted and undirected networks in the work of [5]. We considered the weights in the calculation of subgraph entropy of neighbour nodes on directed networks. The networks are broken down into components and their weighted subgraph centrality computed based on their direct and indirect influences on their neighbours and next nearest neighbours respectively. Using real world weighted and directed network datasets such as Moreno High school network, US Airport network, C. Elegans network and Facebook-like network we implemented the proposed centrality measure and analyzed the node ranking capacity, spreading efficiency and influence on networks in comparison with centrality metrics such as Out-Degree, Betweeness, Out-Closeness, Eigenvector and Pagerank centralities. We noted that PC ranked nodes distinctly just like some of its counterpart centrality metrics.

It is very necessary that we must point out some of the presumptions we made in order to achieve this centrality. In considering length of paths between nodes and its neigbhours, we focused only on two degrees of freedom thereby limiting the number of next neigbhours to 2-hops as research has proven that this gives the best results [5,24]. Also, we did not consider dynamic processes as we focused only on static networks. We hope to enhance this proposed centrality since we considered only the topological structure of nodes in this model. We will further examine its sensitivity to different spreading rates on different networks.

Acknowledgments. C. E. wishes to thank Patrice Monkam who read the first draft of this manuscript and offered some valuable suggestions. C. E. equally thanks Prof. Wei Shan who took time to explain their work clearly.

Funding. This work is partially supported by National Natural Science Foundation of China (61473073, 61433014), Fundamental Research Funds for the Central Universities (N2017009, N182608003, N181706001, N2018008, N161702001, N171706003).

References

1. Kim, H., Anderson, R.: Temporal node centrality in complex networks. Phy. Rev. E, **85**(2), p. 026107 (2012). http://dx.doi.org/10.1103/PhysRevE.85.026107
2. Estrada, E., Rodriguez-Velazquez, J. A.: Subgraph centrality and clustering in complex hyper-networks. Phys. A **364**, pp. 581–594 (2006). https://doi.org/10.1016/j.physa.2005.12.002. ISSN 0378–4371
3. Hu, W.: Finding statistically significant communities in networks with weighted label propagation. SN, **2**, 138–146 (2013). http://dx.doi.org/10.4236/sn.2013.23012
4. Borgatti, S. P.: Centrality and network flow. Soc. Net. **27**(1), 55–71 (2005). https://doi.org/10.1016/j.socnet.2004.11.008. ISSN 0378–8733
5. Qiao, T., Shan, W., Zhou, C.: How to identify the most powerful node in complex networks? A novel entropy centrality approach. Entropy **19**, 614 (2017). https://doi.org/10.3390/e19110614
6. Zareie, A., Sheikhahmadi, A., Fatemi, A.: Influential nodes ranking in complex networks: an entropy-based approach. J. Chaos **104**, 485–494 (2017). https://doi.org/10.1016/j.chaos.2017.09.010. ISSN 0960–0779
7. Wang, W., Street, W.N.: Modeling influence diffusion to uncover influence centrality and community structure in social networks. Soc. Netw. Anal. Min. **5**(1), 1–16 (2015). https://doi.org/10.1007/s13278-015-0254-4
8. Palla, G., Farkas, I. J., Pollner, P., Derényi, I., Vicsek T.: Directed Network Modules. New J. Phys. **9** 186 (2007). https://doi.org/10.1088/1367-2630/9/6/186
9. Lü, L., Zhou, T., Zhang, Q., Stanley H. E.: The H-index of a network node and its relation to degree and coreness. Nat. Commun. **7**, 10168 (2016). https://doi.org/10.1038/ncomms10168
10. Chen, D., Lü, L., Shang, M-S., Zhang, Y-C., Zhou, T.: Identifying influential nodes in complex networks. Phys. A **391**(4), 1777–1787 (2012). https://doi.org/10.1016/j.physa.2011.09.017. ISSN 0378–4371
11. Chen D-B., Gao H, Lü L, Zhou T.: Identifying influential nodes in large-scale directed networks: the role of clustering. PLOS ONE **8**(10), e77455 (2013). https://doi.org/10.1371/journal.pone.0077455

12. Zhao, X-Y., Huang, B., Tang, M., Zhang, H-F., Chen, D-B.: Identifying effective multiple spreaders by coloring complex networks. EPL **108** 68005 (2014). https://doi.org/10.1209/0295-5075/108/68005

13. Min, B., Liljeros, F., Makse, H. A.: Finding influential spreaders from human activity beyond network location. PloS One **10**(8), e0136831 (2015). https://doi.org/10.1371/journal.pone.0136831

14. Coleman, J. S.: Introduction to Mathematical Sociology. Free Press, New York, pp. 450–451 (1964). http://konect.uni-koblenz.de/networks/moreno_highschool

15. Opsahl, T.: Why Anchorage is not (that) Important: Binary Ties and Sample Selection (2011). http://konect.uni-koblenz.de/networks/opsahl-usairport

16. Watts, D. J. Strogatz, S. H.: Collective dynamics of "small-world" networks. Nature **393**, 440–442 (1998). https://doi.org/10.1038/30918

17. Opsahl, T., Panzarasa, P.: Clustering in weighted networks. Soc. Netw. **31**(2), 155–163, (2009). https://doi.org/10.1016/j.socnet.2009.02.002

18. Dehmer, M.: Information processing in complex networks: graph entropy and information functionals. J. AMC **201**(1–2), 82–94 (2008). https://doi.org/10.1016/j.amc.2007.12.010. ISSN 0096–3003

19. Dehmer, M., Varmuza, K., Borgert, S., Emmert-Streib, F.: On entropy-based molecular descriptors: statistical analysis of real and synthetic chemical structures. J. Chem. Inf. Model **49**(7), 1655–1663 (2009).https://doi.org/10.1021/ci900060x

20. Bonchev, D., Trinajstić, N.: Information theory, distance matrix, and molecular branching. J. Chem. Phys. **67**, 4517 (1977). https://doi.org/10.1063/1.434593

21. Cao, S., Dehmer, M., Shi, Y.: Extremality of degree-based graph entropies. J. Inf. Sci. **278**, 22–33 (2014). https://doi.org/10.1016/j.ins.2014.03.133. ISSN 0020–0255

22. Cao, S., Dehmer, M.: Degree-based entropies of networks revisited. J. AMC **261**, 141–147 (2015). https://doi.org/10.1016/j.amc.2015.03.046. ISSN 0096–3003

23. Onnela, J-P., Saramäki, J., Kertész, J., Kaski, K.: Intensity and coherence of motifs in weighted complex networks. J. Phys. Rev. E. **71**(6,), 065103 (2005). https://doi.org/10.1103/PhysRevE.71.06510

24. Easley, D., Kleinberg, J.: Networks, Crowds and Markets: Reasoning about a highly Connected World. Cambridge University Press (2010). https://www.cs.cornell.edu/home/kleinber/networks-book/networks-book.pdf

25. Fei, L., Deng, Y.: A new method to identify influential nodes based on relative entropy. J. Caos **104**, 257–267 (2017). https://doi.org/10.1016/j.chaos.2017.08.010. ISSN 0960–0779

Target Detection in UAV Automatic Landing Based on Artificial Bee Colony Algorithm with Bio-Inspired Strategy Approach

Beiwei Zhang[✉], Jiangtao Cao, and Ying Guo

School of Information and Control Engineering, Liaoning Shihua University,
Fushun 113001, Liaoning, China
xiamilpiggy@126.com

Abstract. To meet the requirements of detecting runway and salient ground targets precisely and rapidly for Unmanned Aerial Vehicles (UAV) during automatic landing, a scenario that integrates our modified Artificial Bee Colony (ABC) algorithm with template matching is proposed in this paper. This study aims at enhancing the traversal search capability and avoid evolutionary stagnation with the Bio-inspired ABC (BABC) algorithm. Firstly, the initial solution optimization strategy of reverse learning is adopted in the initialization phase. Secondly, an improved search strategy with Levy flight characteristics is applied in the phase of updating the population and local searching. Additionally, chaotic searching strategy is introduced into ABC to reinforce the ability of system and get rid of local optimal solution, thus showing advantage of convergence property and robustness when compared with ABC. Series of experimental results demonstrate the feasibility and effectiveness of our presented approach over the standard method. The novel algorithm also meets the requirements of speed and precision in practical application.

Keywords: Unmanned Aerial Vehicles · Swarm intelligence · Chaotic search · Artificial Bee Colony algorithm

1 Introduction

Advancements in Unmanned Aerial Vehicles (UAV) have made it possible to perform dangerous or dull missions without participation of man for military and civilian purposes [1–3]. Multiple scenarios have been designed for the collision avoidance, route planning and guidance of UAV [4–6]. It's of great necessity to equip UAVs with precise target information to enable them to execute various tasks, especially in the procedure of automatic landing [7–9]. With the spread of computer vision methods and developments in digital camera technology, vision-based sensors has been gaining a lot of popularity in recent years [10]. Much attention has been directed forward vision-based methods due to their great potential in both theoretical study and practical applications [11–14]. Vision-based recognition and control on UAVs can be a more challenging task to some extent, considering the higher flight velocity [15]. A volume of research on vision-based control in regions of formation flight and aerial refueling has been conducted by many groups [16–20]. It has been proposed to use only natural

© Springer Nature Singapore Pte Ltd. 2020
J. Qian et al. (Eds.): ICRRI 2020, CCIS 1335, pp. 253–271, 2020.
https://doi.org/10.1007/978-981-33-4929-2_18

features of the scene in vision-based systems for the location or prediction of targets [21]. The applications of algorithms in Machine Vision (MV) contribute a lot in the procedure of UAV landing through the detection of crucial targets undoubtedly [22].

A reliable and efficient process of the runway recognition, for which various methods have been proposed, is greatly desirable for UAV landing safely beyond doubt. For instance, Hough Transform has been utilized for the edge extraction during the runway detection, which fits for UAVs drawing near to the ground. This work applies template matching to solving the problem. Generally the part of runway is cut out as the template, which has been known by us, and it is our target to seek out the exact location of runway or other salient ground targets in the original image for automatic landing. Common template matching methods require searching all the points in the searching region, which are time-consuming and end up with low efficiency. Methods that were proposed in succession for the sake of accelerating matching velocity are mainly classified into two sorts [23]: One is feature-based approach, which shrinks the searching space to enhance the efficiency. The other involves all kinds of evolution algorithms to conduct the matching process, such as the Simulated Annealing algorithm, the Gauss-Newton method [24], the Levenberg-Marquart method and multiple evolution algorithms (EA). EAs are implemented by generating an initial population randomly [4]. Each individual takes values in the searching space and their corresponding fitness functions are evaluated. Various operators are introduced into the population to update the individuals afterwards, simulating the process of natural evolutionary. Optimal solutions can be achieved by repeating the process of evaluating and renovating the population. The disadvantage of being computationally time consuming can be overcome with the process of computer technology. Using the Genetic algorithm (GA) as a substitution to common ergodic searching has been most frequently applied for template matching [25] and feature extraction. However, when the Genetic algorithm evolutes to a certain extent, the accumulation of local optimization turns into the next generation through intersection and mutation, thus increasing the difficulty of changing fitness values, which leads to the so-called "evolutionary stagnation". Swarm intelligence has absorbed points of many scientists in related fields due to its excellent convergence ability. Bonabeau defined swarm intelligence as "any attempt to design algorithms or distributed problem-solving devices inspired by the collective behavior of social insect colonies and other animal societies" [26]. Among various methods, one popular swarm-intelligence-based algorithm, the Particle Swarm Optimization (PSO) algorithm, was introduced by Eberhart and Kennedy in 1995. PSO has already been applied to template matching successfully as one of the meta-heuristic algorithms inspired by biological systems. In 2005, Dervis Karaboga proposed a new swarm intelligence algorithm–the Artificial Bee Colony algorithm (ABC) [27], which simulates the biological mechanism of bee groups that are looking for optimal nectar sources. Equipped with stronger global searching capability, the algorithm obtains a wide range of applications in the field of function optimization, combinatorial optimization and path planning. In [28], ABC is applied to the extraction of the small signal equivalent circuit model parameters of MESFET (Metal Semiconductor Field Effect Transistor) devices by Samrat L. Sabat et al. It's also seen that ABC has been used to resolve the problem of clustering in [29]. However, similar with other global optimization algorithms, ABC is also prone to falling into local optimums. Considering the outstanding performance of the chaos theory, we introduce chaos searching strategy into the ABC in this paper to reinforce the ability of the system to get rid of the local optimal solution, which improves the convergence speed and accuracy in the process of iteration. The Bio-inspired Artificial Bee Colony algorithm (BABC) for template matching in UAV automatic landing is introduced afterwards in this study. Our proposed approach has several key benefits compared with previous works: (1) The initial solution optimization strategy of reverse learning is adopted in the initialization phase. (2) An improved search strategy with Levy flight characteristics is applied in the phase of updating the population and local searching. (3) we introduce chaos theory into swarm-intelligence-based algorithm to reflect the indeterminacy of real swarm-intelligence; (4) BABC can reinforce the ability of the system to get rid of the local optimal solution. Consequently, the search ability of ABC algorithm is obviously enhanced while better results are obtained; (5) It can improve the convergence speed of target detection and the accuracy in the process of iteration. After all, the results demonstrate that the improved algorithm outperforms other algorithms.

The reminder of this paper is organized as follows. Section 2 introduced the standard ABC and proposed BABC algorithm. Section 3 described the principle of template matching approached by BABC, while in Sect. 4, comparative experimental results were shown to demonstrate the advantages of our proposed approach. The final section contained our concluding remarks and future work.

2 Bio-Inspired Artificial Bee Colony Algorithm

2.1 Principles of the ABC Algorithm

Waggle dance of honey bees can be commonly seen in nature. Such a type of dance means the flying path of bees appears "8" character and the bee swings its abdomen simultaneously, which conveys information of food sources to other bees. One study shows that the angle between the axis of the dance and gravity just indicates that between the directions of the sun and food sources. Furthermore, the dance conveys more detailed information on distance and azimuth. The position of food source locates in the direction toward the sun when the bee dances head upward, and away from the sun when head down. Additionally, the duration of bees wagging tails and humming increases as the source gets farther. Bees in the hive make choices between seeking new food sources in the neighborhood and gathering honey according to the information obtained by the dance. Bees exchange messages, communicate and learn mutually through such dances, which ensures the entire bee colony seek out the richest food sources.

ABC was inspired by the interesting phenomenon above, and was initially utilized to solve the extremum problems of continuous multi-peak functions. As a new heuristic bionic intelligent optimization algorithm, it has been applied to various research areas including workshop scheduling and robot path planning. Its more application potential has yet to be explored.

We first introduce three basic parts in the following to give a systematic exposition of the algorithm [30]:

Food Sources: (A and B in Fig. 1) represents the possible solutions within the scope of solution space. Their values depend on multiple factors, such as richness or proximity to the hive. A single quantity can be used to represent the profitability of the source for the sake of simplicity.

Employed Bees: Employed foragers are linked with specific food sources that are currently exploited by them. After recording the location and uploading nectar rapidly, they return to the hive and unload nectar (as presented in the white frames). Then there are several options for the employed bees as follows:

- Abandon the previous source due to low avenues and turn into followers (UF in Fig. 1).
- Dance to recruit mates in the hive (EF1 in Fig. 1).
- Continue to forage without recruiting other bees (EF2 in Fig. 1).

Unemployed Bees: Those searching for food sources to exploit, which can be divided into two types:

- *Scouts* ('S' in Fig. 1): Scout bees search for new sources around the hive spontaneously without any prior knowledge.
- *Onlookers* ('R' in Fig. 1): Onlooker bees wait in the hive and searches for food sources with information obtained from employed bees.

All the bees play the role of scouts without any prior knowledge at the beginning. They turn into any kind above on the basis of the revenue degree after random search according to the principles as follows: Bees whose incomes rank higher than the threshold turn into leaders (otherwise into followers). They continue gathering and recruit more partners. While those with relatively lower income abandon the food sources and continue searching. The bees set for searching new sources when fail to find better sources.

Compared with other methods such as GA and PSO, the most remarkable advantage of ABC shows up as the local search in each iteration, which adds the frequency of searches in the local region. Consequently the probability of finding the optimal parameters and, of course, the volume and time of searching are increased.

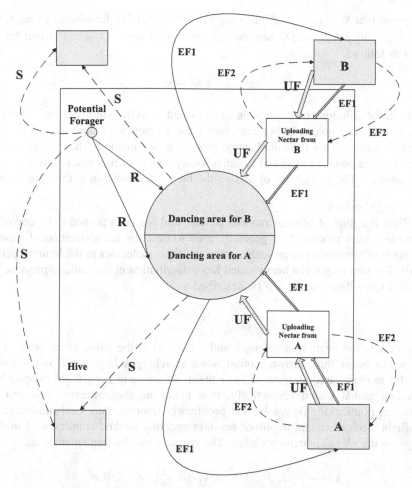

Fig. 1. Behaviors of honey bees seeking for nectar

2.2 Principles of BABC

a). Opposition-based Learning strategy (OBL)

OBL is a new developed method in swarm intelligence study in recent years. It has been applied in multiple optimization algorithms. Rahnamayan et al. proposed the Opposition-Based Differential Evolution method, which put forward that the optimal individual is selected on the basis of population rather than individuals. The opposite point of a feasible solution is also calculated to find the optimal solution. The opposite point is defined as:

Suppose that $X_i = (x_i^1, \ldots, x_i^D)$ is a point in the space of D dimension, where $x_i \in R$ and $x^j \in [a^j, b^j], j = 1, \ldots, D$, then the opposite point $ox_i = (x_i^1, \ldots, x_i^D)$ can be calculated as follows.

$$ox^j = a^j + b^j - x^j \tag{1}$$

The initial solutions of population in the standard ABC are generated randomly, which is easy to implement but cuts down the efficiency of algorithm. The OBL strategy is employed for the initialization process in our method. The current solution and its opposite point are searched simultaneously, and the better one is selected as the initial solution. The probability of finding the best initial solution is thus improved.

b). Levy flight search

Levy flight is a kind of Markov random process and has been proved to be one of the best random walk models. It is generally used to describe the mathematical types of paths made of continuous steps such as motion of gas molecules or the hunting path of animals. The step length is a heavy-tailed Levy distribution in the walking process. The simplified Levy flight model can be described as

$$L(s) \sim |s|^{-1-\delta} \tag{2}$$

where s is the random step length and $\delta \in (0, 2)$ is the value of an index. Levy flight works better than Brown motion when searching a large scale and unknown space due to more rapid increase of Levy flight. Solutions in Levy flight execute local search and global search respectively, thus balancing the convergence speed and diversity of population. To resolve the problem of trapping into local minimum, the Levy flight mechanism and improved position updating method is introduced in ABC to improve the global searching ability. The operator can be implemented as

$$s = \frac{\mu}{|v|^{1/\delta}}, \quad \delta = 1.5, \quad \mu \sim N(0, \sigma_\mu^2), \quad v \sim N(0, \sigma_v^2) \tag{3}$$

where s is random step length and N represents the normal distribution. The Levy flight generates large jumps and changes directions rapidly, thus widening the searching space. Our improved method utilizes Eq. (4) to update the nectar resource positions and adopts nectar resources with Levy flight features to further update the optimal solution in the current population using Eq. (5).

$$v_i^j(t+1) = x_i^j(t) + \phi_i^j(x_i^j(t) - x_k^j(t)) + \varphi_i^j(x_{best}^j(t) - x_i^j(t)) \tag{4}$$

$$x_{best}^{j\prime}(t+1) = x_{best}^j(t) + levy(\lambda) \otimes (x_i^j(t) - x_{best}^j(t)), \quad 1 < \lambda < 3 \tag{5}$$

where $x_i^j(t)$ is the nectar resource position during the t th search by the i th bee, ϕ_i^j denotes a uniformly distributed random number between -1 and 1, and φ_i^j denotes a uniformly distributed random number between 0 and C. We utilize x_{best}^j to represent the optimal individual in the current population, $\lambda = 1 + \delta$ is the scale parameter, and $\otimes$ denotes the vector operator.

c). Chaotic search strategy
Chaos exists widely in nature as a nonlinear phenomenon. In general, chaos refers to the randomness state of motion obtained by deterministic equations. Chaos theory is summarized by the butterfly effect, which was established by Lorenz to simulate the global weather system numerically. He found that subtle changes in initial conditions lead to the radical difference among final results from the subsequent simulation, thus making long-term prediction impossible. We can observe such sensitive dependence on initial conditions not only observed in comparatively complex systems, but also in the simplest logistic equations. These systems possess a delicate internal structure with characteristics of ergodicity, randomness and regularity despite their chaotic appearances. They can travel all the states in accordance with the objective laws of the system itself without any repeating within a certain range. Common Logistic map is a typical chaotic system with the equation as follows:

$$z_{n+1} = \mu z_n (1 - z_n); \quad n = 0, 1, 2, \dots \dots \tag{6}$$

where n is the iteration of the algorithm, and μ stands for the control parameter. Equation (6) can iterate to a fixed time sequence by an arbitrary initial value $z_0 \in [0, 1]$ once the value of μ is determined. If μ equals to 4, and $z_1 \neq 0.25, 0.5, 0.75$, the system will turn into complete-chaos, in which the value of z will visit each neighborhood in a sub-range of $[0, 1]$. Ergodicity of chaos optimization algorithms ensures the evolutionary process able to jump out of local optimal solutions.

ABC is prone to falling into local minima and presents a lower convergence speed in the later period of evolution. Chaotic sequences are applied to resolving the problem in this study. The integration of chaotic sequences with ABC is reflected in the chaotic searching strategy. Chaotic sequences are generated based on the optimal locations observed by each employed bee to achieve random search within a small neighborhood. In this way, the characteristics of the chaotic variables are fully utilized to ensure the individuals of sub-generations distribute ergodically within the solution space in our modified approach, thus avoiding the premature of the solutions and helping find out the optimal solutions more rapidly. It is known that the convergence proof of ABC algorithm has been conducted previously [31]. The chaotic searching strategy is utilized to assist the algorithm jump out of local optimal solutions. It is obvious that optimal solutions can be achieved in our proposed method, that is, the BABC algorithm can be verified convergent.

3 Template Matching Based on the Bio-Inspired Artificial Bee Colony Algorithm

3.1 Template Matching Methods

Template matching is an attractive topic in the field of image processing. It is defined as the procedure of seeking out the right sub-image in line with the known template. As a method that proceeds matching on the basis of the characteristics of the image target, it possesses high accuracy and sensitivity for target positioning even in complicated conditions [32]. Template matching algorithms aim to seek out the target to be recognized by calculating the similarity degree between the template and the matching region and ascertaining the most similar position as the optimum matching point. Generally, template matching is conducted using intensity or features.

The image features achieved from the feature-extraction methods contain a higher level of semantic information. A number of such methods share the characteristics of scale–invariability and affine-invariability, especially the wavelet feature-extraction, which can implement the multi-resolution decomposition and coarse-to-fine matching of images [33]. However, the methods above generally involve a large number of geometric and image morphology calculation without general laws and models to follow. It's needed to extract the respective corresponding characteristics in different applications, which means the increase of difficulty. While the matching methods based on gray scale are equipped with fixed mathematical models and relatively simple. Additional, their matching results and error analysis can be utilized for quantitative analysis. Gray scale method is utilized for template matching in this study due to its simplicity.

Template matching can be used to compute the degree of correlation between the two for seeking out the matching location in the search area. Figure 2 shows the schematic of template matching. Assume that the template (b) stacks upon the original image (a), and the part overrode by the template is recorded as the sub-graph $S_{x,y}$, where (x,y) represent the coordinates of the image point at the upper left corner of this sub-graph. The correlation function of T and the S-graph overrode by T is figured out as the template T slides along the original image I.

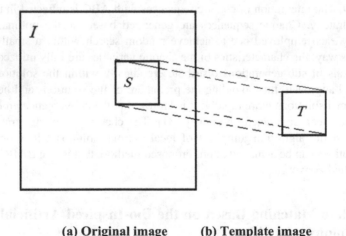

(a) Original image   (b) Template image

Fig. 2. Schematic of the template matching

3.2   Fitness Function

Generally the fitness function is defined for the evaluation of each individual's income under different circumstances [34]. Normalized correlation coefficients (NCC) is generally utilized as the fitness function of template matching. The NCC $R(i,j)$ between the template and the target image area overlapped by the template is presented as:

$$R = \frac{\sum\limits_{n=1}^{N} T(n)S(n)}{\sqrt{\sum\limits_{n=1}^{N} T(n)^2 \sum\limits_{n=1}^{N} S(n)^2}} \tag{7}$$

where $T(n)$ and $S(n)$ are the gray value of the template and the corresponding overlapped target image area (sub-image) respectively. n represents the index of image points and N is the area (in pixels) of template. The exhaustive operation of this function will be extremely time-consuming if the image is relatively huge. To overcome this shortage, we simplify the fitness function as:

$$\hat{R} = \sum\limits_{n=1}^{N} |T(n) - S(n)| \tag{8}$$

Generally, when $\hat{R}$ obtains the minimum $\hat{R}_{min} = 0$ (namely R obtains the maximum $R_{max} = 1$) during the iterative calculation, we confirm the corresponding location as the best matching point.

3.3   Gray Scale Value Template Matching Based on BABC

Template matching aims at finding the best matching point in the original image. Problems of large searching volume and high time complexity emerge in gray scale matching. Therefore, enhancing the searching velocity and accuracy has become the key issue. Although ABC can be utilized to achieve our target matching point rapidly and efficiently, it is prone to getting trapped in local optima. In this paper, the BABC algorithm is introduced to resolve the problem, with the steps as follows:

Step 1: Image pre-processing and set the initial parameters. Obtain the original image and the template and convert them into gray scale. Set a threshold for the fitness function according to different situations. The initial parameters of BABC include the population scale N and the maximum iteration $MAXcycle$.

Step 2: Initialize the solution set in accordance with OBL strategy and calculate the fitness values of population.

Step 3: Implement the iteration process. Jump to Step 9 if the termination condition is satisfied.

Step 4: The employed bees execute global search according to Eq. (4) and generate new positions v_i^j. Select the optimal one as candidate for the next generation on the basis of greedy rule.

Step 5: Calculate the probability of x_i being selected using Eq. (12) and Eq. (13):

$$p_i = fit / \sum_{i=1}^{N} fit_i \qquad (9)$$

$$fit_i = \begin{cases} 1/(1+f_i) & f_i \geq 0 \\ 1 + abs(f) & f_i < 0 \end{cases} \qquad (10)$$

where f_i and fit_i are the objective function value and the fitness value of the ith solution respectively.

Step 6: The onlooker bees choose the individuals with relatively lower fitness values to be leaders according to p_i and update the nectar resource position using Eq. (4).

Step 7: Calculate the current optimal solution and update it using Levy flight strategy.

Step 8: Implement the chaotic search around the best food sources on the basis of Eq. (6) after the parameters' ranges switch into (0, 1). Choose the best one to replace a random employed bee among the series of solutions.

Step 9: If the current iteration T is less than *MAXcycle*, go to Step 3, or else, end the iteration process and deem the searching point corresponding to the final optimal solution to be the best matching point.

4 Experimental Results and Analysis

4.1 Simulation Experiment of Test Functions

The convergence velocity and optimal solution are generally regarded as standard for the judgment of evolution algorithms. In order to investigate the feasibility and effectiveness of our proposed method in this work, series of experiments on five standard test functions are carried out to calculate their minimum and comparative experimental results with ABC are given. The number of onlooker bees is taken equal to the number of employed bees so that we have less parameters to compute. Figure 3 shows the comparative iterative curves by using ABC and BABC.

From Fig. 3, it is obvious that our proposed BABC has shown better convergence velocity and higher accuracy than the standard ABC in terms of *Sphere* function and *Rosenbrock* function. Moreover, the modified method demonstrated better performance in the test of multimoding functions *Schwefel*, *Rastragin*, and *Griewamk* as well. Experimental results show that BABC makes better performance in the optimization of multimodal and multi-variable problems as a robust optimization algorithm due to the utilization of neighbor production mechanism.

4.2 The Simulation Experiments of Image Matching

A series of vision-based experiments are conducted to verify the effectiveness of our proposed method in the template matching. Runways and ground targets are chosen to be templates to show the usefulness of BABC in UAV landing. Simulations are

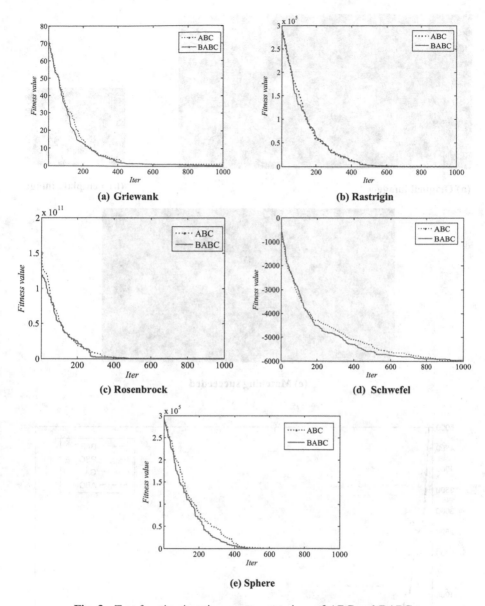

Fig. 3. Test function iterative curve comparison of ABC and BABC

Implemented in Matlab 2012b, and executed by the computer with 2.0 GHZ CPU, 4 GB memory, and operation system of Windows 7. The parameters for the standard ABC, BABC, PSO and GA were set to the following values: $MAXcycle = 200$ $NP = 60$, $limit = 20$. The first two groups simulated the contexts in which the camera (UAV) presses close to and stays far from the runway respectively.

(a) Original image

(b) Template image

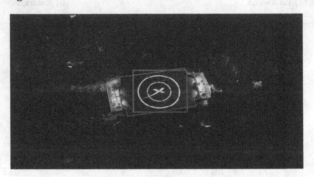

(c) Matching succeeded

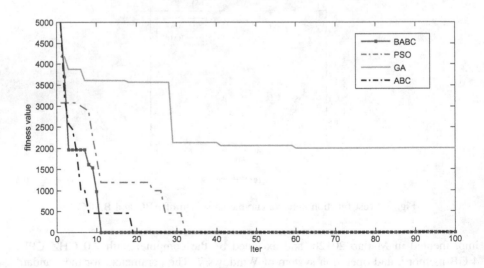

(d) Iterative curve comparison for Case1

Fig. 4. Experimental results of case 1

(a) Original image   **(b)Template image**

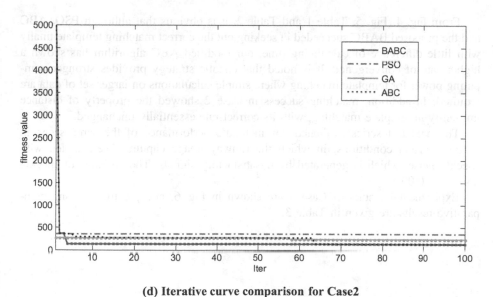

(c)   Matching succeeded

(d) Iterative curve comparison for Case2

Fig. 5. Experimental results of case 2

The results of Case 1 are shown in Fig. 4, and the time-costing comparative results are given in Table 1.

Table 1. Compared time-costing results of Case 1

Algorithm	Total time/s	Converged iteration
ABC	4.039	15
GA	4.457	72
PSO	3.965	39
BABC	4.132	13

The results of Case 2 are shown in Fig. 5, and the time-costing comparative results are given in Table 2.

Table 2. Comparative results of Case 2

Algorithm	Total time/s	Converged iteration
ABC	4.148	15
GA	4.655	80
PSO	4.078	43
BABC	4.263	16

From Fig. 4, Fig. 5, Table 1 and Table 2, it is obvious that although PSO, ABC and the proposed BABC succeeded in seeking out the correct matching template finally with little difference of calculating time, our modified ABC algorithm has shown a higher rate of convergence. It is noted that chaotic strategy provides stronger computing power for template matching where simple calculations on larger set of data are required. In addition, matching success in Case 2 showed the property of distance immunity in template matching, with its correct rate essentially unchanged.

The next test series evaluated our method's performance of the same scene in different visual conditions, in which the runway image captured were added with speckle noise, which is generated by unsatisfactory signals. The variance of speckle noise $v = 0.09$.

Experimental results of Case 3 are shown in Fig. 6, and the time-costing comparative results are given in Table 3.

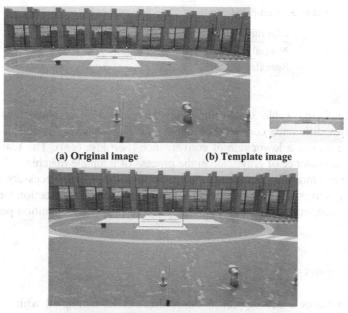

(a) Original image   (b) Template image

(c) Matching of original image succeeded

(d) Matching of speckle-noised image succeeded

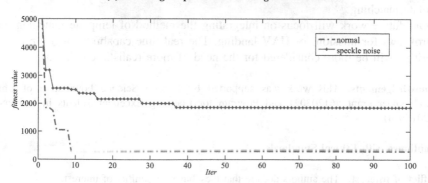

(e) Comparison of iterative curves in different conditions of noise in Case 3

Fig. 6. Experimental results of case 3

Table 3. Compared results of Case 3 in two noise conditions

Algorithm	Total time/s	Converged iteration
Normal	1.727	9
Speckle	1.841	36

From Fig. 6 and Table 3, the outputs of the successful matching results in two situations with difference convergence behaviors, and matching in the condition of speckle noise shows a lower rate of convergence (See Fig. 6(e)). The reason of this result is the dependence of gray scale in the process of template matching. The optimal fitness value is limited in the presence of noise. Therefore, it is necessary to remove noise of images imported from the camera. In addition, the edge detection methods can be applied to cooperate with our method, which makes better recognition performance achievable.

5 Conclusions

This study presents a new approach that utilizes ABC combined with bio-inspired searching strategy to solve the template matching problem, and applies the novel algorithm to UAV automatic landing. The opposition-based learning strategy is employed in the initialization phase to increase the optimization velocity. The Levy flight characteristics is adopted in the phase of updating and local search to widen the searching space. Chaotic search helps ABC avoid local optima and achieve more satisfying results. Numerical simulations were conducted, with experimental comparison results between the proposed method and the traditional methods given to verify the feasibility and effectiveness of it. The simulations have shown that the convergence rate and convergence accuracy of BABC excel that of ABC. We also confirmed the effectiveness of the modified method for resolving template matching problems under various circumstances. It is obvious that our proposed approach is superior to ABC in template matching.

Our future work will focus on integrating the method of template matching with control laws for missions of UAV landing. The real-time capability of our proposed algorithm will be more considered for the need of more realistic conditions.

Acknowledgements. This work was supported by Natural Science Foundation of China (NSFC) under grant #61203021 and Program for Liaoning Excellent Talents in University (LR2015034).

Compliance with Ethical Standards

Conflict of Interest. The authors declare that they have no conflict of interest.

References

1. Millet, P.T., Ready, B.B., Mclain, T.W.: Vision-based precision landings of a tailsitter UAV. In: AIAA Guidance, Navigation, and Control Conference, AIAA, Chicago, Illinois, pp. 1–11 (2009). https://doi.org/10.2514/6.2009-5680
2. Johnson, E., Calise, A., Watanabe, Y., Ha, J., Neidhoefer, J.: Real-time vision-based relative aircraft navigation. J. Aerosp. Comput. Inf. Commun. 4(4), 707–738 (2007). https://doi.org/10.2514/1.23410
3. Fravolini, M., Campa, G., Napolitano, M.: Evaluation of machine vision algorithms for autonomous aerial refueling for unmanned aerial vehicles. J. Aerosp. Comput. Inf. Commun. 4(9), 968–985 (2007). https://doi.org/10.2514/1.17269
4. Lin, C.L., Lee, C.S., Huang, C.H., Kao. T.C.: Unmanned aerial vehicles evolutional flight route planner using the potential field approach. J. Aerosp. Comput. Inf. Commun. 9(3), 92–109 (2012). https://doi.org/10.2514/1.54107
5. Francolin, C.C., Rao, A.V.: Optimal control of a surface vehicle to improve underwater vehicle network connectivity. J. Aerosp. Comput. Inf. Commun. 9(1), 1–13 (2012). https://doi.org/10.2514/1.i010002
6. Clare, A.S., Cummings, M.L., How, J.P.: Operator objective function guidance for a real-time unmanned vehicle scheduling algorithm. J. Aerosp. Comput. Inf. Commun. 9(4), 161–173 (2012). https://doi.org/10.2514/1.I010019
7. Ward, D.G., Monaco, J.F., Schierman, J.D.: Reconfigurable control for VTOL UAV shipboard landing. In: AIAA Guidance, Navigation, and Control Conference and Exhibit, AIAA, Portland, vol. 1, pp. 499–509 (1999). https://doi.org/10.2514/6.1999-4045
8. Schulz, H.W., Buschmann, M., Krüger, L., Winkler, S., Versmann, P.: Towards vision-based autonomous landing for small UAVs - first experimental results of the vision system. J. Aerosp. Comput. Inf. Commun. 4(5), 785–797 (2007). https://doi.org/10.2514/1.26789
9. Wasserman, T., Lennon, J., Atkins, E.: A modal operations paradigm for robust vision-based astronaut following. J. Aerosp. Comput. Inf. Commun. 3(12), 603–618 (2006). https://doi.org/10.2514/1.20201
10. Shen, Y.F., Rahman, Z.: An automatic computer-aided detection system for aircraft emergency landing. In: Infotech@Aerospace 2011, St. Louis, Missouri, pp. 1–10 (2011). https://doi.org/10.2514/6.2011-1465
11. Barber, B., McLain, T., Edwards, B.: Vision-based landing of fixed-wing miniature air vehicles. J. Aerosp. Comput. Inf. Commun. 6(3), 207–226 (2009). https://doi.org/10.2514/1.36201. MIT Lincoln Laboratory
12. Dobrokhodov, V., Kaminer, I., Jones, K., Ghabcheloo, R.: Vision-based tracking and motion estimation for moving targets using unmanned air vehicles. J. Guid. Control Dyn. 31(4), 907–917 (2008). https://doi.org/10.2514/1.33206
13. Campbell, M., Wheeler, M.: Vision-based geolocation tracking system for uninhabited aerial vehicles. J. Guid. Control Dyn. 33(2), 521–532 (2010). https://doi.org/10.2514/1.44013
14. Webb, T., Prazenica, R., Kurdila, A., Lind, K.: Vision-based state estimation for autonomous micro air vehicles. J. Guid. Control Dyn. 30(3), 816–826 (2007). https://doi.org/10.2514/1.22398

15. Effland, J., Seanor, B., Gu, Y., Napolitano, M.: Application of machine vision in unmanned aerial systems for autonomous target tracking. In: AIAA Guidance, Navigation and Control Conference and Exhibit, Honolulu, Hawaii, 18–21 August 2008. https://doi.org/10.2514/6. 2008-7251

16. Choi, H., Kim, Y.D.: Vision-based reactive collision avoidance algorithm for unmanned aerial vehicle. In: AIAA Guidance, Navigation, and Control Conference, AIAA, Portland, Oregon (2011). https://doi.org/10.2514/6.2011-6603

17. Frietsch, N., Seibold, J., Crocoll, P., Wei, M., Trommer, G.F.: Real time implementation of a vision-based UAV detection and tracking system for UAV-navigation aiding. In: AIAA Guidance, Navigation, and Control Conference, AIAA, Portland, Oregon (2011). https://doi. org/10.2514/6.2011-6405

18. Tweddle, B.E.: Relative computer vision based navigation for small inspection spacecraft. In: AIAA Guidance, Navigation, and Control Conference, AIAA, Portland, Oregon (2011). https://doi.org/10.2514/6.2008-7251

19. Yang, J., Rao, D., Chung, S.J., Hutchinson, S.: Monocular vision based navigation in GPS-denied riverine environments. In: Infotech @ Aerospace, St. Louis, Missouri (2011). https:// doi.org/10.2514/6.2011-1403

20. Horvath, T.J., et al.: A vision of quantitative imaging technology for validation of advanced flight technologies. In: 42nd AIAA Thermophsics Conference, Honolulu, Hawaii (2011). https://doi.org/10.2514/6.2011-3325

21. Frew, E., et al.: Vision-based road-following using a small autonomous aircraft. In: Proceedings of IEEE Aerospace Conference, 2004, vol. 5, pp. 3006–3015 (2004). https://doi. org/10.1109/AERO.2004.1368106

22. Stevens, M.R., Reiter, A., DelMarco, S., Vinciguerra, L., Antone, M.: Video aided navigation for small UAVs. In: AIAA Infotech@Aerospace 2007 Conference and Exhibit, Rohnert Park, California, 7–10 May 2007. https://doi.org/10.2514/6.2007-2985

23. Zhang, Y.X., Liu, D., Xin, J.: Research of image correlation matching method based on CPSO. J. Electron. Inf. Technol. **30**(3), 529–530 (2008). https://doi.org/10.3724/sp.j.1146. 2006.01344

24. Kumar, R., Ghosh, A.K.: Nonlinear longitudinal aerodynamic modeling using neural Gauss-Newton method. J. Aircraft **48**(5), 1809–1812 (2011). https://doi.org/10.2514/1.C031253

25. Huang, X.S., Zhang, F.: Multi-modal medical image registration based on gradient of mutual information and hybrid genetic algorithm. In: Third International Symposium on Intelligent Information Technology and Security Informatics, Jinggangshan, China, pp. 125–128 (2001). https://doi.org/10.1109/iitsi.2010.186

26. Bonabeau, E., Dorigo, M., Theraulaz, G.: Swarm intelligence: from natural to artificial systems. Connection Sci. **14**(2), 163–164 (2002). https://doi.org/10.1080/09540090210-144948

27. Karaboga, D., Akay, B.: A comparative study of Artificial Bee Colony algorithm. Appl. Math. Comput. **214**(1), 108–132 (2009). https://doi.org/10.1016/j.amc.2009.03.090

28. Sabat, S.L., Udgata, S.K., Abraham, A.: Artificial Bee Colony algorithm for small signal model parameter extraction of MESFET. Eng. Appl. Artif. Intell. **23**(5), 689–694 (2010). https://doi.org/10.1016/j.engappai.2010.01.020

29. Zhang, C.S., Ouyang, D.T., Ning, J.X.: An Artificial Bee Colony approach for clustering. Expert Syst. Appl. **37**(7), 4761–4767 (2010). https://doi.org/10.1016/j.eswa.2009.11.003

30. Karaboga, D., Basturk, B.: On the performance of Artificial Bee Colony (ABC) algorithm. Appl. Soft Comput. **8**(1), 687–697 (2007). https://doi.org/10.1016/j.asoc.2007.05.007
31. Akay, B.: Synchronous and asynchronous Pareto-based multi-objective artificial bee colony algorithms. J. Global Optim. **57**, 415–445 (2013). https://doi.org/10.1007/s10898-012-9993-1
32. Apalak, M.K., Karaboga, D., Akay, B.: The artificial bee colony algorithm inlayer optimization for the maximum fundamental frequency of symmetrical laminated composite plates. Eng. Optim. **46**, 420–437 (2014)
33. Derrac, J., Garcia, S., Hui, S., Suganthan, P.N., Herrera, F.: Analyzing convergence performance of evolutionary algorithms: a statistical approach. Inf. Sci. **289**, 41–58 (2014)
34. Yi, X., Yuren, Z., Hailin, L.: An elitism based multi-objective artificial bee colony algorithm. Eur. J. Oper. Res. **245**, 168–193 (2015)
35. Phuc, N.H., Chang, W.A.: Fast artificial bee colony and its application to stereo correspondence. Expert Syst. Appl. **45**, 460–470 (2016)

Research on Multi-modal Emotion Recognition Based on Speech, EEG and ECG Signals

Hui Guo, Nan Jiang[(✉)], and Dongmei Shao

College of Public Security Information Technology and Intelligence,
Criminal Investigation Police University of China, Shenyang 110854, China
zgxj_jiangnan@126.com

Abstract. In order to avoid the single-channel emotion recognition feature being easily affected by the interference of subjective ideas, a multi-modal emotion recognition method based on speech signals, EEG signals and physiological signals is proposed. In order to improve the accuracy of sentiment recognition, overcome the limitation of the lack of discrete sentiment recognition information, and combine the continuity of dimensional sentiment recognition, a decision recognition fusion model is proposed. Emotion recognition was performed based on non-invasively collected speech, EEG, and ECG signals, and multiple decision analysis was performed in conjunction with data mining and information fusion technologies to finally obtain the correct emotion category. Experimental results show that the recognition rate of the proposed multi-pattern emotion recognition model is improved by 7.64% compared with the traditional single-modal emotion recognition model, which proves that the system can effectively improve the recognition accuracy and better fit the real human emotions.

Keywords: Speech signals · EEG signals · ECG signals · Multi-modal · Emotion recognition

1 Introduction

With the advent of the era of artificial intelligence, humans have more and more time to communicate with electronic products. People are no longer satisfied with directing cold and rigid electronic products to complete mechanical work, but instead want to have more advanced emotions with intelligent electronic products. communicate with. In order to create a harmonious human-computer interaction environment, new human-computer interaction technology based on emotion recognition is becoming a research hotspot.

Human beings are driven by self-esteem in emotional activities to produce cover-up behavior. Speech signals are generated by vocal cord vibrations, EEG signals are generated by changes in cerebral cortex potentials, and ECG signals are generated by heart beats. These unconditioned reflexes of the human body are governed by the autonomic nervous system, which has obvious nonlinearity, complexity and robustness. Generally, the situation cannot be controlled subjectively. At the same time, because it is collected in a non-invasive way, it reduces the interference caused by the

© Springer Nature Singapore Pte Ltd. 2020
J. Qian et al. (Eds.): ICRRI 2020, CCIS 1335, pp. 272–288, 2020.
https://doi.org/10.1007/978-981-33-4929-2_19

stress caused by the device. Therefore, the emotion recognition research is based on voice signals, EEG signals and ECG signals. However, in special cases, people can control the activities of internal organs and organs through indirect channels. Therefore, emotion recognition based on a single signal will be affected by subjective ideas to a certain extent, leading to a reduction in the accuracy of recognition results. In order to overcome the limitation of single channel emotion recognition error, a multi-modal emotion recognition model based on speech signals, EEG signals and ECG signals is proposed to improve the recognition rate of emotion recognition.

At present, in the field of emotion recognition, there are many scholars who research emotion recognition based on a single signal. Mustaqeem proposed an artificial intelligence-assisted deep stride convolutional neural network architecture using pure net strategy for speech emotion recognition, which improves the accuracy of the model while reducing the model size [1]. Jiefeng Zhao et al. Designed a new long-term and short-term memory neural network for speech emotion recognition with a recognition accuracy of 95% [2]. Literatures [3–5] are based on deep learning methods such as convolutional neural networks and support vector machines for more optimized language emotion recognition research. Literatures [6] and [7] used deep belief neural network and correlation vector machine for brainwave emotion recognition research, respectively. There are also many scholars who research bimodal emotion recognition based on two signals. Most of them are based on auditory signals and visual signals. Yaxiong Ma et al. Proposed a dual-modal discrete emotion recognition of audio and video based on deep weighted fusion of neural network [8]. Juan D. S. Ortega and others based on support vector regression to conduct two-dimensional emotion recognition based on fusion of audio and video features [9]. Yuanyuan Zhang performed emotion recognition on video and audio through full convolutional neural network based on attention mechanism [10]. Scholars who conduct research on multi-modal emotion recognition mainly recognize based on physiological signals. Yang Liu et al. Studied discrete emotion recognition and dimensional emotion recognition based on physiological signals. Experimental results show that the accuracy of dimensional emotion recognition is lower than discrete emotion recognition [11]. Luca Romeo et al. Proposed a method for emotion recognition of physiological signals based on a multi-instance learning framework [12]. Szwoch Wioleta conducted an earlier review of emotion recognition based on physiological signals [13]. Wenping Zhao and Xiaoxiao Zhou also conducted research on multi-modal emotion recognition of physiological signals based on multiple deep learning methods [14, 15].

In order to improve the accuracy of emotion recognition and overcome the limitation that single-channel emotion recognition is easily affected by subjective ideas, this paper proposes a multi-modal emotion recognition method based on speech, EEG and ECG signals. In order to avoid the phenomenon that the effective sentiment feature is easy to be lost when clustering based on discrete sentiment, an emotion recognition model is proposed based on the decision-level voting method for mutual verification of discrete sentiment recognition results and dimensional sentiment recognition results. Emotion recognition is based on speech, EEG and ECG signals, and the correct emotion category is output through information fusion at the decision layer, which can effectively improve the accuracy and robustness of the emotion recognition model, and realize the complementary of information of different emotion characteristics.

2   Multi-modal Emotion Recognition Patterns

Psychologists usually study emotions from two perspectives: basic emotion theory and dimensional emotion theory. Basic emotion theory believes that emotions are divided into specific categories such as happy, angry, sad, and fear; dimensional emotion theory believes that emotions have a multi-dimensional structure and are continuously variable [16]. On the basis of speech emotion-based discrete emotion recognition, in order to overcome the limitation of effective emotion features of discrete emotion recognition, two dimensions, valence and arousal, were selected for two-dimensional emotion space recognition research based on EEG signals and ECG signals. The two dimensions of emotion divide the two-dimensional emotion space into four parts, and quantify 9 points based on each dimension. Each discrete emotion has a unique mapping relationship in the two-dimensional emotion space, and the two are performed at the decision-making level. Mutual evidence can effectively improve the performance of emotion recognition models.

Traditional information fusion technologies are mainly divided into data-level fusion, feature-level fusion, and decision-level fusion. Among them, decision-level fusion is widely used in multi-modal because of its high fault tolerance, high anti-interference, high fusion level, and real-time nature of signal fusion. Pattern recognition in progress [17]. In order to overcome the limitations of single-channel emotion recognition which is easily disturbed by masking behavior and large errors, a voting method for information fusion at the decision-making level is proposed. The specific process is shown in Fig. 1. Pre-processing and feature extraction based on speech, EEG, and ECG signals, and language discrete emotion recognition based on Probabilistic Neural Network (PNN) to obtain predicted emotion categories, and support vector classification (SVC) Emotion recognition in the EEG dimension is used to obtain predicted emotion coordinates, and EEG dimension recognition based on the Extreme Learning Machine (ELM) is used to obtain predicted emotion coordinates. The decision-making layer first decides whether the coordinates obtained by emotional recognition of the EEG and ECG signals are in the same quadrant based on the voting method. If they are different, the input is re-recognized, and if they are the same, the judgment of the emotional labels obtained by the speech signal recognition in the two-dimensional emotional space is continued. Whether the mapping is in the same quadrant as the emotion coordinates obtained from EEG and ECG signal recognition. If not, the input is re-recognized. If it is the same quadrant, the model can output the correct result after double verification, that is, the emotion label obtained by speech emotion recognition.

3   Pretreatment and Emotional Feature Extraction

Affected by recording equipment, glottal excitation, and lip radiation, in order to avoid the lack of effective emotional features in the high-frequency part of the voice signal, the voice signal needs to be processed for noise reduction, pre-emphasis, and framed windowing. Due to the high accuracy of MFCC emotional features and the strong

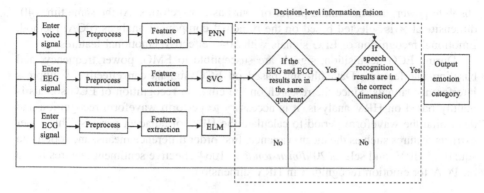

Fig. 1. Flow chart of multi-modal emotion recognition pattern

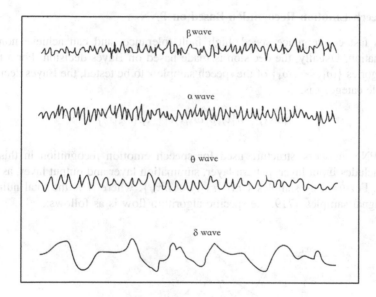

Fig. 2. Four basic waveforms of EEG

robustness of GFCC emotional features, this paper extracts the mixed emotional features of MFCC and GFCC for speech signals for emotion recognition.

Since the potential fluctuations on the human body surface are extremely weak non-stationary signals, in order to suppress the power frequency, eye movement artifacts, and myoelectric interference that EEG signals receive during the process, bandpass filtering is performed based on the EEG signals. Due to the lack of regularity of EEG waveforms, EEG waves are usually separated into four basic bands of β, α, θ and δ for processing, which are distributed at 14–30 Hz, 8–13 Hz, 4--7 Hz, and 0.5–3 Hz [17], which is shown in Fig. 2. In order to perform time-frequency analysis based on brain waves, this paper extracts the central frequency, average power, relative power, absolute power, and maximum power of four basic waveforms. At the same time, 40-dimensional × is selected based on the principal component analysis (PCA) method. Emotional recognition of EEG signals with 5040 effective emotional features.

During ECG acquisition, signals are susceptible to EMG, power frequency, and baseline interference. Data filtering based on ECG signals can effectively suppress interference signals. Since the research on the emotion recognition of ECG signals is mainly based on HRV analysis, it is necessary to perform waveform recognition and determine the waveform period to calculate the HRV of adjacent periods. This paper extracts features such as the mean, variance, first-order difference means, and root mean square of HRV, and selects 20-*dimensional* × 1064 effective sentiment features based on PCA for emotion recognition in HRV dimension.

4   Emotion Recognition Classifiers

4.1   Speech Emotion Recognition Based on PNN

PNN has fast convergence speed, high fault tolerance, and can achieve non-linear approximation. Usually, the decision is made based on Bayes decision. For a total of four categories $\{\omega_1, \cdots, \omega_4\}$ of the speech sample x to be tested, the Bayes decision of the sample category is:

$$\max\{p(\omega_1|x), p(\omega_2|x), \cdots, p(\omega_4|x)\} \tag{1}$$

The PNN network structure used for speech emotion recognition in this paper mainly includes input layer, pattern layer, summation layer and output layer, as shown in Fig. 3. For training samples $\{trx_1, trx_2, \cdots, trx_N\}$, where N is the total number of speech signal samples 1719, the specific algorithm flow is as follows:

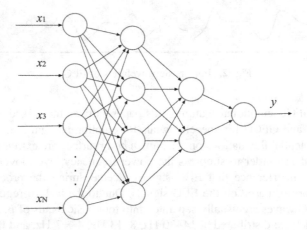

Fig. 3. PNN structure diagram

(1) Enter the test sample, and the number of nodes in the input layer is equal to the sample MFCC and GFCC mixed feature dimension 24;

(2) Calculate the Gauss function of each sample in the test sample and the training sample. The number of nodes in the pattern layer is equal to the number of training samples. The value of the Gauss function between the test sample x and the j th training sample trx_j, that is, for the test sample x, the value output from the j th pattern layer node is:

$$Gauss(x - trx_j) = e^{-\frac{\|x - trx_j\|}{2\delta^2}}$$ (2)

Where δ is the hyperparameter of the j th mode layer node of the model;

(3) Find the sum of the output of the pattern layer nodes corresponding to the test samples of the same sentiment category. The number of nodes in the summation layer is equal to the number of emotion categories in the training sample 4;

(4) Normalize the output of the summation layer to find the probability that the test sample corresponds to different categories, and judge the emotion category of the test sample based on the probability. Number of output layer nodes is 1.

4.2   EEG Emotion Recognition Based on SVC

The SVC classifier used in EEG emotion recognition is a support vector machine with classification function, which non-linearly maps EEG sample data to high-dimensional kernel space is shown in Fig. 4. The optimal emotion classification of EEG sample data is realized by constructing the optimal classification hyperplane in space. The specific algorithm flow is as follows:

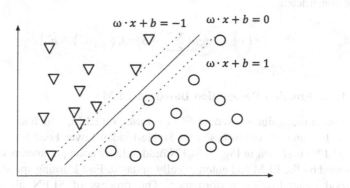

Fig. 4. SVC schematic

(1) For   EEG   sample   set $T = \{(\vec{x_1}, \vec{y_1}), (\vec{x_2}, \vec{y_2}), \cdots, (\vec{x_N}, \vec{y_N})\}$,   Where $\vec{x_i} = (\vec{x_i}^1, \vec{x_i}^2, \cdots, \vec{x_i}^n)$, and Radial Basis Function (RBF) is selected as the

activation function. For the training set (x_i, y_i) which has been given the hyperplane $\vec{w}\vec{x} + b = 0$, solve the optimization problem based on the EEG sample data:

$$\min_{\vec{a}} \frac{1}{2} \sum_{i=1}^{N} \sum_{j=1}^{N} \alpha_i \alpha_j y_i y_j K(\vec{x_i}, \vec{y_j}) - \sum_{i=1}^{N} \alpha_i,$$

$$s.t. \sum_{i=1}^{N} \alpha_i y_i = 0,$$

$$C \geq \alpha_i \geq 0, i = 1, 2, \cdots, N \tag{3}$$

Where C is the penalty coefficient, $K(x_i, y_j)$ is the activation function, N is the total number of EEG samples 1280, and α_i is the Lagrange multiplier. The optimal solution is:

$$\vec{\alpha}^* = (\alpha_1^*, \alpha_2^*, \cdots, \alpha_N^*)^T \tag{4}$$

(2) Based on hyperplane computing $\vec{w}$:

$$\vec{w}^* = \sum_{i=1}^{N} \alpha_i^* y_i \vec{x_i}, \tag{5}$$

Simultaneously select the component of α_j^* to calculate b^*:

$$b^* = y_i - \sum_{i=1}^{N} \alpha_i^* y_i K(\vec{x_i}, \vec{y_j}) \tag{6}$$

(3) After determining the hyperplane, the emotion classification decision function can be constructed:

$$f(\vec{x}) = sign\left(\sum_{i=1}^{N} \alpha_i^* y_i K(\vec{x_i}, \vec{y_j}) + b^*\right) \tag{7}$$

4.3 ECG Emotion Recognition Based on ELM

The emotion recognition based on ECG is based on ELM, which as a new type of fast machine learning algorithm based on Single-hidden Layer Feed Forward Neural Network (SLFN) is shown in Fig. 5. SLFN hidden layer node parameters can be randomly determined by the ELM and automatically updated, Fast learning speed, high precision, and good generalization performance. The process of SLFN algorithm for ECG emotion recognition based on ELM is as follows:

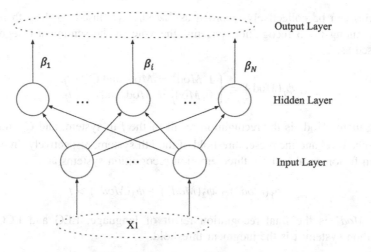

Fig. 5. Structure of SLFN

(1) Determine the number of hidden neurons in the SLFN as the number of ECG training samples 46, randomly generate the connection weight w and the hidden neuron threshold b of the input layer and the hidden layer, and follow the continuous probability distribution;

(2) Select the RBF function as the activation function of the SLFN hidden layer neurons, and calculate the output matrix H of each node of the hidden layer;

(3) Solve the SLFN output layer weight $\widehat{\beta}$:

$$\widehat{\beta} = H^{+}T' \tag{8}$$

Where H is the hidden layer output matrix of SLFN, H^{+} is the Moore-Penrose generalized inverse of the hidden layer output matrix H, T is the matrix of the SLFN output layer, and T' is the transpose of the T matrix.

5 Emotional Information Fusion at Decision Level Based on Voting

When the emotion information fusion at the decision-making level is based on the voting method, the emotion classifiers of each channel give classification suggestions through decision-making, and vote based on emotion categories. The category with the most votes wins, and the final decision result is the emotion category with the most votes.

There are 4 categories of emotion recognition in language in the multi-modal emotion recognition model, and there are 9 categories of valence and arousal degree in EEG and ECG emotion recognition, respectively. Modi represents the i th emotion category of the signal to be identified. A recognition containing a total of C emotion categories can be collectively recorded as $\{Mod_i\}_{i=1}^c$. Based on the j th recognition system among the 3 recognition systems, the vote of the sentiment category can be expressed as:

$$\phi_j(Mod_j) = \begin{cases} 1, Mod_j^* = Mod_t \text{ and } C_j > t_j, \\ 0, Mod_j^* \neq Mod_t \text{ and } C_j \leq t_j, \end{cases} \tag{9}$$

Among them, Mod_j^* is the recognition result of the j th system, and C_j and t_j are the confidence level and the preset threshold of the j th system, respectively. In the end, the decision fusion results of the three emotion recognition systems are:

$$\phi_1(Mod^*) + \phi_2(Mod^*) + \phi_3(Mod^*) > t \tag{10}$$

Where Mod^* is the final recognition result of language, EEG and ECG emotion recognition system; t is the judgment threshold.

6   Experimental Analysis

6.1   Data Selection

Multi-modal emotion recognition pattern based on language, EEG and ECG signals are trained based on CASIA Chinese Affective Corpus, DEP Affective EEG Database and RECOLA Affective Physiological Signal Database respectively. CASIA Chinese Emotional Corpus was recorded by the Institute of Automation of the Chinese Academy of Sciences. Four professional speakers were invited to record six emotions: angry, happy, fear, sad, and surprise. The DEAP emotional EEG database was recorded by Queen Mary University of London in the United Kingdom. It recorded EEGs of 32 participants who watched 40 music videos. Participants according to the degree of arousal, valence, happiness, control and familiarity of each ratings for videos. The RECOLA emotional physiological signal database was recorded by the University of Fribourg in Switzerland. It recorded the ECG of 46 participants to complete the online collaboration task. At the same time, 6 people who were not involved in the project were invited to perform emotions based on the two dimensions of valence and arousal. Evaluation, and finally take the average value as the evaluation result.

In order to verify the feasibility and performance of the emotion recognition model, a volunteer was invited to perform emotion evoking and recording of language, EEG and ECG signals based on four emotions: anger, happy, sad and fear. The presentation of emotion-evoked materials uses a high-performance computer system. The distance between the video presentation screen and the subject is about 50 cm. When the material is played, the participants are required to wear K701 high-fidelity headphones from Austrian Love Technology Co., Ltd. to ensure that the emotions are fully aroused and induced. The subjects continued to play the material for 5 min when different types of emotions were being played, in order to evoke different levels of emotions. The voice signal collection is based on the LS-12 mini digital recorder from Olympus Japan. The sample rate of the corpus is 22050 Hz, the sampling accuracy is 16 bits, and the number of channels is double. The data is saved in the ".wav" format. 40 articles.

The collection of EEG signals is based on the Brain Link smart brain wave collection headband of Shenzhen Macro Intelligence Company. There is a total of 40 EEG samples with different emotional valence and arousal levels. The collection of ECG signals is based on Wuhan Zhongqi Biomedical Electronics Co., Ltd. PM-7000D multi-parameter monitor, a total of 40 ECG samples with different emotional valence and arousal degrees.

6.2 Results and Discussions

The training of the multi-modal emotion recognition model classifier is based on 1,719 voice signal samples of four emotions: angry, happy, sad, and afraid in the CASIA Chinese emotional corpus, 1,280 samples of EEG signals from DEAP emotional EEG database and 1,288 samples of ECG signals from RECOLA emotional physiological signal database. The recognition results of each channel classifier after training are shown in Figs. 6, 7 and 8, and the overall recognition rate is shown in Table 1.

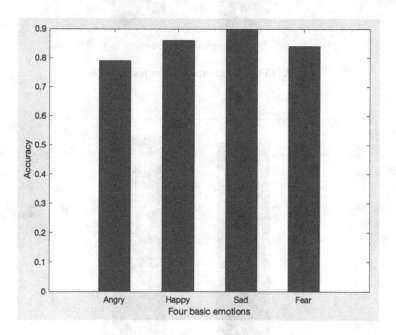

Fig. 6. Speech emotion recognition result graph

The chart shows that the accuracy of angry emotion categories in PNN-based speech emotion recognition is 79.18%, the accuracy of happy emotion categories is 88.50%, the accuracy of sad emotion categories is 90.44%, and the accuracy of fear emotion categories is 84.27%. It can be obtained that the overall average accuracy of the model speech emotion recognition reaches 85.60%. The accuracy of the valence dimension in SVC-based EEG emotion recognition is 81.09%, the arousal accuracy is

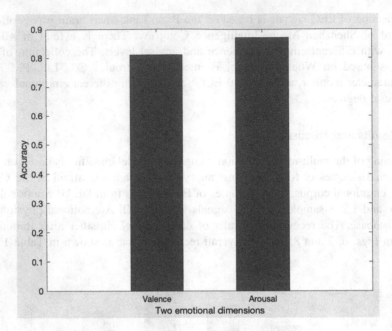

Fig. 7. EEG emotion recognition result graph

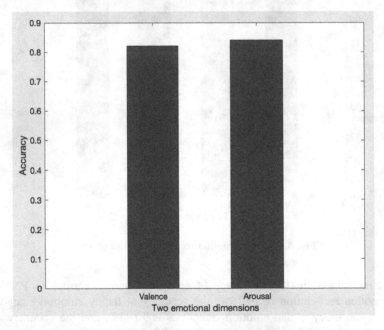

Fig. 8. ECG emotion recognition result graph

Table 1. Comparison of emotion recognition results based on classifier

Emotional signal	Emotion classifier	Affect category/dimension	Accuracy	Average accuracy
Speech	PNN	Angry	79.18%	85.60%
		Happy	88.50%	
		Sad	90.44%	
		Fear	84.27%	
EEG	SVC	Valence	81.09%	84.20%
		Arousal	87.31%	
ECG	ELM	Valence	81.83%	82.88%
		Arousal	83.92%	
Total				84.23%

87.31%, and the valence average accuracy of EEG signal emotion recognition is 84.20%. In the ECG emotion recognition based on ELM, the accuracy of the valence dimension is 81.83%, the arousal dimension is 83.92%, and the overall average accuracy of the emotion recognition of the ECG signal is 82.88%. The overall average accuracy of the model is 84.23%, and the classification effect of each channel classifier after training is better. In the discrete emotion recognition of speech signals, the recognition rate of sad emotion is the highest, and the recognition rate of angry emotion is the lowest. This indicates that the emotional activity of sadness is most obviously reflected on the human vocal cords through the autonomic nervous system, while anger is weaker. In the dimensional emotion recognition of EEG and ECG signals, the recognition rate of the arousal dimension is higher than that of the valence dimension, which shows that the emotional intensity is more obvious in emotional activities and conforms to objective laws (Table 2).

Table 2. Comparison of emotion recognition results

Number of channels	Signal type	Accuracy
Single-Mode	Speech	90.07%
	EEG	85.00%
	ECG	87.48%
	Mean	87.52%
Double-Mode	Speech, EEG	91.50%
	Speech, ECG	92.33%
	EEG, ECG	89.91%
	Mean	91.25%

In addition, bimodal emotion recognition research was conducted based on speech, EEG and ECG signals, and comparison experiments were carried out based on speech and EEG signals, speech and ECG signals, and EEG and ECG signals. The comparison results are presented in Table 3. Among them, the emotion recognition rate of speech and EEG signals is 91.50%, the emotion recognition result of speech signals and ECG signals is 92.33%, the recognition result of EEG signals and ECG signals is 89.91%, and the overall recognition rate is 91.25%. Experimental results show that the three two-channel emotion recognition effectively improve the recognition effect of emotion recognition, and the overall increase is 3.73%. At the same time, the combination of voice signal and ECG signal has the best recognition effect, and the combination of EEG signal and ECG signal has the lowest recognition rate, which shows that the voice signal and ECG signal have a high correlation in emotional activities.

The verification of the multi-modal emotion recognition model is based on the speech, EEG and ECG signals of the same person under four emotions. The single-modal emotion recognition of three single signals and the multi-modal fusion of the three signal recognition results are performed separately. State emotion recognition, the recognition results are shown in Figs. 9, 10, 11 and 12, and the overall recognition rate is shown in Table 3. It can be seen from the chart that when performing single-modal emotion recognition, the accuracy rate of the angry category based on speech emotion recognition is 82.35%, the accuracy rate of the happy category is 95.10%, the accuracy rate of the sad category is 93.81%, and the accuracy rate of the fear category is 89.02%, the overall average accuracy rate is 90.07%; the accuracy rate of the valence dimension based on EEG for emotion recognition is 81.83%, the accuracy rate of the arousal dimension is 88.17%, and the overall average accuracy rate is 85.00%; emotion based on ECG The accuracy rate in identifying the potency dimension is 85.66%, the accuracy in the arousal dimension is 89.29%, and the overall average accuracy is 87.48%. When performing multimodal emotion recognition, the accuracy rate of the

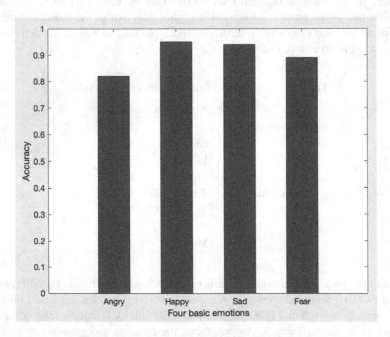

Fig. 9. Speech emotion recognition result graph

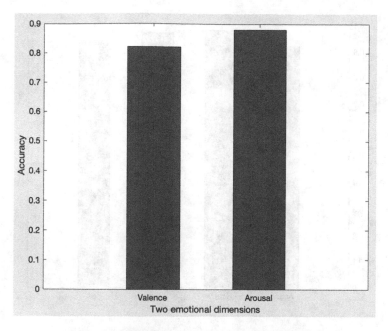

Fig. 10. EEG emotion recognition result graph

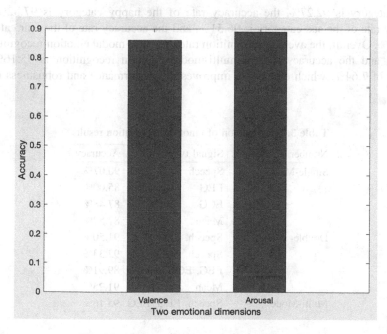

Fig. 11. ECG emotion recognition result graph

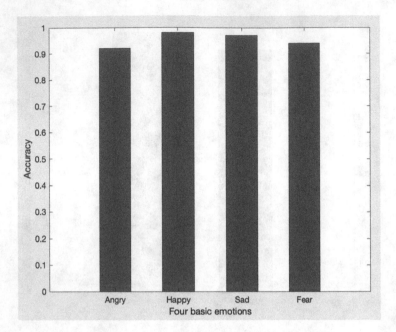

Fig. 12. Multi-modal recognition result graph

angry category is 92.27%, the accuracy rate of the happy category is 97.82%, the accuracy rate of the sad category is 96.51%, and the accuracy rate of the fear category is 94.02%. Overall, the average recognition rate of single-modal emotion recognition is 87.52%, and the accuracy rate of multi-modal emotion recognition is 95.16%, an increase of 7.64%, which effectively improves the performance and robustness of the model.

Table 3. Comparison of emotion recognition results

Number of channels	Signal type	Accuracy
Single-Mode	Speech	90.07%
	EEG	85.00%
	ECG	87.48%
	Mean	87.52%
Double- Mode	Speech, EEG	91.50%
	Speech, ECG	92.33%
	EEG, ECG	89.91%
	Mean	91.25%
Multi-Mode	Speech, EEG, EEG	95.16%

7   Conclusion

This paper proposes a multi-modal emotion recognition model based on speech, EEG and ECG signals. This model avoids single-channel emotion recognition from being susceptible to masking behaviors and big mistakes, and overcomes the limitations of discrete emotions that lack effective emotional features. The decision-level emotion information fusion based on the innovative point voting method effectively improves the performance of model recognition and can better adapt to the real emotions of people. The proposed multi-modal emotion recognition model can provide technical support for mental health monitoring in the medical field, facilitate interrogation of investigators in the field of public safety, and provide data support for a smooth user experience in the field of software development. The breakthrough progress of emotion recognition technology is of great significance to the development of interdisciplinary disciplines of physiology, psychology and computer vision. In future research, other information fusion methods will be explored to study whether information fusion technologies such as data layer and feature layer can optimize the effect of emotion recognition to a greater extent.

Acknowledgements. This work was supported in part by Natural Science Foundation of Liaoning Province 2019-ZD-0168 and 2020-KF-12-11, and Major Training Program of Criminal Investigation Police University of China 3242019010, and Key Research and Development Projects of Ministry of Science and Technology 2017YFC0821005.

References

1. Mustaqeem, S.K.: A CNN-assisted enhanced audio signal processing for speech emotion recognition. Sensors **20**(183), 1–15 (2020)
2. Zhao, J., Mao, X., Chen, L.: Speech emotion recognition using deep 1D& 2D CNN LSTM networks. Biomed. Signal Process. Control **47**, 312–323 (2019)
3. Wang, K., An, N., Li, B.N.: Speech emotion recognition using fourier parameters. IEEE Trans. Affect. Comput. **6**(1), 69–75 (2015)
4. Pan, Y., Shen, P., Shen, L.: Speech emotion recognition using support vector machine. Int. J. Smart Home **6**(2), 101–108 (2012)
5. Li, P.: Research on Speech Emotion Recognition Based on Deep Learning. University of Science and Technology of China, Hefei (2019)
6. Chao, H., Liu, Y.: Emotion recognition from multi-channel EEG signals by exploiting the deep belief-conditional random field framework. IEEE Access **2**(8), 33002–33012 (2020)
7. Zhang, X., Wang, W., Sun, Y.: Improved multi-class RVM model for EEG emotion recognition. Comput. Eng. Appl. **55**(9), 112–117 (2019)
8. Ma, Y., Hao, Y., Chen, M.: Audio-Visual Emotion Fusion (AVEF): a deep efficient weighted approach. Inf. Fusion **46**, 184–192 (2019)
9. Ortega, J.D.S., Cardinal, P., Koerich, A.L.: Emotion recognition using fusion of audio and video features. IEEE Int. Conf. Syst. **62**(23), 6235–6248 (2014)
10. Zhang, Y.: Research on Multi-modal Emotion Recognition Method Based on Deep Learning. University of Science and Technology of China, Hefei (2019)
11. Liu, Y., Gedeon, T., Caldwell, S.: Emotion recognition through observer's physiological signals. arXiv, 1–6 (2020)

12. Romeo, L., Cavallo, A., Pepa, L.: Multiple instance learning for emotion recognition using physiological signals. IEEE Trans. Affect. Comput. 1–18 (2019)
13. Wioleta, S.: Using physiological signals for emotion recognition. IEEE HIS **6**(8), 556–561 (2013)
14. Zhao, W.: Research on Multi-modal Emotion Recognition Based on Physiological Signals. Tianjin Normal University, Tianjin (2019)
15. Zhou, X.: Research on Affective Computing Based on Multi-modal Fusion. Xi'an University of Posts and Telecommunications, Xi'an (2018)
16. Mao, Q.: Research on Speech Emotion Feature Extraction and Recognition Method. Jiangsu University, China (2009)
17. Wu, R.: Research on Key Technologies of Information Fusion in Military Information System. University of Electronic Science and Technology, Chengdu (2016)

Research on Performance Evaluation Method of Fuzzy C-Means Clustering Process Based on PLS Method

Wenhua Tao, Yuying Wang$^{(\boxtimes)}$, and Qiushi Tian

School of Information and Control Engineering,
Liaoning Shihua University, Fushun 113001, LN, China
wyy001200@163.com

Abstract. Aiming at the problem that the principal component analysis method does not consider the quality variables when dimensionality reduction, some process complex process performance evaluation models are difficult to obtain online. In this paper, a fuzzy C-means clustering algorithm based on PLS algorithm is used to implement an online performance evaluation model. This paper first uses the PLS algorithm to establish an industrial process model for online prediction, and then uses the fuzzy C-means clustering algorithm to calculate the membership of the online data processed by the industrial model for each category, and obtains the online evaluation results of quality variables through fuzzy synthesis. Simulation experiments verify that the algorithm of this method has high accuracy, simple calculation, and has certain application value.

Keywords: PLS algorithm · Fuzzy C-means clustering · Performance evaluation

1 Introduction

The industrial process is complicated and has many processes. Excellent industrial system performance is very important for enterprise production. Affected by factors such as aging of the device, improper operation of the operator, and changes in the production environment, the performance status of the industrial process will deviate from the optimal working performance, which may lead to poor economic performance and even production accidents [1]. Therefore, it is meaningful to design an online performance evaluation model for industrial processes.

In recent years, performance evaluation algorithms have developed rapidly. Performance evaluation methods are roughly divided into three types: analytical model-based methods, knowledge-based methods, and data-driven methods. Principal component analysis, independent principal component analysis, and partial least squares are the more widely used algorithms in multivariate statistical analysis algorithms in data-driven methods [2]. The basic idea of principal component analysis is to convert high-dimensional raw data into low-dimensional feature elements for modeling. Although the principal component analysis method can reduce the dimensionality of the data and can obtain the principal components with low correlation, principal component analysis

J. Qian et al. (Eds.): ICRRI 2020, CCIS 1335, pp. 289–303, 2020.
https://doi.org/10.1007/978-981-33-4929-2_20

is difficult to deal with nonlinear problems. Scholkopf et al. proposed kernel principal component analysis solved this problem [3]. The kernel principal component analysis uses the kernel function to project the data into a high-dimensional space, and then uses principal component analysis to reduce the dimension to realize the analysis of the indicator data [4]. The insufficient of above two algorithms is that these algorithms only focus on the internal relationship of process variables, without considering the relationship between quality variables (output variable) and process variables. PLS algorithm can solve this shortcoming. PLS algorithm can focus on the relationship between process variables and quality variables [5].

Therefore, in this paper, the PLS algorithm is used to decompose the process variable data space according to the degree of correlation with industrial process quality related variables, establish an industrial process model, and use the process model, the data is calculated online. When judging the rank of the online calculation result of the process model, the fuzzy C-means clustering analysis method with concise calculation process is used to divide the performance grade of the quality variable data during offline modeling and obtain the membership function of the corresponding grade. The membership function corresponding to each performance level is used to calculate the membership of the quality variable data obtained through the online calculation of the process model, and the performance level corresponding to the quality variable data is judged according to the membership degree corresponding to the quality variable data, to realize the online evaluation of industrial process performance.

2   Partial Least Squares

Partial least squares are commonly used for dimensionality reduction and regression analysis of high-dimensional data. Partial least squares can investigate the correlation between variables and establish regression equations [6]. This algorithm predicts quality variables by observing process variables. Partial least squares can decompose the sample space according to the correlation between the input variables and the output variables, extract the principal components of the sample space to establish a regression model [5]. Partial least squares can handle the case where the matrix dimension is larger than the number of samples. Partial least squares can simultaneously achieve multiple linear regression, principal component analysis and canonical correlation analysis. Partial least squares are also called second-generation regression analysis method [6].

Let the input matrix of sample data be X, which contains m sample data and n indicator variables, Y is the output matrix, which contains m sample data and l variables.

First, standardize the input and output sample data and convert it into standardized data with zero mean and unit variance. Let the standardized input and output matrices be X_0 and Y_0, respectively.

Extract the components t_1 of the input matrix and the components u_1 of the output matrix.

$$\begin{cases} t_1 = X_0\omega_1 \\ u_1 = Y_0c_1 \end{cases} \tag{1}$$

Among them, ω_1 and c_1 are the weight coefficients of the input and output matrices, respectively, both of them are unit vectors, i.e. $\|\omega_1\| = \|c_1\| = 1$. When extracting the above two components, need to meet the following requirements.

1. t_1 and u_1 should have as much variation information as possible in their data.
2. The relationship between t_1 and u_1 can reach the maximum.

It can be seen from the above conditions that t_1 and u_1 can reflect the data of variables X and Y as fully as possible, and the input variable component t_1 has a strong explanatory ability for the output variable u_1.

Partial least squares can be transformed into solving the following optimization problem:

$$\begin{cases} \max(\mathrm{cov}(t_1, u_1)) \\ st \; \|\omega_1\| = \|c_1\| = 1 \end{cases} \tag{2}$$

Using the Lagrange method to solve Eq. (3), we can obtain the unitized eigenvector ω_1 corresponding to the largest eigenvalue of $X_0^T Y_0 Y_0^T X_0$. c_1 is the unitized eigenvector corresponding to the largest eigenvalue of $Y_0^T X_0 X_0^T Y_0$. Obtain the weight vectors of the input and output matrices can get the first component t_1 and u_1. Establish the three regression equations are as follows:

$$\begin{cases} X_0 = t_1 p_1^T + X_1 \\ Y_0 = u_1 q_1^T + Y_1 \\ Y_0 = t_1 r_1^T + Y_1' \end{cases} \tag{3}$$

$$\begin{cases} p_1 = \dfrac{X_0^T t_1}{\|t_1\|^2} \\ q_1 = \dfrac{Y_0^T u_1}{\|u_1\|^2} \\ r_1 = \dfrac{Y_0^T t_1}{\|t_1\|^2} \end{cases} \tag{4}$$

Among them, X_1, Y_1 and Y_1' are the residual matrices of the three regression equations. Then use X_1 and Y_1. instead of X_0 and Y_0 to find the weight coefficient ω_2 and c_2 of the second component and the second component t_2 and u_2.

$$\begin{cases} t_2 = X_1\omega_2 \\ u_2 = Y_1c_2 \end{cases} \tag{5}$$

Repeat the solution until all the components are found or meet the requirements.

After repeated solving, if the rank of X_0 is a, so can get:

$$\begin{cases} X_0 = t_1 p_1^T + \ldots + t_a p_a^T \\ Y_0 = t_1 r_1^T + \ldots + t_a r_a^T + F_a \end{cases} \tag{6}$$

Because both $t_1, \ldots, t_a$ are linear combinations of $E_{01}, \ldots, E_{0p}$, the linear regression equation can be expressed as follows:

$$Y = t_1 r_1^T + t_2 r_2^T + \ldots + t_a r_a^T + F_a = X \left[\sum_{j=1}^{b} \omega_j^* r_j^T \right] + F_a \tag{7}$$

Among them,

$$\omega_j^* = \prod_{i=1}^{j-1} \left(I - \omega_i p_i^T \right) \omega_j \tag{8}$$

In the formula, $\left[\sum_{j=1}^{b} \omega_j^* r_j^T \right]$ is the regression coefficient of partial least squares, F_a represents the column of the residual matrix.

Partial least squares are often used to determine the number of principal components by checking the validity of crossover. The cross-validity test method first uses all sample sets except for a certain sample point i and uses d components to fit a regression equation, and then calculates the sum of squared prediction errors $press_d$ of sample point.

$$press_h = \sum_{j=1}^{p} \sum_{i=1}^{n} \left(y_{ij} - \hat{y}_{dji} \right) \tag{9}$$

In the formula, $\hat{y}_{dji}$ represents the fitted value of Y_j at the sample point i.

Use all sample points to fit a regression equation containing d components.

$$ss_d = \sum_{j=1}^{p} \sum_{i=1}^{n} \left(y_{ij} - \tilde{y}_{dji} \right) \tag{10}$$

In the formula, $\tilde{y}_{dji}$ represents the predicted value of the i-th sample point, where $i = 1, 2, \ldots, n$.

For all output variables, the cross-validity of component e can be expressed as follows.

$$Q_d^2 = 1 - \frac{press_d}{ss_{d-1}} \tag{11}$$

Using cross-validity to measure component e has the following criteria for the marginal contribution of prediction model accuracy:

1. For $k = 1, 2, \ldots, q$, there is at least one k, making $Q_d^2 \geq 0.0975$. Adding component e at this time improves the prediction model of at least one output variable y_k. Therefore, it can be considered that increasing component e is obviously beneficial.

Through the above steps, the input variable X and output variable Y can be decomposed into the following form:

$$\begin{cases} X = TP^T + \tilde{X} \\ Y = TQ^T + \tilde{Y} \end{cases} \tag{12}$$

In the formula, P represents the score matrix, Q is the regression matrix of Y to T, also known as the load matrix, and $\tilde{X}$ and $\tilde{Y}$ are the residual matrix of X and Y, respectively.

3 Fuzzy C-Means Clustering Algorithm

Cluster analysis is a process of classifying objects based on the similarity between things, and classifying things according to certain requirements and laws [7]. Cluster analysis is an important unsupervised learning method, Cluster analysis is an important unsupervised learning method and one of the important means of data recognition and processing in data mining algorithms, it has been widely used in many fields [6]. Cluster analysis algorithms can be divided into hard clustering algorithms and soft clustering methods. Fuzzy C-means clustering algorithm is a flexible fuzzy clustering method [8]. The fuzzy C-means clustering algorithm obtains the membership degree of each sample data to all clustering centers by iterative optimization of the objective function, and then determines the classification of each sample data according to the membership degree to realize the automatic classification of sample data, compared with the ordinary C clustering method, the fuzzy C-means clustering method is more superior to the processing of sample data [9].

Let the sample data be X, which contains n samples, and the sample data can be expressed as $x_i, (i = 1, 2, \ldots, n)$. If the sample data X is divided into g categories, the membership degree of the sample $x_i, (i = 1, 2, \ldots, n)$ in the data set X to the $j, (j = 1, 2, \ldots, g)$ th category is u_{ij}, and the clustering result is expressed by the membership matrix U of $g \times n$ dimension. The elements in the membership matrix satisfy the following three requirements:

$$\begin{cases} \sum_{j=1}^{g} u_{ij} = 1, (\forall i = 1, 2, \ldots, n) \\ 0 < \sum_{i=1}^{n} u_{ij} < n (\forall j = 1, 2, \ldots, g) \\ u_{ij} \in [0, 1] (i = 1, 2, \ldots, n; j = 1, 2, \ldots, g) \end{cases} \tag{13}$$

The following is the objective function of fuzzy C-means clustering algorithm:

$$\min J_m(U, V) = \sum_{j=1}^{n} \sum_{i=1}^{g} u_{ij}^m d_{ij}^2 \tag{14}$$

In the formula, g represents the number of clusters, the value range is $2 \leq g \leq n$, m is the fuzzy weighting coefficient, usually $m = 2$, $V = \{v_1, v_2, \ldots, v_g\}$ represents the set of category centers, and $d_{ij}^2 = \|x_i - v_j\|^2$ represents the Euclidean Distance from the sample x_i to the category center v_j, u_{ij} represents the degree of membership of x_i to category j.

In order to satisfy the constraint condition of $J_m(U, V)$ in formula (14), the Lagrange multiplier is introduced into the objective function, which can be expressed as:

$$\min J_m = \sum_{i=1}^{g} d_{ij}^2 - \lambda \left(\left(\sum_{i=1}^{g} u_{ij} \right) - 1 \right) \tag{15}$$

Find the corresponding partial derivative of all parameters, and make the partial derivative of each parameter equal to 0, and the necessary conditions to obtain the minimum value of the objective function (14) are:

$$u_{ij} = \frac{1}{\sum_{k=1}^{g} \left(\frac{d_{ij}}{d_{ik}} \right)^{2/(m-1)}} \tag{16}$$

$$v_j = \frac{\sum_{i=1}^{n} (u_{ij})^m \bullet x_i}{\sum_{i=1}^{n} (u_{ij})} \tag{17}$$

The fuzzy C-means clustering method uses the above constraint as the basic condition, and iteratively obtains the center v_j and membership u_{ij} of each category, and continuously optimizes the value of the objective function $J_m(U, V)$. The calculation process of the fuzzy C-means clustering method is as follows:

1. Initially set each parameter, including the number of categories g, clustering weighted index m, maximum number of iterations T_{MAX}, iteration threshold $\varepsilon > 0$, and initialize clustering center $V(1)$.
2. Update membership u_{ij} according to formula (16).
3. Update category center v_j according to formula (17).
4. If $\|V(k+1) - V(k)\| < \varepsilon$, then terminate the iteration and the algorithm ends, otherwise let $k = k+1$ and jump to step 2.

Through the iterative update of category centers and membership in steps 2 and 3, the membership of the centers of each category and the sample data can be obtained, and the fuzzy division of the sample data set can be achieved [9].

4 Process Performance Evaluation Method

In this paper, a partial least squares method is used to establish a prediction model for the sample data, and a fuzzy C-means clustering algorithm is used to establish the membership function of each performance level required for performance evaluation. When input data is obtained online, the prediction data of the output variable is first obtained through the prediction model, and then the process performance is evaluated by calculating the membership of the prediction data for each category. The specific evaluation process is as follows.

Due to the large hysteresis in industrial production, the single sample obtained cannot characterize the performance status of the entire industrial production process. Therefore, the data needs to be pre-processed before the process performance evaluation. Use a data window of width H as the basic evaluation unit during data pre-processing [5]. Use the average value of the data in the data window of length H as the pre-processed data for process performance evaluation. The following describes the specific process of process performance evaluation.

1. Preprocess the historical data, fuse the data with the length of time in the historical data, and find the average of each basic data unit as the fused data, namely:

$$\bar{m}_i = \sum m_{ik} \Big/ H \tag{18}$$

In the formula, $\bar{m}_i$ represents the average value of the i indicator, m_{ik} represents the k data in the i indicator, and H represents the data window width.

2. Standardize the data obtained from formula (18).
3. Use formulas (1) to (11) to build a prediction model to obtain the coefficient matrix b and constant terms of the prediction model.
4. Use formulas (13) to (17) to divide the performance level of historical data and get the membership function of each performance level.

After obtaining the prediction model of the industrial process, the specific process of online evaluation is as follows:

1. Using Eq. (18) to preprocess the obtained data and normalize the preprocessed process data to obtain the processed data.
2. Using the coefficient matrix and constant terms obtained by partial least squares to calculate the predicted value of the output variable to the preprocessed data.
3. Using the membership function of each performance level to calculate the membership of the predicted value of the output variable to each performance level.
4. Using fuzzy operators to determine the performance level of the predicted value of the output variable.

5 Simulation

5.1 Industrial Process Description

Blast furnace ironmaking is an industrial production process in which coke, iron-bearing ore and other raw materials are produced in the blast furnace to produce iron, which plays an important role in the modern steel production process [11]. The blast furnace production process is to load iron ore, fuel and flux from the top of the blast furnace, and the hot air is blown in from the tuyere under the blast furnace. The carbon element in the fuel is burned in the hot air to produce reducing gas, which produces a reduction reaction with iron-containing ore The reduced iron is melted and carburized to form pig iron, which is discharged through the iron port. The gangue in the ore is converted into slag and discharged from the blast furnace through the slag port [12]. The blast furnace ironmaking process is very complicated. In blast furnace automatic control, furnace temperature control is the most important control. Furnace temperature control is the core of blast furnace ironmaking, and furnace temperature fluctuation is controlled by furnace temperature. Blast furnace temperature is more sensitive to the reduction rate of silicon in molten iron. Therefore, the silicon content of hot metal is often used to reflect the temperature of the blast furnace. There are many factors that affect the silicon content of hot metal in the blast furnace smelting process. Table 1 lists the process variables that can be measured for online evaluation during the blast furnace smelting process [11].

Table 1. Main variables of blast furnace temperature state evaluation

Num.	Main variable
1	Air volume
2	Wind temperature
3	Wind pressure
4	Breathability
5	Oxygen enrichment
6	Coal injection
7	Feed rate
8	Iron difference

5.2 Establishment of Prediction Model

In order to test the effect of the method proposed in this paper in practical application, this paper selects No.15321 $\sim$ No.15450 data of iron temperature in No.6 blast furnace temperature data of Inner Mongolia Baotou Steel to model and analyze the furnace temperature during the blast furnace smelting process.

Use formulas (1) to (11) to decompose the sample data and use cross-validation to get the number of principal components.

When the second component is extracted, the square sum of the prediction error pre_2 of y is:

$$press_2 = 70.3473 \tag{19}$$

The sum of squared errors ss_2 of y is:

$$ss_1 = 64.4186 \tag{20}$$

Using formula (11) to calculate the cross-validity, we can get:

$$Q_2^2 = 1 - \frac{press_2}{ss_1} = -0.0435 < 0.0975 \tag{21}$$

From formula (21), it can be determined that the number of principal components of the blast furnace temperature prediction model is 2.

The parameters ω and c can be obtained by solving Eq. (2).

$$\omega = \begin{bmatrix} -0.2508 & -0.4038 \\ -0.6356 & 0.1319 \\ 0.1334 & -0.5807 \\ -0.3956 & -0.2801 \\ -0.2687 & -0.4549 \\ 0.2950 & 0.0734 \\ 0.1255 & -0.1608 \\ -0.4288 & 0.4070 \end{bmatrix} \tag{22}$$

$$c = \begin{bmatrix} -0.9680 & -0.4038 \\ 0.1647 & 0.1319 \\ -0.0345 & -0.5807 \\ 0.1025 & -0.2801 \\ 0.0696 & -0.4549 \\ -0.0764 & 0.0734 \\ -0.0325 & -0.1608 \\ 0.111 & 0.4070 \end{bmatrix} \tag{23}$$

Through formula (1), can get the score matrix T, the matrix P and the matrix Q can be obtained by bringing the scoring matrix T into formula (4).

$$P = \begin{bmatrix} -0.4486 & -0.4777 \\ -0.5710 & 0.0683 \\ -0.1511 & -0.5726 \\ -0.5328 & -0.3526 \\ -0.4916 & -0.2964 \\ 0.3310 & 0.0425 \\ 0.0467 & -0.2120 \\ -0.2294 & 0.4784 \end{bmatrix} \tag{24}$$

$$Q = [0.2671 \quad -0.1703] \tag{25}$$

Use formula (8) to get a new weight matrix, let D be the regression coefficient matrix, then:

$$D = \begin{bmatrix} -8.213 \times 10^{-5} \\ -0.043 \\ 1.1737 \\ -1.702 \times 10^{-5} \\ -8.706 \times 10^{-6} \\ 1.596 \times 10^{-6} \\ 0.0013 \\ -2.769 \times 10^{-4} \end{bmatrix} \tag{26}$$

Then, calculate the average value and standard deviation of the input variable and output variable, and calculate the constant term of the original data regression equation by the following method:

$$ch = \bar{Y} - (\bar{X}/\sigma_X \times d) \times \sigma_Y \tag{27}$$

$$ch = 6.2322 \tag{28}$$

In the formula, $\bar{X}$ and $\bar{Y}$ represent the average value of input variable and output variable, σ_X and σ_Y represent the standard deviation of input variable and output variable, respectively.

The partial least squares regression model of the silicon content of blast furnace molten iron from formula (27) and formula (28) can be roughly expressed as:

$$y = -0.043x_2 + 1.1737x_3 + 0.0013x_7 + 6.2322 \tag{29}$$

The 37 sets of process variable data used for prediction are substituted into formula (29) to calculate the predicted value of silicon content in the blast furnace molten iron, and compared with the actual value,the specific comparison results are shown in Fig. 1. The average error of partial least squares prediction is $\Delta = 0.0753$. It can be seen that the partial least squares method can more accurately reflect the trend of silicon content in blast furnace molten iron. However, this algorithm is not very effective in predicting the high silicon content of blast furnace molten iron.

5.3   Performance Status Analysis

After establishing the industrial process model using partial least squares, this paper uses fuzzy C-means clustering algorithm to divide the performance level of historical data and establish the membership function of each performance level. The input data obtained online is calculated by the process model to obtain the output variable prediction data. The output variable prediction data uses the membership function to calculate its membership for each performance level, and finally get the current performance level of the industrial process.

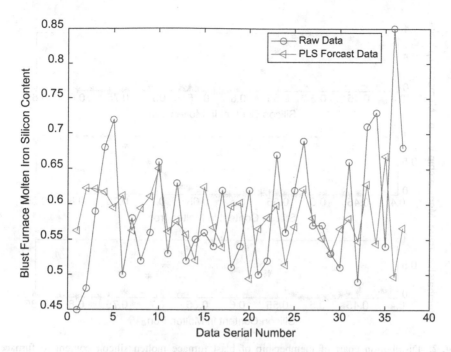

Fig. 1. Predictive model of partial least squares method for silicon content in blast furnace hot metal

The blast furnace temperature state can usually be divided into "high", "low" and "normal". In this paper, the fuzzy C-means clustering algorithm is used to divide the historical data of silicon content in blast furnace molten iron into three categories: "high", "normal" and "low". Because the silicon content data of blast furnace molten iron roughly meets the Gaussian distribution, this article uses the Gaussian function as the membership function. The Gaussian distribution function is shown in Eq. (30). The relevant parameters of the membership function of the three performance levels of the blast furnace temperature are shown in Table 2. The distribution of the membership of the blast furnace molten silicon content to each blast furnace temperature performance level is shown in Fig. 2.

Table 2. Relevant parameters of membership function of each level

Performance level	c	σ
High	0.78	0.026
Normal	0.62	0.026
Low	0.51	0.026

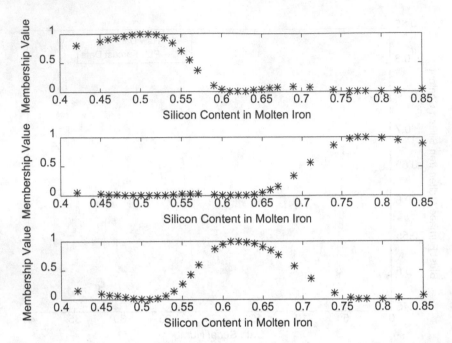

Fig. 2. Distribution chart of membership of blast furnace molten silicon content to furnace temperature grade

$$f(x, \sigma, c) = e^{-\frac{(x-c)^2}{2\sigma^2}} \tag{30}$$

Substituting the parameters in Table 2 into Eq. (30), the membership function of each furnace temperature performance level can be obtained. Using 78 sets of historical data of blast furnace molten iron silicon content to verify membership function, see Table 3 for verification results. It can be seen from Table 3 that the membership function of each performance level of the blast furnace temperature can well reflect the performance status of the blast furnace temperature. Then, use the membership function to calculate the membership of 37 sets of output variable data processed by the prediction model for each level, and finally determine the performance level of each group of data, and compare it with the performance level of the original data. The comparison results are shown in the Table 4. It can be seen from Table 4 that the online monitoring results obtained by the partial least square method are basically consistent with the actual data.

Due to the limitation of the sample size of the modeling data, this prediction model has a certain delay from the sample data in the actual data monitoring. It can more accurately detect the occurrence of abnormal conditions, and then remind engineers to take corresponding measures (Fig. 3).

Table 3. Comparison of classification results of membership function and actual classification results

Furnace temperature status	Serial numbers of various data obtained by fuzzy C-means clustering	Serial numbers of various types of data obtained using membership function
Low	1–9,19,22,25,28–30,33 35,36,41,43,47–49,50, 51,54,55,57–59,60,62, 64,71,73,77	1–9,19,22,25,28–30,33 35,36,41,43,47–49,50, 51,54,55,57–59,60,62, 64,71,73,77
Normal	11–15,17,27,31,32,34, 37–39,40,42,45,46,52 53,56,61,63,65–68,70 72,74,75,76,78	11–15,17,27,31,32,34, 37-39,40,42,45,46,52 53,56,61,63,65–68,70 72,74,75,76,78
High	10,16,18,20,21,23,24,26, 44,69	10,16,18,20,21,23,24,26, 44,69

Table 4. Comparison of online monitoring results and actual results

Furnace temperature status	Sample number of actual classification result	Sample number of online monitoring results
Low	1,2,6,8,9,11,13–15 16,18,19,21,22,24 29,30,32,35	1,7,11,13,14,17,20,24,28 29,32,34,36
Normal	3,4,7,10,12,17,20, 23,25–28,31,37	2–6,8–10,12,15,16,18,19, 21–23,25–27,30,31,33,35, 37
High	5,33,34,36	

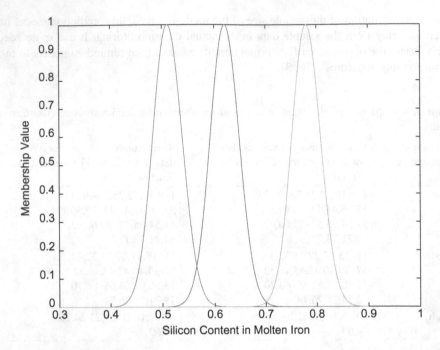

Fig. 3. Membership function graph of each performance level

6 Conclusion

In order to solve the problems that the evaluation results are difficult to obtain online in the evaluation of complex process performance, this paper uses partial least squares method, uses the relevant process data to establish the prediction model, and uses the fuzzy C-means clustering algorithm to determine the performance level of the online calculation result of the prediction model to achieve industrial online evaluation of the production process. The advantage of the industrial process monitoring method is that the process of establishing the prediction model is simple, the calculation complexity is low, and it can accurately reflect the changes of the industrial process. However, this method has errors in the online evaluation of process performance. Therefore, to obtain better modeling results, nonlinear modeling methods can be combined with principal component analysis or partial least squares. In addition, the number of classifications of the FCM clustering method is artificially set, and the FCM clustering method can be improved by analyzing the sample data.

References

1. Tarafder, A., Rangaiah, G.P., Ray, A.K.: A study of finding many desirable solutions in multiobjective optimization of chemical processes. Comput. Chem. Eng. **31**(10), 1257–1271 (2007)
2. Dai, M., Yang, F., Peng, N., Hou, J., Zhang, Z.: Historical data modeling of ammonia synthesis process based on PCA-clustering analysis. Comput. Appl. Chem. **35**(03), 211–219 (2018)
3. Sun, X., Kong, J., Liu, Z.: Power load mid-term forecasting model based on kernel principal component analysis and improved neural network. J. Nanjing Univ. Sci. Technol. (Natural Science Edition) **042**(003), 259–265 (2018)
4. Xu, C., Xiong, X.: Bearing fault diagnosis based on multi-scale feature extraction and KPCA. J. Electron. Measur. Instrum. **33**(11), 22–29 (2019)
5. Kong, X., Cao, Z., An, Q., Xu, Z., Luo, J.: Partial least squares linear model and its nonlinear dynamic expansion model overview. Control Decis. **33**(09), 1537–1548 (2018)
6. Liu, X., Wang, Z., Wang, X.: Online economic performance grading evaluation method based on similarity grid model. Chem. Ind. Eng. **67**(11), 4724–4731 (2016)
7. Ma, X., Chen, H., Ma, X., Zhai, C., Zeng, X.: Research on the influence of high-speed rail construction on highways based on FCM algorithm. Wirel. Internet Technol. **17**(06), 127–128 (2020)
8. Pan, G.: Cluster analysis and research on power load based on FCM. Sci. Technol. Guide **27**(17), 41–42 (2019)
9. Wang, Z., Long, C., Chang, X., Jiang, X.: Random subspace low frequency oscillation modal recognition algorithm based on FCM clustering. J. Pow. Syst. Automat. **32**(04), 69–75 (2020)
10. Wang, W., Yang, G., Ge, W., Liu, P., Qian, C.: Fuzzy C-means clustering algorithm and application under uneven noise conditions. Comput. Eng. Appl. **54**(19), 172–178 (2018)
11. Yuan, D., Cao, F., Li, D.: Model of blast furnace temperature prediction with principal component analysis combined with extreme learning machine. J. Inner Mongolia Univ. Sci. Technol. **36**(04), 327–332 (2017)
12. Wang, Y., Sun, C., Dong, D., Jia, H.: Online blast furnace temperature detection sensor based on lining thickness. Instrum. Technol. Sensor **19**(02), 14–15 (2015)

Intelligent Control and Perception

Path Following Method of a Snake Robot Based on Unilateral Edge Guidance Strategy

Danfeng Zhang[✉]

School of Information and Control Engineering, Liaoning Shihua University,
Fushun 113001, China
Zhangdanfeng9021@sina.cn

Abstract. In order to control the snake robot to follow the desired path with a path edge, on the basis of the direction control method which is based on the angle symmetry, a path edge guidance strategy is proposed. The core idea of this method is that one edge of the desired path can be detected, and the points on the edge are used to guide the snake robot's locomotion direction. When the forward extension line of the head module and the path edge intersect at a point, the forward temporary target point is detected on the path edge. When the reverse extension line of the head module and the path edge intersect at a point, the reverse temporary target point is detected on the path edge. By introducing the temporary target point into the direction control method, the locomotion direction is adjusted by the temporary target points. The simulation results show that the snake robot can adjust the locomotion direction by detecting one edge of the desired path on the ground with unknown friction coefficient. The simulation results show that the snake robot can follow the expected path, and the distance between the snake robot and the path edge can meet the expected value.

Keywords: Snake robot · Creeping · Desired path · Unilateral edge · Edge guidance

1 Introduction

Biological snakes have the special body structure, hundreds of joints and no limbs [1–3]. The body structure endows snakes with the ability to adapt to various environments. The characteristics attract the researches' attentions. A snake robot has these merits by imitating the structure of the biological snake. The researchers hope to apply the snake robot to environmental investigation, disaster rescue, etc. Tracking ability is the key for the snake robot to explore an unknown environment.

The snake robot can move in many gaits. The serpentine is a typical gait. At present, researchers have proposed a variety of path following methods. Based on the simple model of the snake robot [4], a waypoint guidance strategy for steering a snake robot along a path which is defined by waypoints interconnected by straight lines is proposed in [5, 6]. [7] presents a control system which is enable an underwater snake robot to converge toward and follow a straight path in the presence of constant irrotational ocean currents. In [8], the head joint is adjusted by the desired angle, so that the centroid of the snake robot follows the expected straight path. In [9], an integral line-of-sight

© Springer Nature Singapore Pte Ltd. 2020
J. Qian et al. (Eds.): ICRRI 2020, CCIS 1335, pp. 307–318, 2020.
https://doi.org/10.1007/978-981-33-4929-2_21

(LOS) guidance law is presented, which is combined with a sinusoidal gait pattern and a directional controller that steers the robot toward and along the desired path. In [10], the camera has been used to measure the position of the snake robot in real time. According to the position of the snake robot, the desired angle can be calculated, and the snake robot can follow the straight path and the circular path. To control the snake robot to follow the desired circular path, the function of the desired path can be introduced into the control method [11]. In [12], the snake robot follows the circular path by adjusting velocity components. [13] gives the reference trajectory of the head position and the orientation of link 1, and torque is determined to reduce the tracking errors.

In the above researches, the width of the desired path is ignored. When the snake robot follows the desired path, it will swing around the desired line or curve. The desired line or curve can be a line which is representing the direction of a path, between two edges, and at a certain distance from two edges of path. The above methods are suitable for the case where the path direction is known and the distance between robot and path edges is not required. In practical application, it is difficult to obtain the above line or curve. The reasons are shown as follow. In practical applications, the path has a certain width. In most applications, the direction of the path cannot be measured in advance. When the snake robot moves, the trajectory of each module is S-shape, so it is difficult to detect the straight line or curve which is representing the path direction.

In fact, the path edge constrains the moving space, and represents the direction of the path. Therefore, the locomotion direction can be adjusted by the path edge. In some applications, only one edge can be detected, or the control target can be achieved by detecting one edge of the path.

This paper has two contributions. The first contribution is that the snake robot follows the desired path by detecting one edge of the path. The second contribute is that the distance between the snake robot and the path edge can be adjusted.

In order to steer a snake robot along the desired path with unilateral edge, the path edge guidance strategy is proposed. The core idea of this method is that one edge of the desired path can be detected, and the detected points on the edge are used to adjust the snake robot's locomotion direction. When the forward extension line of the head module and the path edge intersect at a point, the forward temporary target point is detected on the path edge. When the reverse extension line of the head module and the path edge intersect at a point, the reverse temporary target point is detected on the path edge. By introducing the temporary target point into the direction control method, the direction of motion is adjusted by the temporary target point.

The paper is organized as follows. Section 2 introduces the influence of joint symmetry on locomotion direction. Section 3 presents a direction control method. Section 2 and Sect. 3 have been proposed in the previous study[14]. Section 4 presents a path edge guidance strategy. Section 5 presents simulation results. Finally, Sect. 6 presents concluding remarks.

2   Analysis of Serpentine Gait

There is no passive wheel between the head module and the ground, and other modules have passive wheels. The number of modules is n. The length of each module is l. The length of the robot is $L = nl$. The coordinate of the i th module is (x_i, y_i), $i = 0 \cdots, n - 1$. The 0 th module is head. The angle between the i th module and x-axis is θ_i. The angle between two adjacent modules is $q_j, j = 1, \cdots, n - 1$. The pose of the snake robot is defined as $\Psi_0 = (x_0, y_0, \theta_0, q_1, \cdots q_{n-1})$. The simplified model of the snake robot is shown in Fig. 1.

According to literature [15], when the curvature function of the snake robot's trajectory is formula (1), the corresponding joint angle is formula (2).

$$\kappa(s) = -\frac{2K_n \pi A}{L} \sin\left(\frac{2K_n \pi}{L} s\right) \tag{1}$$

Where K_n is the number of S-shape of the robot; A is the amplitude parameter of joint angles; s is the arc length from any point on the curve to point O.

$$q_j = -2A \sin(\frac{K_n \pi}{n}) \sin(\frac{2K_n \pi}{L} s_0 + \frac{2K_n \pi}{n} j) \tag{2}$$

Where, s_0 is the arc length from the head to point O.

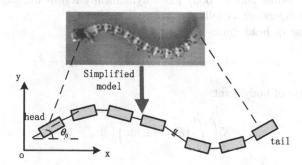

Fig. 1. Simplified model of snake robot

According to literature [16], the locomotion direction can be adjusted by A. When the snake robot tracks a desired path, the direction may be changed at any position. The turning angle is measured by the rotation angle of the head module in a cycle.

$$\varphi = \int_0^{\frac{L}{K_n}} \left[-\frac{2K_n \pi A(s)}{L} \sin(\frac{2K_n \pi}{L} s) \right] ds \tag{3}$$

According to Eq. (3), we can get

$$\varphi = \int_0^{\frac{L}{2K_n}} \left[\frac{2K_n\pi\left(-A(s)+A(s+\frac{L}{2K_n})\right)}{L} \sin(\frac{2K_n\pi}{L}s) \right] ds \qquad (4)$$

Where $L/2K_n$ is the curve length of the trajectory of the head in half cycle. If $A(s) = A(s + L/2K_n)$, the change of A is symmetrical, and $\varphi = 0$. The direction is not changed. If $A(s) \neq A(s + L/2K_n)$, the change of A is asymmetrical, the direction of motion may be changed, and the turning angle is related to the asymmetry degree of A.

According to the above analysis, the symmetry of the joint angle affects the locomotion direction. When the change of the joint angle is symmetrical, the direction is not changed. When snake robot changes the locomotion direction, the change of the joint angle must be asymmetric. When the asymmetry degree of the joint angle is adjusted according to the path edge, the snake robot can be steered along the desired path.

3   Direction Control Method

The direction control method is the basis of the path following. In order to steer the snake robot along the desired path, the snake robot needs to adjust its direction many times. In this section, a direction control method is proposed. The head is used to guide the direction of motion, and the body joints dynamically follow the angle of the head joint. The joint torques are as follows.

Control torque of head joint:

$$\tau_1 = K_1 \cdot (\ddot{q}_d + k_d(\dot{q}_d - \dot{q}_0) + k_p(q_d - q_0) + \delta) \qquad (5)$$

Control torque of body joint:

$$\tau_j = K_j \left(\int_0^t (E_{ref} - E)dt \right) (q_{j-1} - q_j) \qquad (6)$$

$j = 2, \ldots, n-1$. Where δ is a compensating torque, k_d and k_p are two coefficients, $K = [K_1, K_2, \ldots, K_{n-1}]^T$ is a coefficient vector, E_{ref} and E are the expected kinetic energy and the actual kinetic energy [14], respectively. The amplitude of the body torque is adjusted by $\left(\int_0^t (E_{ref} - E)dt \right)$, so that the velocity of the movement cannot be zero. q_d is the reference angle of the head joint. According to Eq. (6), the torque of the jth joint is influenced by the difference between the angle of the jth joint and the $(j$-1)th joint, so the angle of the jth joint must lags behind the $(j$-1)th joint. Due to the continuous swing of the head and the dynamic following of the body joints, the body of the robot appears S- shape, and the serpentine locomotion can be realized.

When $\delta = 0$, there is no path direction information in the torque of the head joint, and the direction of motion is not affected by the path.

When $\delta = (e^{c-a} - 1)q_{d}$, the direction of motion is related to parameter a. The relationship between a and the direction of motion is given as follows. When a is asymmetrical, the angle of the head joint is asymmetric. The body joints follow the head joint in order, so the angle of body joint is asymmetrical. According to the relationship between the joint angles and the direction, the direction of motion may be changed. When a is symmetrical, the change of the head joint is symmetrical. The body joints follow the head joint in order, so the body joints also change symmetrically. When the joint angle changes symmetrically, the direction of motion is not changed.

4 Path Edge Guidance Strategy

The control objective is that the snake robot can track a straight path with unilateral edge, and the distance between the centerline of the snake robot's trajectory and the path edge is h_{e}.

In the torque of the head joint, a is a direction parameter. In order to steer the snake robot along an expected line on the path, the relationship between the torque of the snake robot and the path edge is established by $a(t)$.

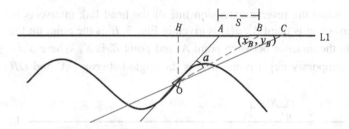

Fig. 2. Schematic diagram of forward temporary target point

Since the trajectory of the snake robot is S-shape, when $t_0 < t < t_0 + T/2$, the forward extension line of the head link intersects the path edge, when $t_0 - T/2 < t < t_0$, the reverse extension line of the head link intersects the path edge.

First, when the forward extension line of the head link intersects the path edge L1, $a(t)$ is shown in Fig. 2. The intersection is point A. B is a point on the edge of the desired path. The distance between point A and point B is S. Point B is called forward temporary target point. S is a constant. O is the mass center of the head link. $a(t)$ is the angle between OA and OB. With the movement of the robot, the forward temporary target point is updated along the edge of the desired path.

The key problem of this method is to obtain point B and angle $a(t)$. As shown in Fig. 2, point C is a point on the edge of the desired path. l_{OA} and l_{OC} can be measured by sensors in real time. $\angle AOC = \chi$. OA is parallel to the head link. According to the cosine formula of triangle, we can get

$$l_{AC} = \sqrt{l_{OA}^2 + l_{OC}^2 - 2l_{OA} \times l_{OC} \times \cos\chi} \tag{7}$$

So, $\angle CAO = \arccos\frac{l_{AC}^2 + l_{OA}^2 - l_{OC}^2}{2l_{AC} \times l_{OA}}$. Then, the distance between point O and point B can be obtained.

$$l_{OB} = \sqrt{l_{OA}^2 + l_{AB}^2 - 2l_{OA}l_{AB}\cos\angle OAB} \tag{8}$$

Therefore, when the straight extension line of head link intersects the path edge L1, $a(t)$ is

$$a(t) = \arccos\frac{l_{OA}^2 + l_{OB}^2 - l_{AB}^2}{2l_{OA} \times l_{OB}} \tag{9}$$

Second, when the reverse extension line of the head link intersects the path edge L1, the intersection is point A'. $a(t)$ is given in Fig. 3. B' is the point on the edge of the desired path, the distance between point A' and point B' is S', Where $S' = \frac{SH'}{2h_d - H'}$. B' is the reverse temporary target point. $a(t)$ is the angle between OA' and OB'.

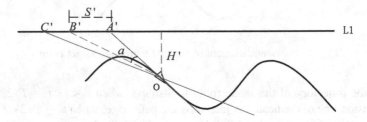

Fig. 3. Schematic diagram of reverse temporary target point

To calculate S', we need to obtain the vertical distance H' between point O and the edge of the path. $l_{OA'}$ and $l_{OC'}$ can be measured by sensors in real time. $\angle A'OC' = \chi$. According to the cosine formula, we can get $\angle A'C'O = \arccos\frac{l_{A'C'}^2 + l_{OC'}^2 - l_{OA'}^2}{2l_{A'C'} \times l_{OC'}}$, where $l_{A'C'} = \sqrt{l_{OA'}^2 + l_{OC'}^2 - 2l_{OA'} \times l_{OC'} \times \cos\chi}$, so $H' = l_{OC'} \times \sin\angle A'C'O$. Then, the distance between point O and point B' can be obtained.

$$l_{OB'} = \sqrt{l_{OA'}^2 + S'^2 - 2l_{OA'}S'\cos(\pi - \chi - \angle OC'A')} \tag{10}$$

Therefore, when the reverse extension line of head link intersects the path edge L1, $a(t)$ is

$$a(t) = \arccos \frac{l^2_{OA'} + l^2_{OB'} - S'^2}{2l_{OA'} \times l_{OB'}} \tag{11}$$

$a(t)$ is the direction control parameter. The relationships between $a(t)$ and the direction of motion are given as follows.

When the trajectory or the extension line of trajectory intersects the edge of the desired path, as shown in Fig. 4, the snake robot cannot follow the desired path. When the snake robot moves along the desired path, but the distance between the centerline of the trajectory and the path edge is not equal to expected value h_e, the trajectory is shown in Fig. 5. It can be proved that $a(t_1) \neq a(t_1 + T)$ in the above two cases, that is, the change of $a(t)$ is asymmetrical. When the change of $a(t)$ is asymmetrical, the direction of motion is changed. With the change of direction, the asymmetry of $a(t)$ decreases until the change of $a(t)$ is symmetrical.

In Fig. 6, the snake robot moves symmetrically along a line on the desired path. The line is parallel to the edge of the desired path, and the distance between the line and the edge of the desired path is h_e. We can get $a(t_1) = a(t_1 + T/2)$. t_1 can be set to any value, so $a(t) = a(t + T/2)$, that is, the change of $a(t)$ is symmetrical. When the change of $a(t)$ is symmetrical, the direction of motion is not changed, so the snake robot can move along this line.

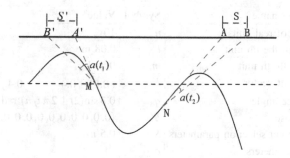

Fig. 4. When the robot does not move along the desired path, the change of a

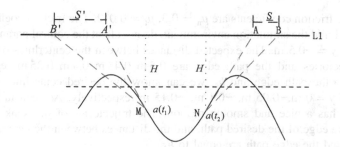

Fig. 5. When the snake robot moves along the desired path, but the distance between the centerline of snake robot's trajectory and the path edge is not equal to h_e, the change of a

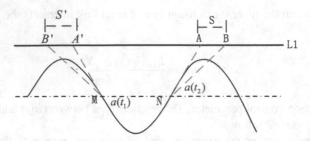

Fig. 6. When the snake robot moves along the desired path, and the distance between the centerline of snake robot's trajectory and the path edge is equal to h_e, the change of a

5 Simulation and Analysis

The simulations are programmed based on Open Dynamics Engine (ODE). The basic fixed parameters of the simulations are presented in Table 1. In the following simulations, the friction coefficient of ground is unknown, and the path edge is not given in advance, but the points on the path edge can be detected in real time. The purpose of simulation is to verify that the snake robot can follow the desired path on the ground with the above conditions, and the distance between the centerline of trajectory and the path edge can be equal to the expected value.

Table 1. Basic parameter of simulation

Parameter name	Symbol	Value
Number of real units	n	10
Length of the ith unit	l	0.08 m
Mass of th ith unit	m_i	0.50 kg
Coefficient vector	K	$(0.01 \quad 0.4 \quad \cdots \quad 0.4)^T$
Reference angle	q_d	$0.3 \sin(2t + 2\pi i/n)$ rad
Initial pose	Ψ_0	$(0,0,0,0,0,0,0,0,0,0,0,0)$
Target point selection parameters	S	0.5 m
Other parameters	c	1

5.1 Trajectory Analysis

In Fig. 7, the friction coefficients are $\mu_n = 0.3$, $\mu_t = 0.012$. The friction coefficients are only used to build the simulation environment, do not affect the control parameters. The path edge is y = −0.5 m. The expected distances between the centerlines of the snake robot's trajectories and the path edge are 0.5 m, 0.45 m, 0.4 m, 0.35 m respectively. According to the path edge and h_e, we can know the desired centerlines of the trajectories are y = 0 m,−0.05 m, −0.1 m, −0.15 m respectively. As seen in Fig. 7, the snake robot has a nice and smooth motion. The trajectories of the snake robot are parallel to the edge of the desired path, and the distances between the centerline of the trajectories and the edge path are equal to h_e.

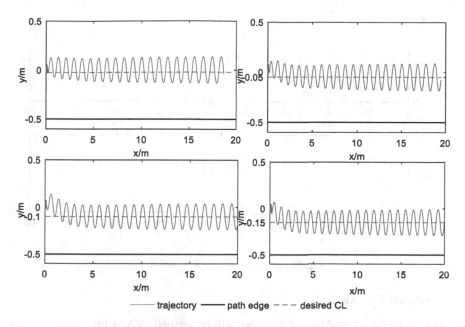

trajectory ——— path edge – – – desired CL

Fig. 7. The trajectories of the snake robot

5.2   Parameter Analysis

In this section, we will take the locomotion process in Fig. 7(d) as an example to analyze the relationship between $a(t)$ and the direction of motion. The change of $a(t)$ in 0 s–20 s and 60 s–80 s are shown in Fig. 8. In 0–20 s, the distance between the centerline of the trajectory and the path edge is not equal to h_e, so $a(t)$ changes asymmetrically, and the direction changes. With the adjustment of direction, the degree of asymmetry gradually decreases. As shown in Fig. 8 and Fig. 7 (d), When $a(t)$ changes symmetrically, the centerline of the trajectory is parallel to the path edge, and the distance between the centerline and the path edge is h_e.

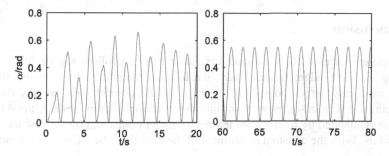

Fig. 8. When the edge of the desired path are $y = -0.5$ m, $h_e = 0.35$ m, the change of $a(t)$

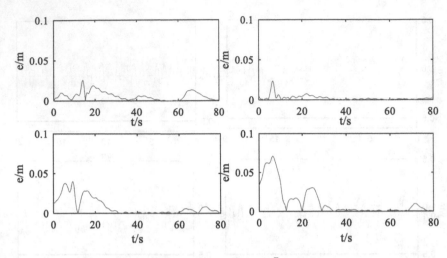

Fig. 9. The error between $\bar{X}$ and X

5.3   Adaptability Analysis

In this section, the adaptability of the robot will be verified. A new line which is called X is given in this section. The distance from the edge of the path to the line X is h_e. $(\overline{x(t)}, \overline{y(t)})$ is the mean value of head link in a period. The trajectory of $(\overline{x(t)}, \overline{y(t)})$ is the centerline of the head trajectory, recorded as $\bar{X}$. The error between $\bar{X}$ and X is $e(t)$.

$$e(t) = \min\{\bar{y} - y(\bar{x}), \bar{x} - x(\bar{y})\} \tag{12}$$

Where $(\bar{x}, \bar{y})$ is a point on $\bar{X}$, $y(\bar{x})$ is the y value when $x = \bar{x}$ on the X, $x(\bar{y})$ is the x value when $y = \bar{y}$ on the X.

We choose the friction coefficients (μ_n, μ_t) of the ground as $(0.4, 0.012)$ $(0.3, 0.012)(0.2, 0.012)(0.1, 0.012)$, respectively. The errors between $\bar{X}$ and X are shown in Fig. 9. According to the errors, we obtain that the distance between the centerline of trajectory and the path edge can meet the expected value.

6   Conclusion

The purpose of this study is to steer the snake robot to follow the straight path by detecting one path edge. In this study, the forward temporary target points and the reverse temporary target points on the edge of the desired path are detected. By introducing the temporary target point into the direction control method, the direction of motion is adjusted by the forward temporary target points and the reverse temporary target points. With the constant adjustment of the direction, the snake robot can follow the desired path. The reverse temporary target points are changed according to the reference distance between the centerline of the snake robot's trajectory and the path edge, so the distance between the centerline of the snake robot's trajectory and the path edge can meet the expected value.

In future work, the author will establish an experimental platform to verify the effectiveness of the method, and improve the method so that the minimum and maximum distance between the trajectory and the path edge can be adjusted according to requirements.

Acknowledgements. This work is supported by PhD Research Startup Foundation of Liaoning Shihua University 2016XJJ-021.

References

1. Zhang, A.F., Ma, S.G., Li, B., Wang, M., Guo, X., Wang, Y.: Adaptive controller design for underwater snake robot with unmatched uncertainties. Sci. China Inf. Sci. **59**(5), 1–15 (2016)
2. Malayjerdi, M., Akbarzadeh, A.: Analytical modeling of a 3-D snake robot based on sidewinding locomotion. Int. J. Dyn. Control, 1–11 (2018)
3. Guo, X., Wang, M.H., Li, B., Ma, S., Wang, Y.: Optimal torque control of a snake-like robot based on the minimum infinity norm. Robot **36**(1), 8–12 (2014)
4. Liljeback, P., Pettersen, K.Y., Stavdahl, Ø., Gravdahl, J.T.: A simplified model of planar snake robot locomotion. In: IEEE/RSJ International Conference on Intelligent Robots & Systems, pp. 2868–2875. IEEE (2010)
5. Liljeback, P., Pettersen, K.Y.: Waypoint guidance control of snake robots. In: IEEE International Conference on Robotics and Automation, pp. 937–944 (2011)
6. Liljeback, P., Haugstuen, I.U., Pettersen, K.Y.: Path following control of planar snake robots using a cascaded approach. IEEE Trans. Control Syst. Technol. **20**(1), 111–126 (2012)
7. Kelasidi, E., Pettersen, K.Y., Kohl, A.M., Gravdahl, J.T.: Planar path following of underwater snake robots in the presence of ocean currents. IEEE Robot. Autom. Lett. **1**(1), 383–390 (2017)
8. Rezapour, E., Liljebäck, P.: Path following control of a planar snake robot with an exponentially stabilizing joint control law. IFAC Proc. Volumes **46**(10), 28–35 (2013)
9. Kelasidi, E., Liljeback, P., Pettersen, K.Y., Gravdahl, J.T.: Integral line-of-sight guidance for path following control of underwater snake robots: theory and experiments. IEEE Trans. Robot. 1–19 (2017)
10. Hasanabadi, E., Mahjoob, M.J.: Trajectory tracking of a planar snake robot using camera feedback. In: International Conference on Control, Instrumentation and Automation, Shiraz, Iran, pp. 894–897. IEEE (2011)
11. Sato, H., Tanaka, M., Matsuno, F.: Trajectory tracking control of snake robots based on dynamic model. In: IEEE International Conference on Robotics & Automation, pp. 3029–3034 (2005)

318   D. Zhang

12. Mohammadi, A., Rezapour, E., Maggiore, M., Pettersen, K.Y.: Maneuvering control of planar snake robots using virtual holonomic constraints. IEEE Trans. Control Syst. Technol. **24**(3), 884–899 (2016)
13. Ariizumi, R., Takahashi, R., Tanaka, M., Asai, T.: Head-trajectory-tracking control of a snake robot and its robustness under actuator failure. IEEE Trans. Control Syst. Technol. 1–9 (2018)
14. Zhang, D.F., Li, B., Chang, J.: The path following method of a snake robot based on the angle symmetry adjustment. Robot. **41**(6), 788–794,833 (2019)
15. Ma, S.G.: Analysis of creeping locomotion of a snake-like robot. Adv. Robot. **15**(2), 205–224 (2001)
16. Ye, C.L.: Mechanism Design and Locomotion Control of Snake-like Robots. Shenyang Institute of Automation, Chinese Academy of Science, Shenyang (2005)

A Novel Method on Probability Evaluation of ZC Handover Scenario Based on SMC

Jia Huang, Jidong Lv$^{(\boxtimes)}$, Yu Feng, Zhengwei Luo, Hongjie Liu,
and Ming Chai

Beijing Jiaotong University, Beijing, China
jdlv@bjtu.edu.cn

Abstract. Zone Controller (ZC) handover scenario is a typical operation function in Communication Based Train Control (CBTC) system. However, due to the nondeterministic communication between the onboard equipment and the ZC, the delay behavior and timeout behavior lead to the complexity of probability evaluation. In this paper, a novel method on probability evaluation of ZC handover scenario has been proposed based on Statistical Model Checking (SMC), which introduces a sequential operator to evaluate the probabilities of all scenarios in CBTC system. In CBTC system, different scenarios have different behaviors and probabilities. Therefore, in ZC handover process, the trigger handover, crossing the demarcation point and logout switching scenarios are modeled by Network Priced timed automata (NPTA) and the whole probability has been evaluated. The result with successful handover is 0.99985, shows that the probability of successful ZC handover is high, which meets the security requirements of CBTC system.

Keywords: ZC handover · SMC · Probability

1 Introduction

As the core control equipment of in CBTC system, ZC mainly responses for computing Movement Authority (MA) within its control area and sending it to the Vehicle On-Board Controller (VOBC) through wireless communication equipment to ensure the safe operation of trains [1]. Duo to the limitation of the control areas, the handover process between two adjacent ZCs is a typical operation scenario to ensure the train safety running from one control area to another. However, since the wireless transmission communication between two different adjacent ZCs may have stochastic nondeterministic behavior, such as delay, timeout or failure and so on, it is important to evaluate the probability of ZC handover in CBTC system [2].

The current research on ZC handover is based on formal modeling and verification. Wang applies the Unified Modeling Language (UML) to model the generation process of ZC in different operational scenarios, and visually demonstrate the structure of the system [3]. In [4], according to the characteristics of the ZC subsystem, the Message Sequence Processes model and the time automaton network model of the ZC handover scenario have been established for the ZC handover scenario function, which mainly focus on one scenario of ZC handover security. Huang defines translations rules of the

© Springer Nature Singapore Pte Ltd. 2020
J. Qian et al. (Eds.): ICRRI 2020, CCIS 1335, pp. 319–333, 2020.
https://doi.org/10.1007/978-981-33-4929-2_22

translations from UML sequence diagram model to linear hybrid automaton model to verify the security of ZC handover in CBTC system [5]. However, it mainly considers the hybrid characteristics not the randomness in communication between ZC and Onboard subsystems. Li uses SCADE to model the ZC handover scenario, in which the controller's train management function and MA calculation function are verified [6]. One good result is [7], in which the concurrent graph theory and state space, the colored Petri net are combined and used to perform layered verification on the ZC subsystem, and the result that the system can be safely handover in the worst state is concluded. None of the above research can overall estimate of the correctness of the ZC handover design, therefore they can also not get overall evaluation property.

Recently, based on NPTA, SMC method is widely used to solve the probability evaluation problems of safety-critical system. One is qualitative analysis through hypothesis testing. It solves whether the probability of satisfying an attribute in NPTA random operation is greater than a certain threshold. The other is quantitative analysis [8], which solves the probability of satisfying an attribute in NPTA random operation.

In this paper, in order to evaluate the probabilities of ZC handover scenarios, a novel method on probability evaluation of ZC handover scenario has been proposed based on SMC, in which the main difference from [9] is that we introduce a sequential operator to evaluate all the scenarios probability of CBTC system. Therefore, according to the ZC handover process, the trigger handover, crossing the demarcation point and logout switching scenarios are modeled by NPTA. The different stages of ZC handover, the probability of successful handover and the consequences of handover failure have been analyzed, and optimization suggestions for MA update time have been proposed. The results show that the probability of successful ZC handover is high, which meets the security requirements of CBTC system.

2   Method

2.1   Priced Timed Automata Model of CBTC

PTAs are derived from time automata [10, 11], whose clocks can be converted to different ratios in different states [12]. Its manifestation is as follows: $C' == N(N \in R \geq 0)$ and the clock has no restrictions on guards and invariants. The stochastic semantics of PTA associate probability distribution with delays in a given state and transitions between states. NPTA consists of a set of PTAs [13]. The time delays in a given state show uniform bounded distribution and unbounded exponential distribution. The PTAs communicate with each other through broadcast channels and shared variables to generate NPTA [14]. Here we can use UPPAAL to model the stochastic behavior of CBTC system.

The stochastic priced timed automata model of CBTC is defined as a tuple $<P, P_0, S, S_i, S_f, A_{tr}, E_{tr} \rightarrow, I>$, For simplify, we only give the syntax of our stochastic priced timed automata model.

- P is the NPTA model of each scenario of CBTC
- P_0 is the initial NPTA model of each scenario of CBTC
- S the states of each scenario NPTA of CBTC

- S_i the initial state of NPTA model of each scenario of CBTC
- S_f is the final state of NPTA model of each scenario of CBTC
- A_{tr} the trigger handover actions, which can trigger from one scenario NPTA model final state S_f to the next scenario NPTA model initial state S_i
- E_{tr} a set of edges between NPTA model P with an action E_{tr}, a guard and invariants to I
- $\rightarrow$ the sequential operator, which denotes the sequential behavior of the two different PTA scenarios of CBTC system when E_{tr} and I are satisfied
- I is the set of invariants of NPTA models

2.2   The Algorithm Stochastic Model Checking of CBTC System

In order to analyze the probability of CBTC, we use SMC technique to get an overall estimate of the correctness of the design. The SMC is based on priced timed automaton model that can be seen as a balance between test and formal verification. It is mainly to simulate the system multiple times and verify that they meet certain specific properties. The result is then used with a statistical algorithm to determine if the system satisfies the attribute with some possibility. SMC can also be used to estimate the probability that a system will satisfy a given attribute. Of course, SMC cannot guarantee 100% confidence compared to detailed statistical methods [15]. However, SMC can limit the probability of errors.

Here, we stick to the algorithm defined in [9]. We use the form $\psi = < > C \leq c\varphi$ to represent an attribute of NPTA of CBTC system during modelling operation scenarios, where C represents an observation clock that will not be reset, $c \in R \geq 0$, and φ represents state predication. When we define NPTA as M, the probability that M runs randomly satisfies ψ is defined as $P_M(\psi)$. Parameters α and β are a set of probabilities used to describe two types of errors in hypothesis testing. The first type of error is a false positive error, that is, the original hypothesis is true, but we reject it, and its probability of occurrence is recorded as α. The second type of error is a false error, that is, the original hypothesis is not correct, but we accept it, and its probability of occurrence is β.

Our SMC algorithm performs a sufficient number of simulations of the CBTC system model at a predetermined level of significance to obtain statistical evidence to check the quantitative nature. The SMC calculates the number of runs N, resulting in a similar interval $[\rho - \varepsilon, \rho + \varepsilon]$ with a confidence of $1 - \alpha$ and a probability of $p = P_r$, where ε and α are defined by the typical CBTC operation scenarios. The runs of the value of N depends on the Chernoff-Hoeffding algorithm [15, 16]. The specific algorithm is shown in Table 1:

Table 1. Algorithm.

Algorithm 1:

Function estimation (P: Scenario Models, p: Scenario model, S: State, S_f :final state,

ψ:attribute,ε:approximation,δ:confidence)

$Pro(P)$

  Choose one operation scenario p,

   If S in p is not ψ

    If P_0 is not in S_f,then choose ε and α, then computes: $Pro(P)$

$$N:= 4 * log(1/\delta)/\varepsilon^2, a:= 0$$

$$for \ i:= 1 \ to \ N \ do$$

        Determine whether each random run x satisfies the attribute ψ

$$a:= a + x$$
      end
      Return $Pro(p):= Pro(p) * a/N$
    else $p:= p \rightarrow$
  else return $Pro(p)$
End

2.3 The Modelling, Verification and Probability Analysis in UPPAAL SMC

UPPAAL can implement model validation through Computation Tree Logic (CTL) language. CTL consists of path formula and state formula. The former quantifies the path of the model, while the latter describes the state. The path expression can classify the validation properties into accessibility, safety and existence. Specifically expressed as:

- $A[]P$: For any reachable state, P is always true
- $E[]P$: There is a path that makes P hold all the time
- $E< >P$: There is a path so that P is finally established
- $A< >P$: For all paths, P is finally established
- $P \rightarrow q$: Only P is established, then q is established

UPPAAL SMC supports checking the quantitative attributes of the model on the basis of UPPAAL. The probability that model M satisfies attribute is expressed by expression $P_M(< >C \leq c\varphi)$. In addition, UPPAAL SMC supports calculating the maximum and minimum values of expressions about clocks or integer values. It is described by the formula $E[bound; N](max(min) : expr)]$. There are three kinds of bound, which are: (1) $\leq n$ specified implicit time limit; (2) $x \leq n$ specified display limits, where x represents clocks; (3) $\# \leq n$ specified discrete steps. N denotes the number of runs and expr denotes the expression to be evaluated [15].

3   The ZC Handover Scenario of CBTC

In CBTC system, each ZC has a certain jurisdiction, and the vehicle can only receive the movement authority within its jurisdiction. When there are multiple ZC sections on the line and train needs to run from one ZC control area to another, VOBC and another ZC are needed to cooperate to complete the train switching process [17, 18]. In the process of handover, when the end point of MA is ZC boundary. The handover ZC requests to takeover ZC to calculate MA for train. The handover process of ZC is completed by VOBC and two ZCs together. According to the sequence of train crossing ZC boundary points in the process of handover, it is divided into three processes: trigger handover, crossing the border and handover cancellation. These three processes will be analyzed and introduced in the following sections [19].

3.1   ZC Triggered Handover

During the normal operation of trains, When the end point of MA provided by the handover ZC for the train is the boundary of the jurisdiction of it. Handover ZC checks the conditions of the handover. When the handover ZC confirms that the train meets the triggered handover conditions, it begins to communicate with the handover ZC and send the running message of the train to be switched and the MA1 information generated for it to the takeover ZC. Takeover ZC generates MA2 information based on the information provided by the handover ZC and track occupancy, as well as other information of the line, and sends MA2 information to the handover ZC. Handover ZC sends mixed MA information to train by integrating MA1 information and MA2 information.

3.2   Crossing Demarcation Point

VOBC runs according to the received MA that is generated by handover ZC, until the head of the train crosses the demarcation point for the handover of ZC. VOBC begins to communicate with handover ZC, sends the train position to take over ZC and applies for MA. At the same time, it retains communication information with the handover ZC. Takeover ZC generates mixed MA based on MA1 generated by handover ZC and MA2 generated by itself. When VOBC registers successfully and receives the mixed MA information generated by the takeover ZC, VOBC begins to be controlled by takeover ZC.

3.3   Logout Switching

VOBC operates according to the takeover ZC's MA. When the train tail crosses the ZC demarcation point of the handover control range, VOBC cancelled and the line resources occupied in the handover ZC are released. The ZC handover process is shown in Fig. 1.

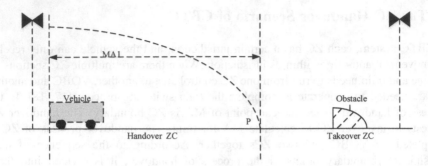

Fig. 1. ZC handover process

4 The Modeling and Probability Evaluation of ZC Handover

In this section, the SMC method is used to establish the NPTA model of trigger handover, crossing the border and handover cancellation process of ZC. And the handover process is integrated and modeled, and the model is verified by simulation and probability calculation.

The NPTA of the trigger handover, the NPTA of crossing the demarcation point and the NPTA of the logout switching are regarded as three processes, which are represented by *TRIGGER*, *CROSSE*, and *LOGOUT* respectively. The handover process is triggered by different positions, that is, the VOBC receives the MA that reaches the demarcation to trigger the process of trigger handover, and the front of the train crosses the demarcation point to trigger the process of crossing the demarcation point, and the rear of the train crosses the demarcation point to trigger the process of logout switching. Therefore, the location information is used as an event that triggers the process, which is represented by *boundary*, *front*, and *after*, respectively, expressed in sequential operator as follows: $NPTA(TRIGGER) \rightarrow boundary \rightarrow NPTA(CROSSE)$ $front \rightarrow NPTA(LOGOUT) \rightarrow after \rightarrow END$.

It should be noted here that if the train does not receive the updated MA within 5−6 s, emergency braking will be carried out. In this paper, the time for emergency braking of trains is set to be 6 s. In CBTC system, three handshakes are needed for vehicle-ground communication and ZCs communication, and the time of each handshake is set to 1 s. In ZC handover, the probability of each process is referenced to the value in [13]. In the process of building the model, the distance between the ZC demarcation point and the initial position of the train equipped with VOBC equipment is referred to as [5], the value is 700 m, and the maximum speed of the train is 80 km/s. At the same time, in the process of establishing the model, only the probability at the time of transmission is considered, and reception is not considered. When evaluating the probability of NPTA model, the parameters are set to $(-\delta, +\delta) : 0.001$, $(\alpha, \beta) : 0.005$, $\varepsilon : 5.0^{-5}$ [13].

4.1 Trigger Handover Modeling and Verification

In trigger handover process, four PTAs, namely handover ZC, takeover ZC VOBC and result, are established to describe the whole process of trigger handover and to evaluate the handover probability.

Modeling of Trigger Handover. As shown in Fig. 2, When the VOBC receives the *MA1*, it sends the *ApplyMA* to the handover ZC by probability selection. When the handover ZC receives the *ApplyMA*, the *TakeoverMA* is sent to the takeover ZC by probability selection. When the handover ZC receives the *TakeoverMA*, it generates *MA2* according to the current line condition, and sends the *MA2* to the handover ZC by probability selection. After the handover ZC receives the *MA2*, it integrates *MA1* and *MA2* to generate the *MixMA*, and sends it to the train through probability selection. After receiving the *MixMA*, the train sends the *PositionMessage* message. When the vehicle crosses the boundary point, the next phase of the trigger is triggered. If it is not reached, the handover ZC system sends an *ApplyMessage* to the takeover ZC system, and the process is repeated until the front of the car crossed the switching point. It should be noted that during the information exchange process, if the three-way handshake fails, the stop command will be sent and the train will stop. And if the information is not received for more than 6 s during the communication, it will reach the *Stop* state. The main information interaction process in the trigger switching phase is shown in Fig. 3.

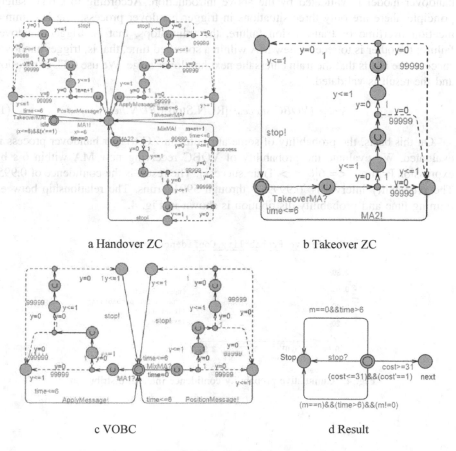

a Handover ZC   b Takeover ZC

c VOBC   d Result

Fig. 2. Trigger handover

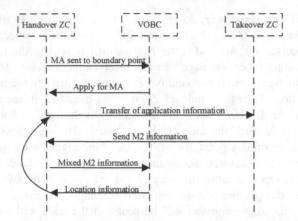

Fig. 3. Trigger handover process information interaction

Trigger Handover ZC Verification and Probability Computation. The trigger handover model is validated by the above introduction. According to CBTC safety principle, there are only three situations in trigger handover process: one is communication overtime or transmission failure, the train stops, that is, trigger handover failure; the other is to receive new MA within a specified time, that is, trigger handover success; the last is that the train enters the next handover stage. We use CTL expression and the result is validated.

$$A <> (VOBC.success||\text{RUN}.Stop\ ||RUN.next) \tag{1}$$

On this basis, the probability of some attributes in ZC trigger handover process is evaluated. We evaluate the probability of VOBC receiving new MA within 6 s by expression $Pr[\text{time} <= 6](<> \text{Train.success})$. The result is the confidence of 0.995. The probability interval is [0.9999, 1] through 59912 runs. The relationship between running time and probability distribution is shown in Fig. 4.

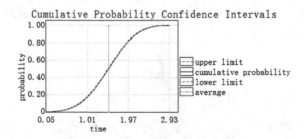

Fig. 4. Cumulative probability confidence interval distribution

Similarly, the probability of switching failure is evaluated by expression: $\Pr[< \ = 6]$ ($< \ >$ Train.Stop). The result is the confidence of 0.995, and the probability interval is [0,9.99994e−005] by 59912 runs. According to the expression: $\Pr[< \ = 31]$ ($< \ >$ RUN.next) evaluate the probability of reaching the next state within 31 s. The result is the confidence of 0.995, and the probability interval is [0.9999, 1].

Based on the above evaluation of the probability of trigger handover process, we have reason to believe that ZC handover has high reliability and accuracy without deceleration.

Finally, we use the expression: $E[< \ = 12; 10000](max : RUN.x)$ to evaluate the time of the highest success rate between the first vehicle-ground communication, and the result is $E(\max) = 11.6639$, that is, the time when the train receives the new MA for the first time in the trigger handover phase is 5.6639.

4.2 Crossing Demarcation Point Modeling and Verification

When the front of the train crosses the demarcation point, the next sub-process of the handoff process will be triggered. In this process, the final state of the handover ZC, the takeover ZC, the VOBC, and crossing the demarcation point are modeled separately. Probabilistic assessment is achieved through different state transitions and information interactions.

Modeling of Crossing Demarcation Point. As can be seen from Fig. 5, when the train crosses the demarcation point, the VOBC sends the location information to the handover ZC and the takeover ZC and applies to the takeover ZC for entry control. Since this process is a trigger condition, the three handshake is not considered. After that, the takeover ZC generates the *MixMA* and sends it to the VOBC according to the *MA1* of the handover ZC and the route conditions. If the VOBC does not receive the mixed MA for more than 6 s, emergency braking will be performed. If VOBC receives the *MixMA*, it sends a *PositionMessage* to the handover ZC and the takeover ZC. After receiving the *PositionMessage*, the takeover ZC sends the takeover message *Completetakeover* to the handover ZC. After the handover ZC receives the message, it sends the updated *UpdateMA1* to the takeover ZC. After receiving the *UpdateMA1*, the takeover ZC updates the *MixMA* information and the rear of the train passes over the demarcation point to trigger the next stage. The specific information interaction is shown in Fig. 6.

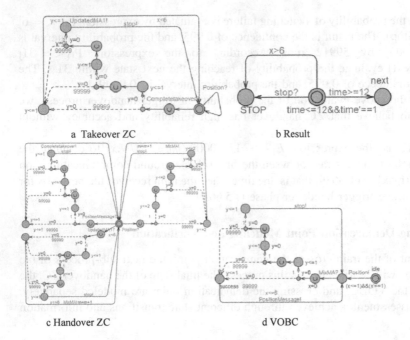

a Takeover ZC   b Result

c Handover ZC   d VOBC

Fig. 5. Crossing the demarcation point

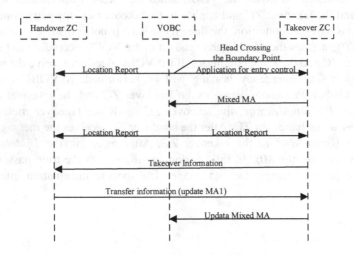

Fig. 6. Information interaction of crossing the demarcation point

Verification and Probability Calculation of Crossing Demarcation Point. In the NPTA model of crossing the demarcation point, the final state is the same as the triggering handover, that is, the handover succeeds, the handover fails, and the next state is successfully entered. The existence of the model is verified by the following CTL expression, and the verification result is passed.

$$A < > (\text{Result.next} \parallel \text{Result.STOP} \parallel \text{VOBC.success}) \tag{2}$$

On this basis, the expression $\text{Pr}[< = 6](< > \text{Train.success})$ is used to evaluate the probability that the train successfully receives a new MA within 6 s. At a confidence level of 0.995, the result is that the probability interval is obtained by running 59,912 times. The relationship between running time, probability distribution and frequency is shown in Fig. 7.

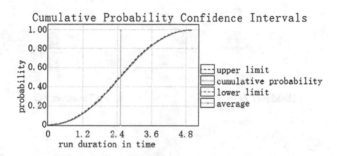

Fig. 7. The relationship of probability distribution

In addition, the probability of a switch failure can be evaluated by the expression $\text{Pr}[< = 6](< > \text{Result.STOP})$. The result is that under the suppose of 0.995 confidence, the probability interval [0,9.99994e−005] s obtained by running 59,912 times. According to the expression $\text{Pr}[< = 12](< > \text{Result.next})$, we can evaluate the probability of reaching the next state within 12 s. The result is that under the premise of 0.995 confidence, the probability interval [0.9999, 1] is obtained by running 59,912 times. Finally, by the expression $E[< = 6; 10000](\text{max:Result.time})$, we can evaluate the maximum probability that VOBC will receive the new MA for the first time at this stage. And the result is $E(\text{max}) = 5.66872$.

4.3 Logout Switching Modeling and Verification

After the rear of the train passes over the boundary point of the handover ZC, VOBC will apply for cancellation of the communication with the handover ZC. In this process, the final states of the handover ZC, takeover ZC, VOBC, and logout handover are modeled separately. The existence verification of the logout switching process and the evaluation of the probability of a certain attribute are realized by information interaction between different PTA models.

Logout Switching Modeling. It can be seen from Fig. 6 that when the rear of train crosses the demarcation point, the VOBC sends a logout request to the handover ZC to apply for *LogoutMessage*, and at the same time, it takes a position report to the takeover ZC. After receiving the position report, the takeover ZC sends *Confirm-Takeover* message to the handover ZC. The handover ZC sends *ConfirmLogout* to the VOBC, and if the VOBC does not receive it, the handover ZC directly clears the train data. The VOBC sends Position to the takeover ZC. The takeover ZC generates a new MA1 based on the location information. The specific information interaction is shown in Fig. 9.

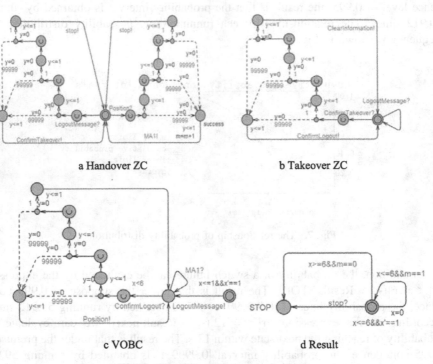

a Handover ZC   b Takeover ZC

c VOBC   d Result

Fig. 8. Modeling of logout switching

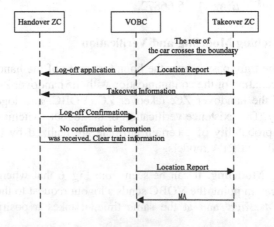

Fig. 9. Information interaction diagram of logout switching

Verification and Probability Calculation of Logout Switching. Through the scenario analysis of the logout switching, there are only two final states in the logout switching process, that is, the handover is successful and a new MA is received within 6 s and no new MA is received within 6 s. The existence of the model is verified by the following CTL expression, and the verification result is passed.

$$A < > (\text{HandOverZC.success} \parallel \text{Result.STOP}) \tag{3}$$

Then, the probability of receiving a new MA within 6 s can be evaluated by the expression $\Pr[< = 6](< > \text{Result.STOP})$. The result is that under the premise of a confidence of 0.995, the probability interval [0.9999, 1] is obtained by running 59,912 times.

In addition, the expression $\Pr[< = 6](< > \text{Result.STOP})$ is used to evaluate the probability that the train does not receive a new MA within 6 s. At a confidence level of 0.995, the result is that the probability interval [0,9.99994e−005] is obtained by running 59,912 times. According to the expression $E[< = 6; 10000](\text{max:Result.time})$, we can evaluate the maximum probability that VOBC will receive the new MA for the first time at this stage. And the result is $E(\text{max}) = 4.6549$.

4.4 Probability Evaluation of ZC Handover

Finally, the overall probability is evaluated by the expression $P = \prod_{n}^{i=1} p_i$, where p_i represents the probability of the process i executing sequentially in a sequential process, i represents the number of processes executing sequentially, and P represents the total probability of sequential processes.

The probability of successful ZC handover can be evaluated by the expression:

$$P_{\text{ZC}} = P_{\text{TRIGGER}} * P_{\text{CROSSE}} * P_{\text{LOGOUT}} \tag{4}$$

In the foregoing, the SMC models have been built for the three processes of ZC handover, and the probability of each process is evaluated by the SMC models. We can conclude that the probability of successful ZC handover is: $P_{zc} = 0.99995 * 0.99995 * 0.99995 = 0.99985$.

5 Conclusion

This paper focus on probability evaluation of ZC handover in CBTC system due to the wireless transmission communication between two different adjacent ZCs, which may have stochastic nondeterministic behavior, such as delay, timeout or failure and so on. We propose a novel method on probability evaluation of ZC handover scenario based on SMC to evaluate all the scenarios probability of CBTC system. According to the ZC handover process, the trigger handover, crossing the demarcation point and logout switching scenarios are modeled by NPTA. The different stages of ZC handover, the probability of successful handover and the consequences of handover failure have been analyzed, and optimization suggestions for MA update time have been proposed. As the successful handover is 0.99985, we can conclude that the probability of successful ZC handover meets the security requirements of CBTC system.

Acknowledgments. The research work reported here was supported by "the Fundamental Research Funds for the Central Universities" (2019JBM009).

References

1. Zhang, L.: Software development of CBTC area controller based on scade. Beijing Jiaotong University (2010)
2. Zhu, L., Yu, F.R., Ning, B., et al.: Cross-layer handoff design in MIMO-enabled WLANs for communication-based train control (CBTC) systems. IEEE J. Sel. Areas Commun. **30**(4), 719–728 (2012)
3. Wang, L., Wang, C.L.: Unified modeling language (UML) modeling and verification of regional controller mobile authorization. Urban Rail Transit Res. **07**, 54–57 (2014)
4. Yang, L., Chen, Y.G.: Scene modeling and verification of regional controller switching based on MSC and UPPAAL. Railway Stan. Des. **5**, 171–179 (2018)
5. Huang, Y.N., et al.: Modeling and verification method of urban rail transit ZC subsystem based on hybrid automata. China Railway Sci. **37**(2), 114–121 (2016)
6. Li, R.: SCADE-based CBTC regional controller modeling and verification. Southwest Jiaotong University (2015)
7. Zhu, D.: Modeling of regional controller subsystem switching function based on colored petri nets. Beijing Jiaotong University (2008)
8. Bulychev, P., et al.: UPPAAL-SMC: Statistical model checking for priced timed automata. In: Electronic Proceedings in Theoretical Computer Science, 85(Proc. QAPL 2012), pp. 1–16 (2012)
9. Bulychev, P., David, A., Guldstrand Larsen, K., Legay, A., Mikučionis, M., Bøgsted Poulsen, D.: Checking and distributing statistical model checking. In: Goodloe, A.E., Person, S. (eds.) NFM 2012. LNCS, vol. 7226, pp. 449–463. Springer, Heidelberg (2012). https://doi.org/10.1007/978-3-642-28891-3_39
10. David, A., Larsen, K.G., Legay, A., Mikučionis, M., Poulsen, D.B.: UPPAAL SMC tutorial. Int. J. Softw. Tools Technol. Transfer **17**(4), 397–415 (2015). https://doi.org/10.1007/s10009-014-0361-y
11. Legay, A., Delahaye, B., Bensalem, S.: Statistical model checking: an overview. In: Barringer, H., et al. (eds.) RV 2010. LNCS, vol. 6418, pp. 122–135. Springer, Heidelberg (2010). https://doi.org/10.1007/978-3-642-16612-9_11
12. Kwiatkowska, M., Norman, G., Parker, D.: PRISM 2.0: a tool for probabilistic model checking. In: First International Conference on the Quantitative Evaluation of Systems, 2004. QEST 2004. Proceedings, pp. 322–323. IEEE (2004)
13. Basile, D., ter Beek, M.H., Ciancia, V.: Statistical model checking of a moving block railway signalling scenario with UPPAAL SMC. In: Margaria, T., Steffen, B. (eds.) ISoLA 2018. LNCS, vol. 11245, pp. 372–391. Springer, Cham (2018). https://doi.org/10.1007/978-3-030-03421-4_24
14. Guldstrand Larsen, K.: Priced timed automata and statistical model checking. In: Johnsen, E. B., Petre, L. (eds.) IFM 2013. LNCS, vol. 7940, pp. 154–161. Springer, Heidelberg (2013). https://doi.org/10.1007/978-3-642-38613-8_11

15. Hoeffding, W.: Probability inequalities for sums of bounded random variables. J. Am. Stat. Assoc. **58**(301), 13–30 (1963)
16. Pearson, C.J.C.S.: The use of confidence or fiducial limits illustrated in the case of the binomial. Biometrika **26**(4), 404–413 (1934)
17. Gao, C.H.H.: CBTC system based on communication. China Railway Science (2018)
18. Lu, L.: Talking about the technology of cross district switching. Guide Becoming Rich Sci. Technol. **23**, 63–63 (2010)
19. Park, H.-D., Lee, K.-W., Lee, S.-H., Cho, Y.-Z., An, Y.-Y., Kim, D.-H.: Fast IP handover for multimedia services in wireless train networks. In: Chong, I., Kawahara, K. (eds.) ICOIN 2006. LNCS, vol. 3961, pp. 102–111. Springer, Heidelberg (2006). https://doi.org/10.1007/11919568_11

Smart Remanufacturing and Industrial Intelligence

Dynamic Adaptive Impedance Matching in Magnetic Coupling Resonant Wireless Power Transfer System

Qiang Zhao$^{(\boxtimes)}$ ⓘ and Chang Cui

School of Information and Control Engineering, Liaoning Shihua University,
Fushun 113001, China
lnshzq@126.com

Abstract. The circuit and environment of Wireless Power Transfer (WPT) system are always complex, which makes it difficult to design the magnetic coupler and reduce the transmission efficiency. Recently, a few advanced matching strategies and structures have been able to improve transmission efficiency in the WPT system when the magnetic coupling system is detuning. But the current compensation and matching system suffers from difficulties in optimization and generalization. To address this issue, in the present work, an external coupling modeling and adaptive dynamic impedance matching method are applied to the energy transmission optimization for WPT system. Firstly, an external coupling network modeling method is designed in a four-coil magnetic coupling resonance system, and a scattering parameter expression method of T-type network transmission efficiency is defined. Secondly, we add frequency independent compensation reactance to the circuit, and use external coupling coefficient to match the impedance of the system to eliminate the detuning and improve the transmission efficiency. Tertiary, the frequency independent impedance is realized by using controllable capacitance, and the value range of compensation capacitance is determined. Finally, experiments and simulations show that the optimized system has good impedance matching and transmission efficiency.

Keywords: Wireless energy transfer · Impedance matching · External coupling · Scattering parameters · Dynamic compensation

1 Introduction

Wireless power transfer (WPT) mainly realizes the energy transmission through non-contact way, which is a new mode for electrical equipment to obtain electric energy from the fixed grid system [1–3]. Its appearance has completely changed the mode that people only rely on the contact type electric energy conduction mode of electrical equipment for hundreds of years. It is also a hot topic of human research. At present, wireless power supply technology has been widely used in many fields, such as household appliances, convenient equipment, vehicles, aerospace, medical devices, industrial robots, oil fields and mines, underwater operations, etc. [4–6]. Compared with the traditional power supply technology, wireless energy transmission technology has the advantages of high

© Springer Nature Singapore Pte Ltd. 2020
J. Qian et al. (Eds.): ICRRI 2020, CCIS 1335, pp. 337–352, 2020.
https://doi.org/10.1007/978-981-33-4929-2_23

security, stable operation, low failure rate, suitable for various environments, conducive to the standardization of the interface, easy to achieve automation and unmanned operation, convenient maintenance and other advantages [7].

WPT technology is the most active research direction in the field of electrical engineering at present. It is an interdisciplinary research field integrating electromagnetic field, power electronic technology, physics, materials science, circuit and other disciplines. It has a broad development prospect [8]. The research of this subject has important scientific significance, high practical value and broad application prospect, which can bring huge economic and social benefits. Therefore, WPT will become the latest significant research direction in the field of modern industrial automation, which is important for accelerating the development of high-end equipment manufacturing industry, energy conservation and environmental protection, promoting industrial upgrading Meaning [9].

With the in-depth study of the magnetic coupling resonance WPT technology, there are still many problems in the development process of the technology, including coupling coefficient [10], frequency stability in the process of energy transmission and frequency bifurcation phenomenon [11], system structure optimization [12], multi-resonance network compensation [13], electro-magnetic compatibility [14], impedance matching and transmission power [15, 16], which are urgent to be solved.

For the WPT system, the influence of circuit parameters on the resonance frequency of the system and how to solve the influence caused by the relative position change between coils are all the problems faced by the research on the mechanism of magnetic coupling resonance WPT. It mainly involves resonance compensation, frequency splitting, impedance matching, dislocation coupling and other key issues [17]. Impedance matching is the core problem in the magnetic coupling resonance system. The research on system structure, transmission characteristics and frequency splitting are focused on impedance matching [18]. In the radio energy transmission system, the reflection impedance, the parasitic resistance of the coil, the source impedance and the load impedance constitute a complex impedance network, which has an important impact on the transmission characteristics of the system [19].

In order to improve the transmission power of the system, a new matching structure is proposed in [20]. In the transmitter, the series/parallel capacitor matrix can be dynamically matched by changing the combination matrix impedance of the series and parallel capacitors of the capacitor Tracking the best impedance matching point at different distances. In [21], according to the load change of multi coil radio energy transmission system, the system network is impedance matched by capacitance. The influence of different matching structures on the system performance is analyzed when the load and cross coupling relationship change. In [22], multi capacitor topology is adopted to reduce the influence of distance variation on the input impedance of the system. According to the input return loss, an adjustable impedance matching circuit and an optimization algorithm are designed. The experimental results show that the designed impedance matching circuit can achieve high-efficiency power transmission when the transmission distance changes in a large range. Multiple impedance matching networks of shunt switchable capacitors are added to the transmitter, and the coupling relationship is adjusted by adjusting each shunt capacitor. The performance of multi load power transmission is improved in [23]. A new adaptive impedance matching network is proposed in [24], which is automatically reconfigured to keep the matching with the coil and adjust the output power to adapt to the change of coil distance The closed-loop control algorithm is used to continuously change the capacitor, which can compensate the mismatch and adjust the output power at the same time. Compared with the system without matching, when the coupling coefficient is 0.05-0.8, the power transfer efficiency of the system is improved by 31.79% and 60%. [25] discussed a new design method of matching circuit based on genetic algorithm and random measurement of S parameters of moving coil. It can further simplify the simulation of matching circuit design and derive the optimal matching circuit.

In this work, we investigate a method of dynamic impedance matching in the wireless power transfer. A matching control method with controllable capacitance compensation is introduced to reduce the transmission efficiency caused by the system detuning. A two-port network analysis model based on external coupling network is established, and a frequency independent controllable compensation reactance is introduced into the resonant circuit. The half bridge circuit is used to realize the controllable compensation capacitance, and determined the value range of the compensation capacitance.

2 Magnetically Coupled Resonators Model

In the four-coil magnetically coupled resonant system shown in Fig. 1.

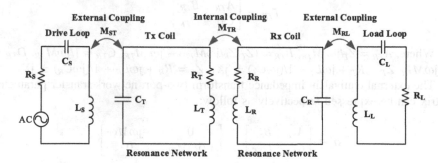

Fig. 1. Equivalent circuit model of four coils wireless power transmission.

In Fig. 1, M_{ST} and M_{RL} are the mutual inductance between the external and the internal resonant network, and the M_{TR} is the mutual inductance of the internal resonance. The excitation and the pick coil constitute an external resonant network. The transmit and receive coils constitute an internal resonant network. The high-frequency electric power source drives the excitation coil to generate a high-frequency oscillating magnetic field, which can be transmitted to the internal LC network by external coupling to resonate. The internal transmit coil and receive coil are internally coupled with each other to form contactless electrical energy transmission. As the internal resonant network and external resonant network using magnetic coupling method for energy transmission, so the use of coil mutual inductance as a parameter to establish T-type equivalent impedance conversion network circuit model, as shown in Fig. 2.

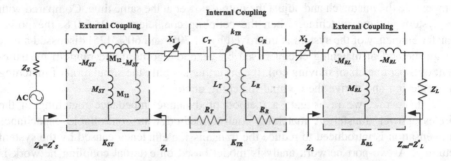

Fig. 2. Equivalent impedance conversion two-port network circuit model.

In Fig. 2, $Z_S = R_S + j\omega L_S + 1/j\omega C_S$, $Z_L = R_L + j\omega L_L + 1/j\omega C_L$, K_{ST}, K_{RL} are the equivalent impedance transform network of the external resonant network, and K_{TR} is the equivalent impedance transformation network of the internal resonant network, which is only related to the coupling coefficient k_{TR} between the two coils. The internally coupled two-port network transfer parameter matrix can be expressed as follow:

$$T_{TR} = \begin{bmatrix} A_{TR} & B_{TR} \\ C_{TR} & D_{TR} \end{bmatrix} \tag{1}$$

Where, $A_{TR} = Z_T/j\omega M_{TR}$, $B_{TR} = (Z_T Z_R/j\omega M_{TR}) - j\omega M_{TR}$, $C_{TR} = 1/j\omega M_{TR}$, $D_{TR} = Z_R/j\omega M_{TR}$, $Z_T = R_T + j\omega L_T + 1/(j\omega C_T) \pm jX_1$, $Z_R = R_R + j\omega L_R + 1/(j\omega C_R) \pm jX_2$

The external equivalent impedance transform two-port network transfer parameter matrix can be expressed respectively as follow:

$$T_{ST} = \begin{bmatrix} A_{ST} & B_{ST} \\ C_{ST} & D_{ST} \end{bmatrix} = \begin{bmatrix} 0 & -j\omega M_{ST} \\ 1/j\omega M_{ST} & 0 \end{bmatrix} \tag{2}$$

$$T_{RL} = \begin{bmatrix} A_{RL} & B_{RL} \\ C_{RL} & D_{RL} \end{bmatrix} = \begin{bmatrix} 0 & -j\omega M_{RL} \\ 1/j\omega M_{RL} & 0 \end{bmatrix} \tag{3}$$

In Fig. 2, the external and internal coupling networks are cascaded to form a new two-port network. According to the cascade characteristic of the two-port network, the total transmission parameters after cascade are shown in (4).

$$T = T_{ST} \times T_{IR} \times T_{RL} = \begin{bmatrix} A & B \\ C & D \end{bmatrix} \tag{4}$$

Where, the total transmission parameters are
$A = (A_{ST}A_{TR} + B_{ST}C_{TR})A_{RL} + (A_{ST}B_{TR} + B_{ST}D_{TR})\,C_{RL}$

$$B = (A_{ST}A_{TR} + B_{ST}C_{TR})B_{RL} + (A_{ST}B_{TR} + B_{ST}D_{TR})\, D_{RL}$$
$$C = (C_{ST}A_{TR} + D_{ST}C_{TR})A_{RL} + (C_{ST}B_{TR} + D_{ST}B_{TR})\, C_{RL}$$
$$D = (C_{ST}A_{TR} + D_{ST}C_{TR})B_{RL} + (C_{ST}B_{TR} + D_{ST}B_{TR})\, D_{RL}$$

The WPT system as the two-port network with one port being input fed by the source and the other being output fed by the load. Based on the scattering parameter, the transfer efficiency will be represented in terms of the linear magnitude of the scattering parameter $|S_{21}|$, the function of power transfer efficiency is defined as $\eta = |S_{21}|^2$ when the network is matching at both ports, in which $|S_{21}|$ is given by

$$S_{21} = 2\frac{V_S}{V_L}\sqrt{\frac{Z_S}{Z_L}} \tag{5}$$

Where, V_S power voltage, V_L load voltage, Z_S power impedance, Z_L load impedance.

Assuming that the port impedance $Z_S = Z_L = Z_P$, S_{21} can be expressed as a T parameter.

$$S_{21} = \frac{2}{A + B/Z_P + CZ_P + D} \tag{6}$$

$$\eta = |S_{21}|^2 = \frac{4}{(A + B/Z_P + CZ_P + D)^2} \tag{7}$$

From the system efficiency in (7), the system efficiency is a function of the port impedances Z_S, Z_L and the transmission parameters A, B, C, D, when the system port impedance value is fixed, the system efficiency is only related to the transmission parameters.

3   Optimization of Magnetic Coupled Resonant Network

Only two levels of headings should be numbered. Input impedance Z_{in} of external coupling network after adding impedance matching circuit is

$$Z_{in} = R_{in} + jX_{in} = \frac{K_{ip}^2}{Z_p} \tag{8}$$

In (8), Z_P is the impedance of power or load port, K_{ip} is the coupling impedance transformation network connecting internal resonance circuit and external port.

It can be seen from (8) that the port impedance Z_P of the equivalent impedance transform network is represented by the input impedance Z_{in} and the characteristic impedance K_{ip} causes the phase shift $\pm 90°$ between the two impedances. The corresponding two-port transfer parameter matrix of (8) can be expressed as

$$\begin{bmatrix} A_{ip} & B_{ip} \\ C_{ip} & D_{ip} \end{bmatrix} = \begin{bmatrix} 0 & \pm jK_{ip} \\ \mp \frac{j}{K_{ip}} & 0 \end{bmatrix} \tag{9}$$

Comparing (8) and (9) can be obtained $K_{ip} = \omega M_{ip}$, therefore, the characteristic impedance K_{ip} can be used to adjust the equivalent impedance of the two-port equivalent circuit of T-type mutual inductance, and the impedance transformation process of the network is shown in Fig. 3.

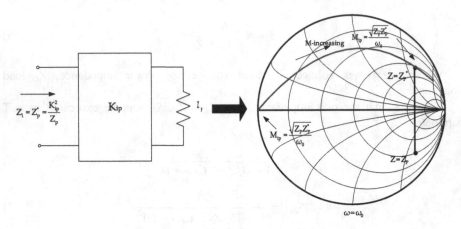

Fig. 3. Input impedance tracking process for coil inductance change in resonant state.

It can be seen from the Smith chart in Fig. 3 that the input impedance and the load impedance are symmetrical about the real axis when $M_{ip} = \sqrt{Z_p Z_p^*}/\omega_0$.

Substituting the port impedance Z_P into (8) can be rewritten as

$$Z_{in} = \frac{K_{ip}^2}{Z_p} = \frac{K_{ip}^2 R_p}{|Z_p|^2} - j\frac{K_{ip}^2 X_p}{|Z_p|^2} \tag{10}$$

As the resonance frequency ω_{0i} is a function of reactance, the reactance part K_{ip}^2/X_p $|Z_p|^2$ in the coupling impedance transformation network will make the working frequency of the circuit deviate from the resonance frequency, which will cause the system detuning and reduce the power transmission efficiency. In order to counteract the reactance that causes the system to be detuned, it is necessary to introduce a frequency-independent controllable reactive reactance X_i in the resonant circuit as shown in (11).

$$X_i = -\mathrm{Im}(Z_i) = \frac{K_{ip}^2 X_p}{|Z_p|^2} = \frac{\omega^2 M_{ip}^2 X_p}{|Z_p|^2} \tag{11}$$

From (11), when the port parameters and ω are constant, the controllable compensating reactance is a function of the externally coupled mutual inductance M. The equivalent impedance K_{ip} of transmission network changes with external coupling. By adjusting the controllable compensation reactance X_i to the matching value, even if it meets the formula (11), for a given port impedance, the network parameters can be configured to prevent the occurrence of detuning. By substituting the parameters related to the transmission efficiency of the system into (6), we can get

$$
S_{21}|_{\omega=\omega_0} = 2j \frac{k_{TR}\frac{k_p^2}{\omega_0 L_i R_p}}{[-k_{TR}^2+(\frac{k_p^2 X_p}{\omega_0 L_i |Z_p|^2}-\frac{j}{Q_i})^2](-j+\frac{X_p}{R_p})^2 - 2\frac{k_p^2}{\omega_0 L_i R_p}(\frac{k_p^2 X_p}{\omega_0 L_i |Z_p|^2}-\frac{j}{Q_i})(-j+\frac{X_p}{R_p})+\frac{k_p^4}{(\omega_0 L_i R_p)^2}}
$$

$$
= 2j \frac{k_{TR}\frac{k_p^2}{\omega_0 L_i R_p}}{-k_{TR}^2(-j+\frac{X_p}{R_p})^2+[\frac{k_p^2}{\omega_0 L_i R_p}-(\frac{k_p^2 X_p}{\omega_0 L_i |Z_p|^2}-\frac{j}{Q_i})(-j+\frac{X_p}{R_p})]^2}
$$

$$(12)$$

Under the given parameters, the (13) is derived as follow.

$$
\frac{\partial S_{21}}{\partial k_p}\bigg|_{\omega=\omega_0} = 0 \tag{13}
$$

The optimal value of k_p when the maximum transmission efficiency is obtained is

$$
k_{p-opt} = \left(1+k_{TR-opt}^2 Q_i^2\right)^{1/4}|z_p|\sqrt{\frac{R_i}{R_p}} \tag{14}
$$

In (14), the port impedance and resonance quality factor are fixed values, and the optimized value of internal coupling factor K_{p-opt} is determined by the distance and position of resonance coil.

The impedance matching and resonance compensation of the system can be easily realized by using the impedance variation characteristic and the transmission response and power transmission efficiency of the system can be optimized.

Half-bridge controllable compensation reactance model shown in Fig. 4, the circuit consists of two full-control switches (V_{T1}, V_{T2}) with capacitor in parallel, and the capacitance voltage is

$$
u_{CT} = U_{CT}\sin(\omega t) \tag{15}
$$

The initial voltage on the capacitor is zero, V_{T1} and V_{T2} are in inverse series and the trigger signal is the same. Trigger angle range is $90° < a < 180°$, while the trigger signal on the voltage zero-point symmetry, conduction angle is $\pi-\alpha$.

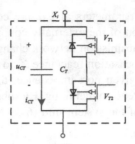

Fig. 4. Half-bridge controllable compensation reactance model.

The voltage of capacitor branch during conduction of V_{T1} or V_{T2} is as follows.

$$u_{CT}(t) = \int\limits_{\alpha}^{\omega t} \frac{1}{\omega C_T} i_{CT}(t)dt = \int\limits_{\alpha}^{\omega t} \frac{1}{\omega C_T}(\sqrt{2}I\cos\omega t)dt \tag{16}$$

$$\omega t = \alpha + k\pi(\ k = 0,1,2,\cdots)$$

The u_{CT} is expanded by Fourier series, and only the fundamental wave term is retained as shown in (17).

$$u_{CT} = \frac{2}{\sqrt{2}\pi} \int_0^{\pi} [u_{CT}(\omega t) * \sin\omega t]dt$$

$$= \frac{\sqrt{2}}{\pi} \int_{\alpha}^{\pi-\alpha} \left\{ \left[\sqrt{2}I\frac{1}{\omega c_T}(\sin\omega t - \sin\alpha) \right] \sin\omega t \right\}dt \tag{17}$$

$$= I\frac{1}{\omega c_T}(1 - \frac{2\alpha}{\pi} - \frac{2\sin 2\alpha}{\pi})$$

From (17), the equivalent capacitor can be solved as follow:

$$C_{eq} = \frac{\pi C_T}{\pi - 2\alpha - 2\sin 2\alpha} \tag{18}$$

Therefore, the controllable compensation reactance value can be expressed as:

$$X_i = \frac{1}{\omega C_{eq}} = \frac{\pi - 2\alpha - 2\sin 2\alpha}{\pi\omega C_T} \tag{19}$$

The controllable compensation reactance can be adjusted continuously in a certain range by changing the trigger angle α. The adjustable reactance is used to compensate the reactance which causes the system detuning in the coupling impedance transformation network, and a variable reactance and reactive power compensation with a variation range of 0-X_c is realized.

In the WPT system, the main compensation in the circuit is the port impedance and the inductive reactive power component in the impedance transformation network. Therefore, the range of values for C_{eq} is as follows.

$$\min\left(\frac{|Z_p|^2}{\omega K_{ip}^2 X_p}\right) < C_{eq} < \max\left(\frac{|Z_p|^2}{\omega K_{ip}^2 X_p}\right) \tag{20}$$

According to the adjustment range $(\alpha_{min}, \alpha_{max})$ of trigger angle, the range of controllable compensation reactance can be calculated as $(X_{i\text{-min}}, X_{i\text{-max}})$.

4 Experimental and Result Analysis

In order to verify the feasibility of the above adaptive tuning method, simulation and experimental verification are carried out. The transmitter and receiver of the system are tuned in tow-sides. The system parameters are shown in Table 1.

Table 1. Detailed parameter values of coils.

Parameters	Symbol	Value
Turns of Tx and Rx coils	N_i	5
Turns of excitation and pickup coils	N_p	1
Outer diameter of Tx and Rx coils	$D_{i\text{-max}}$	300 mm
Inner diameter of Tx and Rx coils	$D_{i\text{-min}}$	200 mm
Diameter of excitation and pickup coils	D_p	180 mm
Coil turn spacing	p	17 mm
Wire diameter	$r_i = r_p$	3 mm
Coil distance	d_{23}	0 ~ 1500 mm

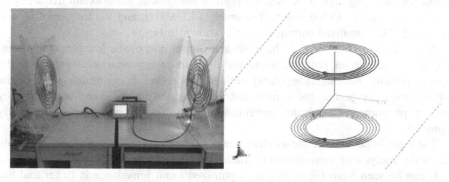

Fig. 5. Simulation model and experimental device of magnetically coupled resonant coil.

Simulation model and experimental device of magnetically coupled resonant coil as shown in Fig. 5.

In the experimental device, the plane spiral coils are made of hollow copper tube. The excitation and the transmitting coil have the same center, and the receiving and picking coils have the same structure. Class E amplifier is used as high frequency inverter to connect excitation coil and programmable electronic load to connect pick-up coil. The operating frequency of the system is 12−15 MHz.The simulation analysis of the plane spiral coil system is carried out. When the transmission distance is 600 mm, the scattering parameter curve of the system is shown in Fig. 6.

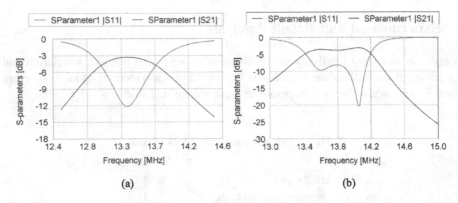

(a)   (b)

Fig. 6. S Parameter before and after optimization of 400 mm transmission distance (a) The original WPT system (b) The optimized WPT system.

As shown in Fig. 6, after optimization, the forward transmission coefficient of the system is larger and has two resonance frequency points, while the original WPT system is only one, and the maximum value of −3 dB is not reached. After optimization, the reflection coefficient of the input port of the system is obviously reduced, which shows that the transmission capacity of the system is significantly improved. The impedance matching method is used to optimize the system, the resonant frequency of the coil is rematches to the power frequency (14.1 MHz), and the frequency split is eliminated. The optimized current of port is shown in Fig. 7.

As can be seen from Fig. 7, the input and output port of the low frequency point have the same current polarity, while the high frequency resonance point has the opposite polarity. Although the polarity of the two ports before optimization is opposite at both frequency points, the current distribution does not have obvious frequency division phenomenon, and the current after optimization has been significantly improved.

The input impedance of the wireless power transmission system plays an important role in the analysis of transmission characteristics, as shown in Fig. 8.

It can be seen from Fig. 8 that the optimized input impedance is larger and has better load performance in general, so that the system has better transmission characteristics.

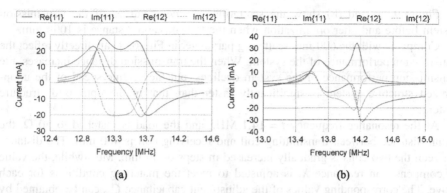

Fig. 7. The optimized current of input and output port (a) The original WPT system (b) The optimized WPT system.

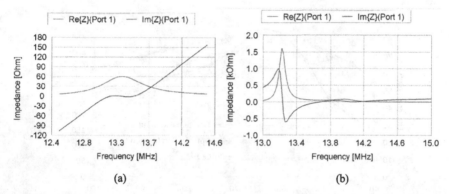

Fig. 8. Input impedance before and after optimization (a) The original WPT system (b) The optimized WPT system.

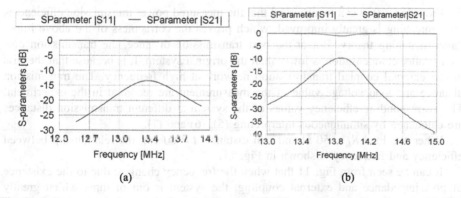

Fig. 9. S Parameter before and after optimization of 800 mm transmission distance (a) The original WPT system (b) The optimized WPT system.

Figure 9 (a) and (b) show the scattering parameter curves of the transmission system before and after optimization when the transmission distance is 1000 mm.

Compared with the 600 mm scattering parameter in Fig. 6, it can directly reflect the transmission performance of the system. When the transmission distance increases, the transmission performance of the system declines, but the decline speed of the unoptimized structure system is significantly faster than that of the optimized structure system.

At the resonance frequency $f = 14.1$ MHz and the load is matched to 50 Ω, the initial distance between transmitting coil and receiving coil is 400 mm. The distance between the two coils is gradually increased in steps of 50 mm. Meanwhile, the value of compensation reactance X_i is adjusted to meet the matching conditions for each move. The corresponding values of the adjustment capacitance C_{eq} can be obtained by the Eqs. (18) and (21), and it varies curve with the transmission distance as shown in Fig. 10.

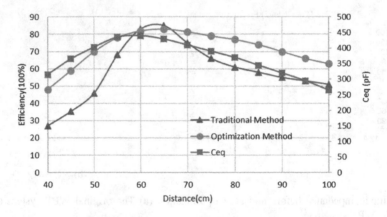

Fig. 10. The relationship between transmission efficiency and distance when $R_L = 50$ Ω.

From Fig. 10, it can be observed that the measured power transmission efficiencies after matching is greatly improved, which proves the correctness of the above impedance matching theory. But at the best transmission distance, the transmission efficiency after compensation is lower than the original system. It is because that the loss of the external coupled transformation network at high frequency. The transmission distance and load voltage value of the measurement are recorded in the experiment. The corresponding efficiency values of the system at different transmission distances are calculated by simultaneous interpreting (5), (6) and (7).

When the load $R_L = 50$ Ω and coil distance $d = 60$ cm, the relationship between efficiency and frequency is shown in Fig. 11.

It can be seen from Fig. 11 that when the frequency changes, due to the existence of port impedance and external coupling, the system is out of tune, which greatly reduces the transmission efficiency of the system. By adding compensation reactance to the system for optimization, the matching capacitance C_{eq} is adjusted according to the impedance matching conditions when the frequency changes, so that it meets the

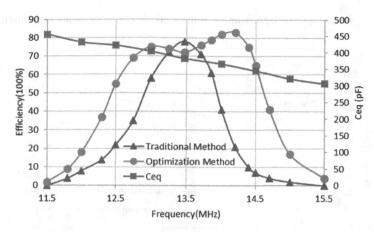

Fig. 11. The relationship between transmission efficiency and frequency when $R_L = 50\ \Omega$.

relationship (12) and (13), the transmission efficiency of the optimized system is significantly improved, and the detuning is greatly improved. The efficiency between the two resonance frequency points of the optimized system will not decrease significantly, and it is far greater than the transmission efficiency of the system before optimization.

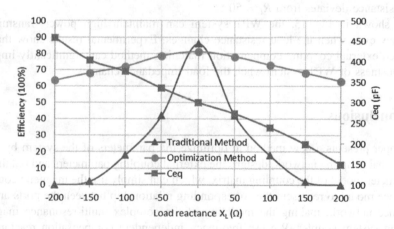

Fig. 12. The relationship between transmission efficiency and load reactance when $R_L = 50\ \Omega$.

At the resonance frequency $f = 14.1$ MHz, load $R_L = 50\ \Omega$ and coil distance $d = 60$ cm, the transmission efficiency of the WPT system decrease dramatically when the load reactance deviates from $X_L = 0\Omega$ is shown in Fig. 12.

From Fig. 12, it can be observed that as the load reactance deviates from $X_i = 0$, the transmission efficiency of the system decreased rapidly with the inherent parameters of the system deviate from the resonant working frequency. By Adjusting the compensation capacitor C_{eq} to match the port impedance, the power transmission efficiency of the WPT system can be significantly improved, but the resonance frequency will change.

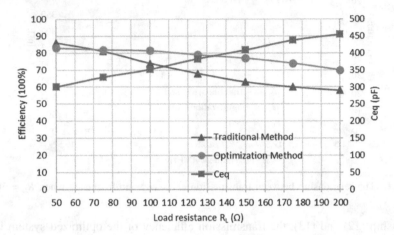

Fig. 13. The relationship between transmission efficiency and load resistance when $X_L = 0\ \Omega$.

In Fig. 13, same operating conditions for transmission efficiency analysis when the load resistance deviates from $R_L = 50\ \Omega$.

As shown in Fig. 13, the WPT system can maintain high power transmission efficiency even when the load resistance changes. Experimental results show that the proposed external coupling compensation matching method can significantly improve the robustness of WPT system when the load impedance changes.

5 Conclusions

This paper presents a new method to optimize the parameters of the system by using the external coupling network, designs the external coupling parameters and obtains the S-parameter model of the coupling matrix, which can simply use the magnetic coupling resonance model to represent the corresponding relationship between the ports and the resonance network, making the modeling of the complex multi-resonance magnetic coupling system simpler. We use frequency independent compensation reactance to match the detuning system, and get the conclusion that the transmission efficiency, resonance compensation and impedance matching of the system can be optimized by external coupling impedance transformation network. Simulation and experimental results show that the existence of external coupling network makes the port impedance match well, and the optimal value of external coupling at the maximum transmission efficiency is obtained.

Acknowledgments. This work was supported in part by the Natural Science Foundation of Liaoning Provincial under Grant 20180551056.

References

1. Shinohara, N.: Power without wires. IEEE Microwave Mag. **12**(7), 64–73 (2011)
2. Kurs, A.B., Karalis, A., Moffatt, R., Joannopoulos, J.D., Fisher, P., Soljacic, M.: Wireless power transfer via strongly coupled magnetic resonances. Science **317**(5834), 83–86 (2007)
3. Brown, W.C.: The history of power transmission by radio waves. IEEE Trans. Microw. Theory Tech. **32**(9), 1230–1242 (1984)
4. Ho, J.S., Kim, S., Poon, A.S.Y.: Midfield wireless powering for implantable systems. Proce. IEEE **101**(6), 1369–1378 (2013)
5. Waters, B.H., Smith, J.R., Bonde, P.: Innovative free-range resonant electrical energy delivery system (free-d system) for a ventricular assist device using wireless power. ASAIO J. **60**(1), 31–37 (2014)
6. Asgari, S.S., Bonde, P.: Implantable physiologic controller for left ventricular assist devices with telemetry capability. J. Thoracic Cardiovasc. Surg. **147**(1), 192–202 (2014)
7. Johari, R., Krogmeier, J.V., Love, D.J.: Analysis and practical considerations in implementing multiple transmitters for wireless power transfer via coupled magnetic resonance. IEEE Trans. Ind. Electron. **61**(4), 1774–1783 (2014)
8. Si, P., Hu, A.P., Malpas, S.C., Budgett, D.: A frequency control method for regulating wireless power to implantable devices. IEEE Trans. Biomed. Circ. Syst. **2**(1), 22–29 (2008)
9. Lee, K., Chae, S.H.: Effect of Quality Factor on Determining the Optimal Position of a Transmitter in Wireless Power Transfer Using a Relay. IEEE Microw. Wirel. Compon. Lett. **27**(5), 521–523 (2017)
10. Wei, X., Wang, Z., Dai, H.: A critical review of wireless power transfer via strongly coupled magnetic resonances. Energies **7**(7), 1–26 (2014)
11. Lin, Z., Wang, J., Fang, Z., Hu, M., Cai, C., Zhang, J.: Accurate maximum power tracking of wireless power transfer system based on simulated annealing algorithm. IEEE Access **6**, 60881–60890 (2018)
12. Li, W., Zhao, H., Li, S., Deng, J., Kan, T., Mi, C.C.: Integrated LCC compensation topology for wireless charger in electric and plug-in electric vehicles. IEEE Trans. Ind. Electron. **62**(7), 4215–4225 (2015)
13. Zhang, Y., Lu, T., Zhao, Z., Chen, K., He, F., Yuan, L.: Wireless power transfer to multiple loads over various distances using relay resonators. IEEE Microw. Wirel. Compon. Lett. **25**(5), 337–339 (2015)
14. Na, K., Jang, H., Ma, H., Bien, F.: Tracking optimal efficiency of magnetic resonance wireless power transfer system for biomedical capsule endoscopy. IEEE Trans. Microw. Theory Tech. **63**(1), 295–304 (2015)
15. Anowar, T.I., Barman, S.D., Wasif Reza, A., Kumar, N.: High-efficiency resonant coupled wireless power transfer via tunable impedance matching. Int. J. Electron. **104**(10), 1607–1625 (2017)
16. Miao, Z., Liu, D., Gong, C.: Efficiency enhancement for an inductive wireless power transfer system by optimizing the impedance matching networks. IEEE Trans. Biomed. Circ. Syst. **11**(5), 1160–1170 (2017)
17. Nguyen, H., Agbinya, J.I.: Splitting frequency diversity in wireless power transmission. IEEE Trans. Power Electron. **30**(11), 6088–6096 (2015)
18. Berger, A., Agostinelli, M., Vesti, S., Oliver, J.A., Cobos, J.A., Huemer, M.: A wireless charging system applying phase-shift and amplitude control to maximize efficiency and extractable power. IEEE Trans. Power Electron. **30**(11), 6338–6348 (2015)

19. Lim, Y., Tang, H., Lim, S., Park, J.: An adaptive impedance-matching network based on a novel capacitor matrix for wireless power transfer. IEEE Trans. Power Electron. **29**(8), 4403–4413 (2014)
20. Beh, T.C., Kato, M., Imura, T., Oh, S., Hori, Y.: Automated impedance matching system for robust wireless power transfer via magnetic resonance coupling. IEEE Trans. Ind. Electron. **60**(9), 3689–3698 (2013)
21. Kim, J., Kim, D.H., Park, Y.J.: Analysis of capacitive impedance matching networks for simultaneous wireless power transfer to multiple devices. Ind. Electron. IEEE Trans. **62**(5), 2807–2813 (2015)
22. Kim, J., Jeong, J.: Range-adaptive wireless power transfer using multiloop and tunable matching techniques. Ind. Electron. IEEE Trans. **62**(10), 6233–6241 (2015)
23. Kim, J., Kim, D.H., Park, Y.J.: Free-positioning wireless power transfer to multiple devices using a planar transmitting coil and switchable impedance matching networks. IEEE Trans. Microw. Theory Tech. **64**(11), 3714–3722 (2016)
24. Miao, Z., Liu, D., Gong, C.: An adaptive impedance matching network with closed loop control algorithm for inductive wireless power transfer. Sensors **17**(8), 1759–1778 (2017)
25. Bito, J., Jeong, S., Tentzeris, M.M.: A novel heuristic passive and active matching circuit design method for wireless power transfer to moving objects. Microw. Theory Tech. IEEE Trans. on **65**(4), 1094–1102 (2017)

Text Recognition for Automated Test Execution in Interlocking: A Deep Learning Approach

Dong Xie[1], Ming Chai[2], Hongjie Liu[2(✉)], Chen Bai[1], Qi Wang[1], and Jidong Lv[2]

[1] Department of Electronic Engineering,
Beijing Jiaotong University, Beijing, China
{16211355,19125083,16211408}@bjtu.edu.cn
[2] The National Engineering Research Center of Rail Transportation Operation
and Control System, Beijing Jiaotong University, Beijing, China
{chaiming,hjliu2,jdlv}@bjtu.edu.cn

Abstract. In this paper, we present a deep learning character recognition algorithm based on multi-level segmentation. It can improve the accuracy of recognition of button characters in the interlocked upper computer interface, which is of great significance to the design of the automatic test execution in the interlocking system. We first analyze the characteristics of characters in the interlocked upper computer interface and combine various methods to segment characters at multiple levels. Then, we build a deep learning character recognition model based on CNN to accurately recognize characters. Finally, we applied our algorithm to recognize the characters of the interfaces in a real interlocked upper computer software and made a comparison with other algorithms. The obtained results show the feasibility and advantage of our approach.

Keywords: Interlock automated testing · Deep learning · Character recognition · Multilevel character segmentation algorithm

1 Introduction

A computer-based interlocking system is the core control equipment of railway operation, which has the characteristics of high efficiency and safety. The interlock system generally consists of two parts – an upper computer and an interlock machine. The upper computer is used for man-machine interaction, generation of issuing operation commands, receiving and processing information of the interlock machine. The interlock machine is used for processing logic of signal equipment and automatic unlocking of access [1].

The computer-based interlocking system is one of the important infrastructures of railway modernization and automation, which plays a very important role in the safe operation of the railway. Therefore, the interlocking software must be fully tested before being put into actual operation to minimize potential hazards. This is also the effective method to ensure the quality and performance of the software [2].

© Springer Nature Singapore Pte Ltd. 2020
J. Qian et al. (Eds.): ICRRI 2020, CCIS 1335, pp. 353–367, 2020.
https://doi.org/10.1007/978-981-33-4929-2_24

Most of the traditional testing methods are to test interlocking software manually by dedicated testers. However, with the continuous upgradation of the computer-based interlocking system, interlocking software has become more and more complex, the traditional manual testing methods can no longer meet the requirements of the inter-locking test.

The model-based automated testing is now a more effective testing method [3]. But this approach generates a large number of test cases, which makes automating these test cases an urgent problem to solve. And there are some difficulties in solving this problem. Since this method generates abstract test cases based on an interlocking table, there is no provision in the interlocking machine to execute the tests directly as test scripts, so they need to be converted into interlocking engineering data. However, the interlocking system is large in scale and complex in structure, interlocking systems at different stations are customized and there is no uniform standard. Additionally, the conversion process is heavy and error-prone. These difficulties lead to the result that there is no unified algorithm to convert abstract cases into executable test scripts based on engineering data.

To solve the above problems, we propose a method to perform tests in the upper computer. The main challenge in this method is how to build a unified model to automatically recognize the interfaces of different stations in the interlocking upper computer. Since deep learning performs well in the field of image recognition, we apply it to the identification task of the interlock interface. However, if directly applied to the task, the deep learning will face problem in identifying the buttons' name of the interfaces, that is, the problem of character recognition. Due to different resolutions, the quality of some characters may be low and there are often some adhesion problems. Therefore, accurate identification of these characters becomes a difficult problem.

In order to solve this problem, we propose a deep learning character recognition algorithm based on multi-level segmentations. We first adopts a three-level character segmentation algorithm to segment the continuous character images of the interfaces, and then feeds the single character images into the CNN-based character recognition model to complete the recognition. This algorithm has a good effect in segmenting adhered characters and can accurately recognize the content of characters. Hence, it could effectively solve the problem of character recognition for the interfaces in the interlocked upper computer. Accordingly, the proposed algorithm has important sig-nificance in realizing the execution of the whole interlocked test.

Related Works

The problem of automated test execution in an interlocking machine has a long history.

Yang et al. [4] proposed a method for generating an optimal test sequence based on an improved Firefly Algorithm. Yan used the time automaton model to automatically generate test sequences, and then use the Firefly Algorithm to automatically optimize them, finally combined the actual station data to automatically execute the test sequences [5]. Xie used the UPPML software to model the interlock control module according to the logic process of the interlock system and the generated test cases [6], and tested the model according to real approaches. Khurana et al. [7] proposed a method to optimize the test sequence using the UML model and genetic algorithm, where the sequence and state diagrams are transformed into a system diagram to cover the maximum number of test cases and then it is optimized with genetic algorithm. Using the Petri net tool to model the interlocking system and computer interlocking software, Ba studied a method for automatic generation of test cases with genetic algorithm, too [8]. The above works focused only on the generation of test cases. However, a large-size interlocking system has a complicated structure and different stations which need to be customized, and no unified algorithm can make an abstract case executable based on engineering data of the test scripts, which led to those studies not to solve the problem of executing generalized test cases.

For the test in the upper computer, the interfaces need to be recognized. The deep learning can recognize images. In terms of character recognition, Jaderberg et al. [9] presented a model for deep learning training that can recognize a large number of English words by introducing different words in batches. Shi et al. [10] constructed an encoder-decoder in a deep continuous neural network, and realized a recognition model with an automatic correction function that can accurately identify Chinese and English words in irregular text areas. In the field of handwritten character recognition, Ciresan et al. [11] proposed a multi-column deep neural network, which achieved certain success in practical application. However, if the above methods are directly applied to the interfaces in the interlocked upper computer, they will not be able to accurately recognize low quality characters.

The rest part of the paper is organized as follows. Section 2 introduces the definition of Computer-Based Interlocking System. Section 3 presents the character recognition model based on CNN, including its basic principle and construction process. Section 4 introduces a three-level character segmentation algorithm. Section 5 presents some experimental results of the CNN character recognition model and three-level character segmentation algorithm compared with other existing algorithms. Section 6 contains the conclusion and scopes for future work.

2 Computer-Based Interlocking System

Computer-Based Interlocking (CBI) system is an integral part of the railway signal system, which has the functions of directing trains, ensuring the safety of the trains, improving the transportation efficiency, etc. The interlocking function in the railway signal is completed through the application of computer technology [12].

The CBI system realizes interlock control among switches, signal machine and track circuits in a station under the operation of the signal operator or the ATS system, which is indispensable guaranteed equipment in railway for safe and efficient driving. The CBI system is mainly composed of the upper computer and the interlocking unit. The relationship among its internal logics is shown in Fig. 1.

The upper computer performs mainly the following functions: (1) to receive operational commands from operators in real time, (2) to issue the operational commands to the interlocking host, (3) to receive real-time information from the interlocking host, and (4) to realize the dynamic display and alarm function of the station.

The interlock machine mainly completes the following functions: (1) Real-time communication with upper computer, (2) Real-time communication between the acquisition board and the driver board in the acquisition drive layer, (3) completion of

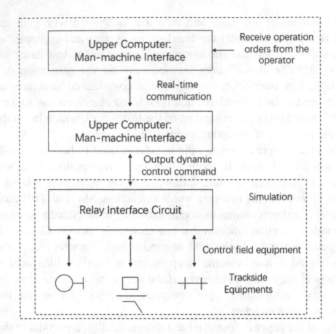

Fig. 1. The internal structure of the computer-based interlocking system.

the operation of the interlocking logic, and (4) the redundancy management function such as synchronization, fault detection and switching between multiple CPU boards.

3   Character Recognition Model Based on CNN

The convolutional neural network is an efficient deep learning neural network, which was proposed by Hubel et al. [13] in 1960s inspired by cortical neurons present in cat's brain whose special structure can reduce the complexity of neural networks. In recent years, the rapid development of the convolutional neural net-work has attracted a large number of scholars to carry out related researches on it, and as a result it has been widely applied in many fields.

In order to realize the recognition of characters (English letters and numbers) in the interlock interface, we build our CNN character recognition model according to the specific situation.

3.1   Making Training Set

The training data set used in the character recognition model mainly contains a part of the data obtained from the MINIST image data set. We supplemented, deleted and cleaned it according to the patterns of the characters in the software used for simulating the interlocked interfaces.

Besides using MINIST data set of some pictures, we also obtain the letters from "A" to "Z" (except the letter "O" as it is not involved by default in the button of an interface), the numbers from "0" to "9" and the symbol "\" (switch) images by taking screen-shots of characters in the operating interface of the interlocking upper computer. After adding and replacing the data set, we get a total of $32 \times 50 = 4800$ pictures in the character training set.

3.2 Model Building and Training

The neural network of the system considered in this paper is constructed based on the Tensorflow deep learning framework, and it is trained on Tensorflow with the help of the GPU acceleration [14].

Firstly, we set up the neural network of the model. In this model, one convolutional layer and one pooling layer are adopted as a composite layer structure. The complete model has 2 composite layers, a fully connected layer and an output layer (Fig. 2).

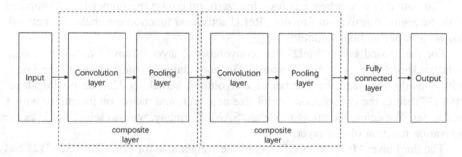

Fig. 2. The structural composition of our model based on convolutional neural network.

In the convolutional layer, the mathematical expression of the convolution process is shown in formula 3-1, where n_in is the number of input matrices, X_k is the k-th input matrix, W_k is the k-th sub-convolution kernel, and $s(i,j)$ is the value of the element in the W output matrix.

$$s(i,j) = (X * W)(i,j) + b = \sum_{k-1}^{n\_in} (X_k * W_k)(i,j) + b \qquad (3\text{-}1)$$

Secondly, we configured the learning rate, decay rate, and other major constants used in the model. We set various values as follows: initial value of the learning rate is 0.001, total number of categories is 35, and total number of iterations is 3000.

The definitions of the main constants used in the model are presented in Table 1.

Then we set the parameters of each layer.

For the first layer "Mul1", the convolutional layer "Conv1" defines the convolution kernel size as an 8×8 matrix. The convolution kernels move on the image with 1 as the step, and the number of convolution kernels is 16. The pooling layer "Pool1" defines the convolution kernel size as 2×2, and moves on the image with 2 as the

Table 1. Main constant definitions of the model.

Constant	Value	Annotation
NUM_CLASSES	35	The number of categories in the character library
PIXEL_THRESHOLD	230	The threshold of binarization
BATCH_SIZE	60	The number of data samples captured in one training session
ITERATIONS	3000	Total number of iterations

step. We define the filling mode as "SAME", which fills the picture with "0" and keeps the image size unchanged before and after the convolution. Finally, we compare and analyze the performances of various activation functions, and then select ReLU as the activation function of this layer whose mathematical formula is as follows:

$$f(x) = \max(0, x) \tag{3-2}$$

The output is zero when x is less than zero, otherwise the output is x. Compared with the Sigmoid activation function, ReLU activation function can make the network converge faster without saturation.

For the second layer "Mul2", the convolutional layer "Conv2" defines the convolution kernel size as an 5×5 matrix. The convolution kernels move on the image with 1 as the step, and the number of convolution kernels is 32. The pooling layer "Pool2" defines the convolution kernel size as 1×1, and moves on the image with 1 as the step. We define the filling mode as "SAME". Finally, we also select ReLU as the activation function of this layer.

The third layer "Full_connect" defines the weight size for [16 * 20 * 32, 512] and sets the initial value of bias as 0.1 with the final size [−1, 16 * 20 * 32] to reshape the output of the tensor.

The fourth layer "Readout" defines the weight size as [512, NUM_CLASSES], sets the initial value of bias as 0.1, and defines the number of categories as NUM_-CLASSES, which is 35 as shown in Table 1.

After constructing the character recognition model, we first use the completed character image data set to carry out the model training by relying on the Tensorflow deep learning framework. Then, we use GPU to accelerate the training, and finally get a model with high accuracy.

4 A Three-Level Character Segmentation Method

Although we have obtained a character recognition model with good accuracy, it does not mean that we can successfully complete the task of recognizing characters on the interlocked upper computer interface.

Before being fed into the character recognition model, it is important to ensure that these characters are single characters rather than consecutive characters, which means that we have to do character segmentation. If an error occurs in the segmentation, the correctness of subsequent character recognition shall have to be greatly compromised [15].

A variety of character segmentation methods have been reported in the specialized literature [16–18]. However, these methods have some limitations, and still there is no cutting method that can be widely used.

In fact, the characters in the interlocked upper computer interface are special. They are not as clear and easy to separate as those in printed documents or license plates. In particular, some low-quality character images often have dirty areas or other interference, low character resolution and fuzzy boundary between adjacent characters. For example, the following figure is some typical examples (Fig. 3):

Fig. 3. Typical adhered characters in the interlocked upper computer interfaces.

If one of the above methods is used alone to those particular cases, the character segmentation ability will be extremely limited and the effect will be poor.

As per above observations, we propose a three-level character segmentation method based on the analysis of the connected domain of characters, the cumulative projection of vertical pixels, and the projection based on the contour difference. The specific process of character segmentation is shown in Fig. 4.

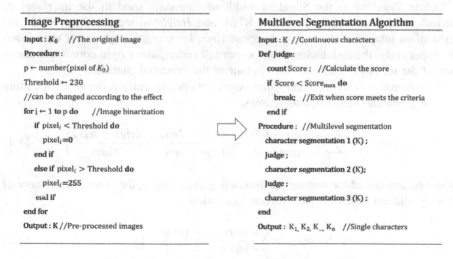

Fig. 4. Three-level character segmentation method.

Based on Fig. 4, the process of character segmentation is explained in the following Sub-Sections.

4.1   Image Preprocessing and Binarization

Some images may have too low basic pixel and small size after cutting. Hence, in order to facilitate the subsequent processing, we firstly use the PIL library built in Python to enlarge an image. Besides, the *Cv2.threshold* function of the OpenCV image processing is used for processing the image binarization. That is, when the pixel value of a certain point of the image is less than the set threshold, the point is set as black, otherwise it is set as white [19]. In this way, after image processing, there will be only two colors for detecting subsequent contours and project-based cutting.

4.2   Criterion for Evaluating Segmentation Effect——Score

Since the binary edge cannot be completely ruled out, the residual interference, such as block, stain, and some low-quality pixels of character images, may often conglutinate the phenomenon by making the effects of single partitions uneven. So, we need to define a criterion for assessing the effect of character segmentation, and to use it as the condition to determine whether the character image needs to be segmented in the next level.

The experimental object in this study is the upper computer interface of the interlocking system. The names of the interfaces are produced by printed characters. And the same group of characters are arranged in order, their sizes are almost the same (the width of different letters and Numbers is slightly different, and the height is basically the same). Therefore, we propose the following evaluation criteria according to the prior law:

Define *Standard* as the Standard width of characters used in the interfaces of interlock system. In addition, we use *Width* and *Height* to represent the width and height of an image before cutting. Suppose (*lefx*, *lefy*), (*rigx*, *rigy*) and (*Cenx*, *Ceny*) are, respectively, the coordinates of the upper left corner, lower right corner and center point of the rectangle enclosing the contour of the truncated character.

For consecutive external rectangular boxes obtained by cutting, the score of cutting effect of a box can be defined as follows:

$$Score_{self} = \sum_{i=1}^{n} \left(\frac{(rigx - lefx)}{Standard} + \frac{(rigx - lefx)}{(rigy - lefy)} + \frac{(rigx - lefx)}{Width} \right) \quad (4\text{-}1)$$

If the total amount of the rectangular boxes is greater than 1, the comparative score of defining adjacent boxes can be expressed as follows

$$Score_{mul} = \sum_{i=1}^{n} \left| \frac{rigx_{i+1} - cenx_i}{cenx_{i+1} - rigx_i} - 1 \right| \quad (4\text{-}2)$$

The final evaluation score will be obtained after comprehensive evaluation as follows:

$$Score = Score_{self} + Score_{mul} \qquad (4\text{-}3)$$

By analyzing the above formula, we could know that the larger the Score calculated, the larger the ratio of width to height, width to height and width of each enclosing rectangle after segmentation is, and the more likely that it's not completely segmented. By comparing the *Score* with the value *Scoremax* we set before, the obtained results can be used as the basis for determining where to enter the next level of segmentation.

4.3 First Level Segmentation Based on Analysis of Character Connected Domain

A large number of practical experiences show that the first analysis of the domain of connected characters in the segmentation process of low-quality characters can achieve better results in eliminating the interference, such as defaced areas. Therefore, we adopt the method of analysis of the domain of connected characters in the first-level character segmentation [20].

In this paper, we use the *cv2.findContours* function provided by OpenCV to detect the contour of a character image. Considering hollow characters, such as "0", "9", "8" and "6", we use the method identifying only the most external contour in detecting contour cut. The function parameter is set as "*cv2.retr_external*" so that the return values are the coordinates of the four vertices of the smallest rectangle enclosing the outer contour.

As shown in Fig. 5, the segmentation method in this level can segment high-quality characters according to their outlines.

Fig. 5. The successful example of the segmentation algorithm based on the analysis of the character connected domain.

However, in some character images of low quality, two or even more characters are not completely separated. So, the connected domain-based segmentation method will mistakenly recognize them as a single character. Therefore, they will score higher than the norm, which will lead to the next level of segmentation (Fig. 6).

4.4 Second Level Character Segmentation Based on Analysis of Cumulative Projection of Vertical Pixels

After the first level of segmentation, if the score remains higher than the given standard score, which indicates that there may be adhesion of characters, we analyze the pixel points of the picture and calculate their cumulative value in the vertical direction. Then,

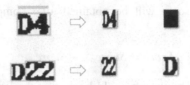

Fig. 6. The failure example of the segmentation algorithm based on the analysis of the character connected domain.

we draw curve diagrams to analyze their projected characteristics [21]. If there is any sudden drastic trough in a continuous curve and the cumulative value of pixels on its both sides is much higher than that at the trough, it indicates that the part of two characters is most likely adhered together. So, the characters can be separated by cutting at this position.

As shown in Fig. 7, this method can segment some characters having relatively simple adhesion.

Fig. 7. The successful example of the segmentation algorithm based on the analysis of the cumulative projection of vertical pixels.

However, in some complicated cases of adhesion, this method cannot accurately find the adhered part, which will also lead to the failure of segmentation. Therefore, they will score higher than the norm, which will lead to the next level of segmentation (Fig. 8).

Fig. 8. The failure example of the segmentation algorithm based on the analysis of the cumulative projection of vertical pixels.

4.5   Third Level Character Segmentation Based on Analysis of Contour Projection

After the second level of segmentation, if the score still remains higher than the given standard score, we adopt the projection analysis method based on contour difference. Firstly, we project the upper and lower contours of the character image, and draw the upper and lower boundary curves of the character image. Then, we subtract the upper and lower contours to obtain the curve with the new contour difference. Finally, we calculate and obtain all the minimum points of the curve and analyze them. After filtering out the interference values which obviously do not conform to the actual situation, we get the final minimum points. These minimum points can be considered as the positions of the character adhesion. Hence, they are selected as the cutting points for segmentation, thus we get the final result.

For example, we split the character "D22" in the three-level segmentation. Firstly, the upper and lower contours are scanned and obtained as shown in Figs. 9(a) and 9(b), respectively. We draw the upper and lower outline curves of the character "D22" in the same figure as shown in Fig. 10(a). Finally, the upper and lower contours are subtracted to obtain the curve as shown in Fig. 10(b).

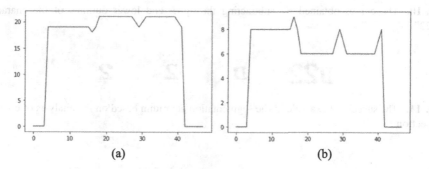

(a)   (b)

Fig. 9. The upper and lower contour curves of the character "D22".

Based on the analysis as shown in Fig. 10(b), the minimum point that meets the condition is found as the cutting point, and the single character images "D", "2" and "2" that are finally segmented are obtained as shown in Fig. 11.

So far, the whole three-level character segmentation algorithm process is completed.

5   Experimental Result

In order to evaluate the performance of our CNN model and three-level character segmentation algorithm on interlocking upper computer interface, we conducted some experiments.

Firstly, we studied the relationship between the iteration number and the loss of the CNN model, and then we got the following curve (Fig. 12):

According to the data curve, the loss of the model decreases with the increase of the number of iterations. However, after the number of iterations exceeds 3000, the loss of the model basically converges to about 0.027. Therefore, 3000 times is selected as the maximum number of iterations of the training.

We also studied the relationship between the number of training samples and the model error rate as presented in Table 2.

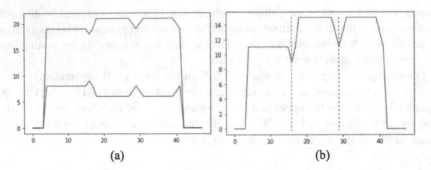

Fig. 10. The curves obtained by subtracting the upper and lower outlines of the character "D22".

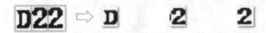

Fig. 11. The successful example of the segmentation algorithm based on the analysis of contour projection.

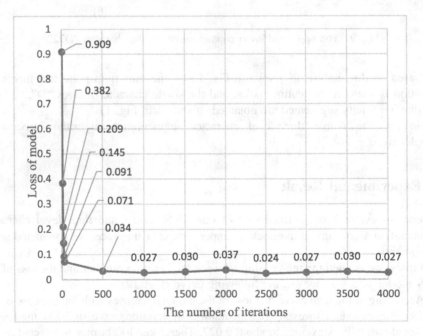

Fig. 12. The model losses with different iterations.

It can be seen in Table 2 that, for this model, the larger the size of the training sample, the higher the accuracy of the model. Due to the limitation of data volume, we select 4000 as the total number of data samples in this paper.

Table 2. The error rates of models with different sizes of training samples.

Training sample size	1000	1500	2000	2500	3000	3500	4000
Error rate		14.56%	10.43%	7.77%	6.79%	5.83%	4.81%	3.88%

In addition to the longitudinal comparison, we also made a horizontal comparison of the CNN model with other character recognition models by adopting the same sample to calculate their recognition accuracy (Table 3).

Table 3. Comparison of the CNN algorithm with other algorithms.

Recognition algorithm	Recognition rate
Template matching recognition	92.50%
KNN	94.08%
CNN	96.12%

It can be seen that our CNN model performs better than other algorithms in the recognition of characters in the interlocked upper computer interface.

After testing the CNN character recognition model, we also conducted several comparative experiments in order to test the effect of our proposed three-layer segmentation algorithm.

We selected 200 buttons of text images taken from the interfaces of the interlocked upper computer as test data and carried out four groups of experiments for comparison.

The first group adopts the segmentation algorithm based on the analysis of the connected domain's features to segment the test data and send it to the CNN character recognition model for recognition. The second group adopts the segmentation algorithm based on the analysis of vertical projection. The third group adopts the segmentation algorithm based on the analysis of the connected domain's features as the first-level cutting algorithm and the segmentation algorithm based on the analysis of vertical projection as the second-level cutting algorithm. The test data are segmented into second-level characters and sent to the model for recognition. The fourth group uses the proposed three-level segmentation algorithm to segment the test data and feed them into the model for recognition.

Then we calculate the segmentation accuracy and recognition accuracy of the above four groups. The comparative results of the segmentation accuracy and recognition accuracy of the four groups of experiments are as follows (Fig. 13):

It can be seen from the experimental results that the accuracy of only one segmentation algorithm is very low, and the accuracy of the combination of two algorithms is improved, while the accuracy of the proposed three-level segmentation algorithm is the highest among the three algorithms. Compared with the single algorithm, the accuracy is increased by more than 20%.

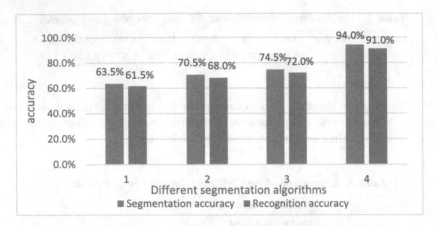

Fig. 13. The accuracy of different character segmentation algorithms.

6 Conclusion

In this paper, we propose a deep learning character recognition algorithm based on multilevel segmentation. This algorithm combines three existing mature character segmentation methods and makes special modification to the design object of this paper. We made some comparative experiments, where the results showed that the algorithm has a relatively better performance in segmenting characters in the inter-locked upper computer interface. The proposed algorithm can improve the accuracy in recognizing the button characters in the interlocked upper computer interface, which is of great significance to the design of the automatic test execution in the interlocking system.

There are several topics worthy for further study in future. Firstly, to find or propose more character segmentation algorithms, to try to replace them with the algorithms in each level of the existing three-level segmentation algorithms, and to investigate whether they will have better effects. Secondly, we will try to build a lager deep learning model to train a more intelligent character segmentation model through a large amount of data. The long-term goal is to develop a fully automated character segmentation and recognition tool with higher accuracy.

Acknowledgment. The work was supported by Beijing Natural Science Foundation (L181005), the subject of the science and technology research development plan of China Railway Corporation (N2018G064), Fundamental Research Funds for the Central Universities under grant 2019JBM002, National Key R&D Program of China (2018YFB1201500) and the Beijing Laboratory of Urban Rail Transit.

References

1. Qiqi, W.: Research and design of computer interlocking simulation system for railway stations. Lanzhou Jiaotong University (2018)
2. Jianhua, L.: Research and application of interlocking software automatic testing technology. Lanzhou Jiaotong University (2013)
3. Lin, X., Yang, Y.: Model-based automatic generation of test case of route establishment process. Ralway Stand. Des. **61**(02), 109–116 (2017)
4. Srivatsava, P.R., Mallikarjun, B., Yang, X.S.: Optimal test sequence generation using firefly algorithm. Swarm Evol. Comput. **8**, 44–53 (2013)
5. Xu, Y.: Research on simulation test method of computer interlocking software. Beijing Jiaotong University (2019)
6. Lin, X.: Automatic generation of test cases for computer interlocking route control process based on UPPAAL. Southwest Jiaotong University (2017)
7. Khurana, N., Chillar, R.S.: Test case generation and optimization using UML models and genetic algorithm. Procedia Comput. Sci. **57**, 996–1004 (2015)
8. Baolian, B.: Research on automatic test technology of station computer interlocking software. Lanzhou Jiaotong University (2017)
9. Jaderberg, M., et al.: Synthetic data and artificial neural networks for natural scene text Recognition. arXiv preprint arXiv:1406.2227 (2014)
10. Shi, B., et al.: Robust scene text recognition with automatic rectification. In: Proceedings of the IEEE Conference on Computer Vision and Pattern Recognition, pp. 4168–4176 (2016)
11. Dan, C., Meier, U., Schmidhuber, J.: Multi-column deep neural networks for image classification. In: IEEE Conference on Computer Vision and Pattern Recognition(CVPR), pp. 3642–3649 (2012)
12. Yan, C.: Research on the application of DSL in software development of computer based interlocking. Beijing Jiaotong University (2012)
13. Hubel, D.H., et al.: Brain mechanisms of vision. Sci. Am. **241**(3), 150–163 (1979)
14. Gong, T., et al.: GPU-based parallel optimization of immune convolutional neural network and embedded system. Eng. Appl. Artif. Intell. **62**, 384–395 (2017)
15. Qiang, S.: Research on segmentation and recognition of low quality printed characters. Nanjing University of Science and Technology (2014)
16. Yun-Jing, X.U., et al.: Character segmentation approach for poor license plate image. Computer Engineering and Design (2019)
17. Zahan, T., et al.: Connected component analysis based two zone approach for bangla character segmentation. In: International Conference on Bangla Speech & Language Processing (2018)
18. Ray, A., Chaudhury, S.: Character recognition using SVM-HMM in a multi-hypotheses architecture. In: 12th IAPR International Workshop on Document Analysis and Systems DAS (2016)
19. Jianliang, X., Yumeng, Q.: Simple license plate recognition algorithm and implementation. Comput. Telecommun. **10**, 50–53 (2011)
20. Chunman, Z., et al.: Multiple vehicle license plate location techniques in complex background. J. Highw. Transp. Res. Dev. **07**, 151–157 (2010)
21. Jagannathan, J., et al.: License plate character segmentation using horizontal and vertical projection with dynamic thresholding. In: International Conference on IEEE Xplore: Emerging Trends in Computing, Communication & Nanotechnology, pp. 700–705. IEEE (2013)

A LSTM-Based Passenger Volume Forecasting Method for Urban Railway Systems

Liya Kang, Hongjie Liu[✉], Ming Chai, and Jidong Lv

School of Electronics and Information Engineering,
Beijing Jiaotong University, Beijing 100044, China
hjliu2@bjtu.edu.cn

Abstract. Accurate and high-resolution passenger volume forecasting is very important for the scheduling and regulation of urban railway trains, both to satisfy the quality of railway service and reduce operation cost. This work proposes a Long Short-Term Memory (LSTM) based method to forecast passenger flow in a railway line. By using LSTM time series forecast model, it not only can adapt to the non-linearity and randomness of subway passenger flow, but also prevent the consequences of gradient explosion. A LSTM network is formulated in this work to forecast the passenger volume for workdays and rest days separately. Actual data from Beijing Metro Line 13 is used to train and test the proposed LSTM network, where 90% of the data are used for training and the others are used for testing. Experimental results prove the efficiency of the proposed method. In this experiment, the forecasted time window is as small as 15 min, the maximum error of forecast is 25.2%, the average error is 10.45%, and the final root mean square error is less than 30. The proposed method and forecast results are very helpful for timetabling and rescheduling of urban railway systems.

Keywords: Urban railway system · Passenger volume forecasting · LSTM network

1 Introduction

1.1 A Subsection Sample

With the rapid development of economy and the significant improvement of urbanization level, the passenger flow of urban travel is also continuously increasing, which has formed a great pressure on the allocation and management of urban public transport. Subway is increasingly becoming one of the main ways of transportation for urban residents in the world. Therefore, as a major artery transportation system of urbanization, an efficient and stable operation environment is conducive to the development of the city. Therefore, the accurate forecast of the passenger volume of the metro line will help to realize the optimization of the operation and dispatching of the metro company, which is an important basis for the planning and scientific decision-making of the metro system [1].

Accurate forecast of the passenger volume is an important index to measure the development of urban public transport. It is helpful for Metro Group to grasp all aspects

© Springer Nature Singapore Pte Ltd. 2020
J. Qian et al. (Eds.): ICRRI 2020, CCIS 1335, pp. 368–380, 2020.
https://doi.org/10.1007/978-981-33-4929-2_25

of passenger volume accurately, and it is also helpful for the managers of Metro Company to work and manage efficiently, which is the basis for the development of urban rail transit. In addition, the accurate forecast of urban rail transit passenger flow can provide subway operators with scientific line passenger volume data, and help managers make scientific decisions based on their own experience [2]. By studying the change and distribution law of passenger volume distribution of urban rail transit lines, we can measure and judge the capacity allocation, vehicle scheduling management, passenger volume change law and vehicle service level of urban rail transit lines, which is conducive to further improving the public transport system management [3].

To sum up, the accurate forecast of passenger volume of urban rail transit is of great significance to the development of the city. In the process of forecasting, researchers need to plan according to the distribution rule and change of urban subway line passenger volume, combined with the production task, line capacity configuration and service level of subway itself.

The main work of this paper are as follows:

(1) A LSTM network is proposed to forecast the passenger volume entering and exiting an urban railway station, the detailed procedure to apply this method is presented.
(2) Actual data from a Beijing metro line is used to verify the effectiveness of the proposed forecasting method. Passenger volume in and out of the station is forecasted separately.
(3) The model is used for weekdays and rest days respectively. The forecasting time window is as small as 15 min. In the process of forecast, researchers need to rely on the distribution law and change of passenger traffic volume on the urban subway line, combined with the production task, line capacity configuration and service level of the subway. Researchers can effectively forecast and accurately grasp the passenger volume of urban rail transit by adopting reasonable forecast scheme and practical forecast means, so as to improve the public management level of urban rail transit and the efficiency and ability of public transport lines to transport passengers.

The rest of this paper is organized as follows. We review the related work on improvement of the forecast method in Sect. 2. Section 3 introduces the construction steps of the prediction model. Section 4 analyzes the prediction results. Finally, Sect. 5 concludes this paper and points out some future research directions.

2 Related Works

According to the existing subway passenger volume forecast literature, there are many technologies and means of public transport passenger volume forecasting, but they can be summarized into two categories: one is regression analysis; the other is time series forecast. For regression analysis method, there are four stages method, such as Kalman filter method and multiple linear regression model method and so on; For time series forecast method, there are discrete forecast model, such as gray neural network model, gray model and time series model method of periodic fluctuation term and so on [4, 5].

Regression analysis is often used in the early stage of urbanization, such as Chicago area transportation in 1962. With the publication of the article "Chicago Area Transportation Study", a four-stage forecast method is proposed [6]. The biggest advantage of this forecast model is that the theoretical basis of the construction model is fully reasonable, which can reflect the outgoing situation of urban residents and the data situation of urban land use, and realize the impact on the passenger volume by the way of feedback on the cross operation between various transportation modes [7]. Another example is multivariate regression, in the process of analyzing the passenger volume of urban rail transit, the paper comprehensively considers the average salary of urban residents, the development of local income and the proportion of urban residents' expenditure. The regression analysis method of "many to many" is adopted, but this forecast method is more suitable for long-term forecast research and railway passenger volume [8]. On the issue of industrial product recycling, a lexicographic multiple target scatter search (SS) method is proposed to solve the proposed multiple target optimization problem [9, 10]. In a word, the regression analysis method is too dependent on the linear causality, which is no longer suitable for the strong randomness and uncertainty of modern subway passenger volume, so the focus of this paper is to analyze a variety of time series forecast models.

Because of its self-adaptive, self-organizing and self-learning ability, neural network has won the favor of many researchers. In 1943, American psychologist W.S. McCullough and mathematician W. Pitts proposed MP model and first proposed the research of artificial neural network. In 1986, Rumelhart and M.Clelland proposed BP neural network learning algorithm [11], and then in 1987, the international neural network society was established. After decades of development, in 2009, Tsung-Hsien Tsai etc. proposed a multivariate time series unit neural network model, then applied it to the forecast of railway passenger demand and achieved good prediction results [12]. Then neural network forecast method is widely used in the field of transportation. Tian and pan have established LSTM RNN model to forecast the 5-min granularity working day road passenger volume. Compared with support vector machine, LSTM RNN has the highest forecast accuracy, and the memory ability of this method is stable. It is also suitable for 15 min, 30 min and 45 min of working days road passenger volume forecasting [13, 14].

In this paper, the time series forecast method of the recurrent neural network (RNN) is not involved in the forecast process, but for the LSTM, in essence, is the development and deformation of RNN model. In order to fully explain the support and analysis of this model for forecast methods, the following two neural network principles will be introduced respectively.

The processing object of RNN is a host of time series data, which is a neural network forecast method based on time series to time series. The data here refers to the historical data obtained, taking the time interval as the granularity division, and then using neural network to make short-term forecast of the future data change trend. The data of research should be able to show the development and change of a certain trend of regular change of something or phenomenon over time [15, 16].

In the early years of this century, it gradually developed into a new algorithm model of deep learning, which can be roughly divided into two types [17]. One is bidirectional RNN, Bi-RNN, the other is the forecast model method that this paper focuses on, that is, Long Short-Term Memory networks (LSTM) [18]. By analyzing the structure of RNN neural network, we can find that there are three layers: input layer, hidden layer and output layer. Before forecast, we can determine a set of activation functions to make the forecasted output pass through the activation function, and then connect the weight of input layer, output layer and hidden layer through the function. This method can connect the historical number in our neural network According to the training situation, we can find the rule and apply it to the weight of the next time automatically [19, 20]. The traditional neural network forecast model only constructs the weight connection relationship between three layers through the activation function, but the special feature of RNN is that the connection function of activation function is applied to the specific neurons in each layer to further improve the effectiveness and precision of training [21, 22] (Fig. 1).

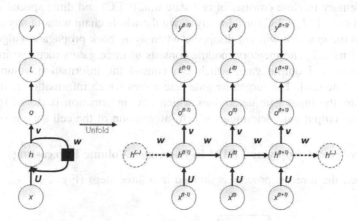

Fig. 1. In RNN network structure.

Each arrow represents a transformation, that is to say, the arrow connection has a weight. The left side is folded, the right side is expanded, and the arrow next to h in the left side indicates that the "loop" in the secondary structure is reflected in the hidden layer.

With the increase of order, the final differential results will be taken as 0, which will cause the gradient explosion. The previous several layers may not have improved the weight, and will not completely eliminate the error [23].

It is easy to cause extreme nonlinear behavior after the error gradient propagates back in multiple time steps, including gradient vanishing and gradient explosion.

In practice, although the serious consequences of gradient explosion may occur, the probability of occurrence is relatively small. In practice, gradient truncation is often used to prevent and solve.

3   LSTM-Based Passenger Volume Forecasting Method

3.1   LSTM Neural Network Model

LSTM model is a branch and module of neural network. It is a chain closed neural network based on the input of the moment and the recursive summary of the previous moment, which has complex nonlinear learning ability. It is specially designed to solve the long-term dependence problem of general RNN. This new method can prevent the gradient explosion in RNN model due to reverse transfer [24].

The main goal of LSTM model is to solve the problem that RNN gradient disappears and it is difficult to process long sequence data [25]. LSTM model consists of input layer, loop hidden layer and output layer [26]. Different from the traditional neural network, the basic unit of the hidden layer of the LSTM model is the memory module. Memory module consists of cell state unit (CEC) and three special operation units called gates [27]. CEC runs directly down the whole chain without any activation function, so the gradient will not disappear when using back propagation algorithm to train LSTM model. The memory module consists of three gate structures: input gate, forgetting gate and output gate, which can control the information volume in the memory module [28]. The forgetting gate determines which information is thrown by the cell state, the input gate determines which new information is stored in the cell state, and the output gate determines which information of the cell state is output.

3.2   Procedure of Applying LSTM in Passenger Volume Forecasting

In this paper, the forecast process is divided into three steps (Fig. 2).

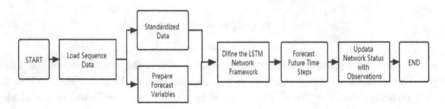

Fig. 2. Procedure of the prediction algorithm

The first part is the standardization of data and the preparation of training work. we set up the corresponding forecast variables and load the original data, and classify the original data: seven days a week as a forecast cycle, and then divided into Monday to Friday as working days and Saturday to Sunday as rest days. And put the data into two files respectively to prepare for the subsequent call training and forecast. In the time sequence, the time step corresponds to the time interval with a certain time interval as a granularity, and the value corresponds to the outbound passenger volume. Each element is a single time step, which reconstructs the data into row vectors.

The second part is to define the network architecture of LSTM and train the historical data to ensure the accuracy of the forecast results, the data is divided into training set and test value, and the training network is constructed. The time series values forecasted by neural network are compared with the known reference test comparison values. The graphs are displayed in the same coordinate axis with different color graphs respectively, and the coincidence degree and error of predicted values and real values can be seen intuitively.

The third part is to forecast the passenger volume in the future. In order to achieve accurate forecast of future passenger volume for a period of time, we require the training function to train the passenger volume data under each time granularity, and forecast the passenger volume data of the next time granularity at the same time, and follow the new network state in real time after each training forecast. In the RNN function forecast above, the previous training data will be used as the historical data of the next training for the current predicted passenger volume data, which will be superposed backward in turn, and the error level will be improved step by step. By using the LSTM time series forecast model, the previous forecast data can be used as the input value of the current function and compared with the known test data to find the error coefficient.

(1) In order to obtain better quasi merging and prevent training divergence, the training data is standardized to have zero mean and unit variance.
(2) In order to prevent the gradient explosion, the gradient threshold value is set to 1, and after 30 rounds of training, the learning rate is reduced by multiplying the factor by 0.2.
(3) For the previous model settings, when following the new network state, for each forecast, the forecast value of the previous time step is used to forecast the next time step. This method is easy to cause large errors with the continuous super-position of data. The improvement used in this paper is: using the observation value instead of the prediction value to follow the new network state.

4 Experimental Results

4.1 Data-Set

This paper selects Xizhimen subway station of Beijing Metro Line 13 from May 1, 2017 to August 31, 2017, and takes 15 min as a particle to carry out the statistics of passenger volume from 5 a.m. to 23:15 p.m. And the date is divided into two kinds: working days and rest days. A total of 120 effective passenger volume analysis days are obtained, i.e. three days from May 1 to May 3 and Dragon Boat Festival on May 30, 2017 are deleted; 86 working days and 34 rest days are obtained respectively. In addition, Xizhimen subway station in Beijing is an office and entertainment center. There are large shopping malls, scenic spots and commercial areas nearby. It is an important transportation hub. In addition, it is also an important transfer station. There are three lines of transfer. The main body of passenger volume is commuters on duty. The volume and trend of personnel are more complex and changeable (Table 1).

Table 1. Xizhimen passenger volume data

Working day date	Passenger volume (person/day)	Rest day date	Passenger volume (person/day)
May 4, 2017	13336	May 5, 2017	13672
May 7, 2017	7026	May 6, 2017	8897
May 8, 2017	17365	May 12, 2017	13772
May 9, 2017	13617	May 13, 2017	8779
May 10, 2017	13660	May 19, 2017	11410
May 11, 2017	13716	May 20, 2017	7847
May 12, 2017	8019	May 26, 2017	13036
May 15, 2017	13793	May 27, 2017	8551
......	......	......	......

This paper makes a statistic on the daily passenger volume of Xizhimen subway station of Beijing line 13 in normal times without special circumstances (too many data, not all listed).

Based on the unit of days, the passenger volume of rail transit changes periodically with every seven days, that is, the distribution of passenger volume changes with the cycle shows a strong correlation (Fig. 3).

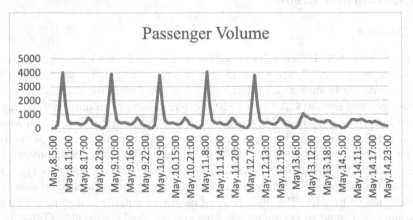

Fig. 3. From May 8, 2017 to May 14, 2017, the daily passenger volume of takes one hour as a granularity.

This week's characteristics are: there is no interference from special events, which belongs to typical normal passenger volume.

The passenger volume of this station is characterized by a significant increase in the morning and evening peak passenger volume on weekdays, among which the morning peak is between 6:30 and 9:00, the evening peak is between 16:00 and 20:00, and the rest of the time is relatively stable; the passenger volume on the rest days is relatively stable, without obvious peak. The station is one of the busiest in Beijing rail transit.

The distribution of passenger volume on Monday, Tuesday, Wednesday and Thursday is similar, and the tidal characteristics of early peak and late peak are obvious. Friday is quite different from the rest of the working day. The difference was mainly between 16 p.m. and 23 p.m., which was later than the usual time of riding. So, using LSTM time series forecast scheme to predict can make a better forecast of the characteristics of the subtle place.

In the rest days of Saturday and Sunday, there are obvious differences compared with the passenger volume on weekdays. The tide volume of morning peak and evening peak does not exist, and the change of passenger volume throughout the day is basically stable without special peak. And the passenger volume from 21:00 to 23:00 on the rest day is significantly higher than that on the working days.

The main reason for this phenomenon is that the proportion of commuter is very high, and for these passengers, work days and rest days are usually distributed in unit cycle. So in the case of the later resume LSTM neural network forecast model, the historical data need to be divided into two types: working days and rest days, which can be forecasted and summarized separately.

4.2 Short-Term Forecast Results

This paper uses 9:1 training forecast ratio, that is, to train the first 90% of the historical passenger volume data, so that the LSTM network model can learn the change of development trend, forecast 10% of the historical data, and compare the results of forecast and observation.

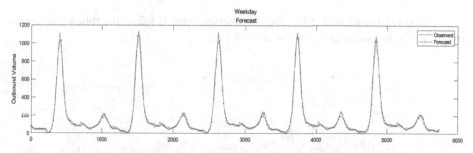

Fig. 4. Comparison between actual outbound passenger volume and forecasted outbound passenger volume. (Color figure online)

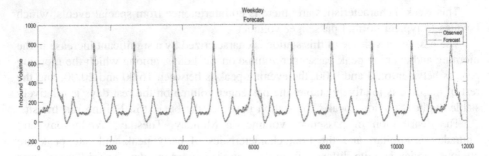

Fig. 5. Comparison between actual inbound passenger volume and forecasted inbound passenger volume. (Color figure online)

Weekday Forecast

The comparison between the actual observed volume change curve (blue) representing passenger volume and the passenger volume change curve (red) after time series forecast shows that the change trend of the two results is basically the same. So, the time series forecast method is very desirable (Figs. 4, 5 and 6).

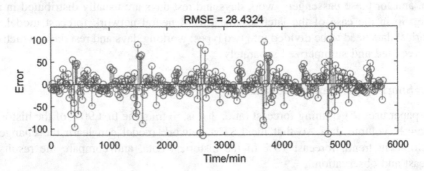

Fig. 6. Working day forecast (outbound passenger). (Color figure online)

The maximum error is 26.54%, and the average error is 8.2%; the root mean square error of outbound passenger volume is 28.4 (Fig. 7).

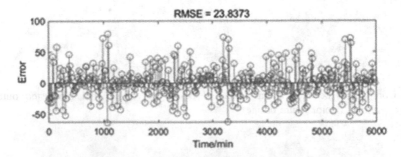

Fig. 7. Working day forecast error (inbound passenger).

The maximum error is 24.43%, and the average error is 12.6%; the root mean square error of outbound passenger volume is 23.8.

Weekend Forecast

In the same way as working days, the figure below shows the passenger volume forecast results corresponding to rest days (Figs. 8 and 9).

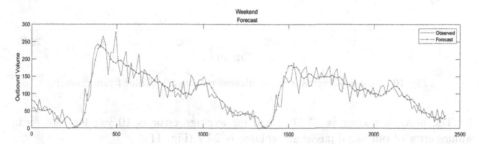

Fig. 8. Comparison between actual outbound passenger volume and forecasted outbound passenger volume.

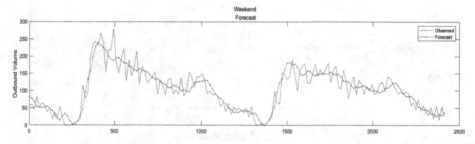

Fig. 9. Comparison between actual inbound passenger volume and forecasted inbound passenger volume.

The blue curve represents the actual trend of passenger volume change, and the red curve is the trend of passenger volume change after the model forecast. It can be seen that there are obvious differences between weekend and weekday passenger volume changes, which is also the reason for the separate forecast of the two in this paper (Fig. 10).

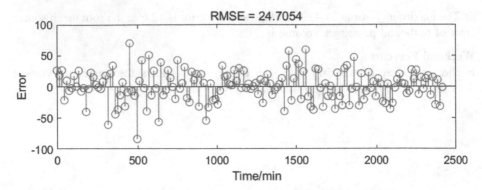

Fig. 10. Rest day forecast error (outbound passenger). (Color figure online)

The maximum error is 24.17%, and the average error is 10.3%; the root mean square error of outbound passenger volume is 24.7 (Fig. 11).

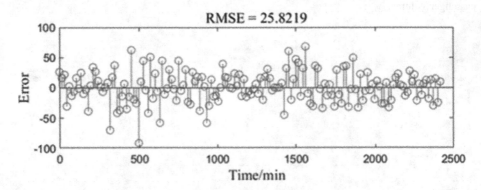

Fig. 11. Rest day forecast error (inbound passenger).

The maximum error is 25.67%, and the average error is 10.7%; the root mean square error of outbound passenger volume is 25.8 (Table 2).

Table 2. error analysis

		Maximum error	Average error
Working day	Outbound passenger volume	26.54%	8.2%
	Inbound passenger volume	24.43%	12.6%
Resting day	Outbound passenger volume	24.17%	10.3%
	Inbound passenger volume	25.67%	10.7%

5   Conclusion

In this paper, the LSTM network model is applied to the passenger volume forecasting of urban rail transit, focusing on the outbound passenger volume of Xizhimen Station of Beijing Metro Line 13. The first 90% data is used for training, the last 10% data is used for forecast, and the forecast results of working days and rest days are obtained.

According to the table, the average maximum error of outbound passenger volume and inbound passenger volume is 25.2%, the average error is 10.45%, and the root mean square error of both is less than 30. the prediction result is good. It is applicable to passenger volume forecast scheme. For urban rail transit, the forecast method of this model can be used to study the change trend of passenger volume without special events, which provides an effective scheme for the subway operators to improve the management system.

In addition, the LSTM model maintains a quantitative error flow in the memory training unit. This stable error flow will not cause gradient explosion and gradient disappearance with the increase of information in the process of model forecast. It can be applied to long-term memory and save a large amount of useful information.

Acknowledgement. The work was partially supported by the Fundamental Research Funds for the Central Universities under grant 2019JBM006, the National Key R&D Program of China (2018YFB1201500), the Beijing Natural Science Foundation (L181005) and the Beijing Laboratory of Urban Rail Transit.

References

1. Zhao, L., Yao, E., Liu, S., et al.: Passenger flow analysis for the surrounding subway stations of large special event site during dissipation. In: The International Conference on Information Systems for Crisis Response and Management (2017)
2. Cervero, R., Pattnaik, J.M.: Rail+Property Development: A Mode of Sustainable Transit Finance and Urbanism (2008)
3. McCulloch, W., Pitts, W.: A logical calculus of ideas immanent in nervous activity. Bull. Math. Biophys. **5**, 115–133 (1943)
4. Cong, Z.: Research on demand forecast of passenger transfer of rail transit and intercity railway-taking Beijing as an example. In: Proceedings of 2017 5th International Conference on Mechatronics, Materials, Chemistry and Computer Engineering (ICMMCCE 2017). Computer Science and Electronic Technology International Society, Chongqing (2017)
5. Jin, F., Li, Y.W., Sun, S.L.: Forecasting air passenger demand with a new hybrid ensemble approach. J. Air Transp. Manage. (2020)
6. McDonald, J.F.: The first Chicago area transportation study projections and plans for metropolitan Chicago in retrospect. Plann. Perspect. **3**(3), 245–268 (1988)
7. Xue, Y.F., Ma, S.: Routing plan of subway line as it's extended based on results of passenger flow forecast. J. Transp. Eng. Inf. (2018)
8. Lu, T.W., Yao, E.J., Liu, S.S., Zhou, W.H.: Short-time forecast of entrance and exit passenger flow for new line of urban rail transit during growth period. J. China Rail. Soc. (2020)

9. Guo, X.W., Zhou, M.C., Liu, S.X., Qi, L.: Lexicographic multi-objective scatter search for the optimization of sequence-dependent selective disassembly subject to multi-resource constraints. IEEE Trans. Cybern. (2019). https://doi.org/10.1109/tcyb.2019.2901834

10. Guo, X.W., Zhou, M.C., Liu, S.X., Qi, L.: Multi-resource constrained selective disassembly with maximal profit and minimal energy consumption. IEEE Trans. Autom. Sci. Eng. (2020, in press)

11. Rumellhart, D.E., Hinton, G.E., Williams, R.J.: Learing representations by back-propagation errors. Nature **323**, 533–536 (1986)

12. Tsai, T.-H.: Neural network based temporal feature models for short-term railway passenger demand forecasting. Expert Syst. Appl. **36**, 3728–3736 (2009)

13. Hou, C.Y., Sun, H., Zhou, Y.F., Cao, B., Fan, J.: Prediction service of subway short-term passenger flow based on neural network. J. Chin. Comput. Syst. (2019)

14. Describe Recent Advances in Applied Sciences (Impacts of Weather on Short-Term Metro Passenger Flow Forecasting Using a Deep LSTM Neural Network). Science Letter (2020)

15. Lint, J.W.C.V.: Online learning solutions for freeway travel time prediction. IEEE Trans. Intell. Transp. Syst. (2008, in press)

16. He, K., Zhang, X., Ren, S., et al.: Going deeper with convolutions. In: Proceedings of the 2017 IEEE Conference on Computer Vision and Pattern Recognition. IEEE Computer Society, Washington, DC (2017)

17. Wu, Y., Tan, H.: Short-term Traffic Flow Forecasting with Spatial-temporal Correlation in a Hybrid Deep Learning Framework, vol. 47, pp. 112–157 (2016)

18. Liu, Y.L., Lv, X.Y., Wang, Y., Kong, D.Y.: Research on the Application of Radial Basis Function Neural Network Model in Beijing-TianJin Inter-city Passenger Flow Forecast. China Academic Journal Electronic Publishing House (2019)

19. Fu, R., Zhang, Z., Li, L.: Using LSTM and GRU neural network for traffic speed prediction using remote microwave sensor data. Transp. Res. Part C **54**, 187–197 (2015)

20. Kuremoto, T., Kimura, S., Kobayashi, K., et al.: Time series forecasting using a deep belief network with restricted Boltzmann machines. IEEE Trans. Intell. Transp. Syst. **137**, 47–56 (2014)

21. Tang, Q.S., Cheng, P., Li, N.: Short time forecasting of passenger flow in urban railway using GSO-BPNN method. Technol. Econ. Areas Commun (2018)

22. Liu, Z.Q., Sun, Y.S.: Driver's lane change intention recognition based on LSTM and D-S theory. Hebei J. Ind. Sci. Technol. (2020)

23. Fu, R., Zhang, Z., Li, L.: Using LSTM and GUR Neural Network Methods for Traffic Flow Prediction. Chinese Association of Automation. IEEE (2017)

24. Zhong, G., Wan, X., Zhang, J., et al.: Characterizing passenger flow for a transportation hub based on mobile phone data. IEEE Trans. Intell. Transp. Syst. **79**, 105–137 (2018)

25. Liu, L., Chen, R.C.: A Novel Passenger Flow Prediction Model Using Deep Learning Approach for Traffic Flow Prediction Research Part C: Emerging Technologies (2018)

26. Huang, J., Levinson, D.M., Wang, Y., et al.: Tracking job and housing dynamics with smartcard data. Proc. Natl. Acad. U.S.A. **24**, 90–108 (2018)

27. Wen, H.Y., Chen, C.W.: Short-term passenger traffic forecast based on deep learning. J. Guangxi Univ. (2020)

28. Roos, J., Gavin, G., Bonnevay, S.: A dynamic Bayesian network approach to forecast short-term urban rail passenger flows with incomplete data. Transp. Res. Procedia **43**, 112–135 (2017)

Intelligent Control of Integrated Energy System

Multi-objective Coordination and Optimal Scheduling of Integrated Energy System for Improving Wind Power Consumption

Ziyang Liu[1(✉)], Yan Zhao[2], and Kaijin Xue[3]

[1] School of Electric Power,
Shenyang Institute of Engineering, Shenyang 110136, Liaoning, China
ziyang_liu@163.com
[2] School of Renewable Energy,
Shenyang Institute of Engineering, Shenyang 110136, Liaoning, China
[3] Information and Communication Branch, Liaoning Electric Power Co., Ltd.,
Shenyang 110006, Liaoning, China

Abstract. At present, China is facing a severe test of the problem of waste wind absorption. As a new technology emerging in recent years, electricity-to-gas technology provides a new method for wind abandonment and will become one of the important technologies in integrated energy systems. This paper first establishes a model of electricity to gas and its operating cost, and builds an energy hub model with electricity to gas. Then, in view of the contradiction between the operating economy and the wind power acceptance capacity of the system when the cost of electricity to gas is high, an improved multi-objective particle swarm optimization algorithm is used for multi-objective coordination and optimization. Finally, the 9-node energy hub system is used as an example to simulate the proposed model. The analysis of the results proves the necessity of taking into account the operation cost of electricity to gas in the integrated energy system, and the feasibility of considering both economics and wind power capacity. By using the algorithm in this paper, while improving the wind power acceptance capacity of the system, it can also effectively ensure the economical operation of the system.

Keywords: Integrated energy system · Wind power dissipation · Power to Gas · Multi-objective optimization

1 Introduction

At present, the output of wind power has the characteristics of intermittent, uncertainty, peak inversion, and large amount of abandoned wind. Constrained by these characteristics, a large amount of wind power is difficult to effectively absorb, and the form of wind power abandonment is severe [1]. The integrated energy system can realize multi-energy co-generation by coordinating and optimizing different forms of energy such as natural gas and electricity, which is an important means to improve energy efficiency [2]. Natural gas has strong storage and transmission characteristics. It can overcome the shortcomings of small storage capacity and high economic cost of electric energy storage systems. It has similar energy flow with the electric power network and has the closest connection. Therefore, the electric-gas integrated energy system has received widespread attention [3].

Power to Gas (P2G) as a hot technology that has emerged in recent years. However, although P2G has a good wind-absorptive effect, it is difficult to significantly reduce its current operating costs. Since P2G operation cost is relatively expensive, it must be reasonably calculated in practical applications. Reference [4] introduced each link of P2G technology, and analyzed the operating cost. Reference [5] introduced the economic feasibility of P2G technology participating in the energy market through the purchase of electricity and the sale of natural gas. Reference [6] used the energy hub model to model P2G technology and analyzed its economics. Reference [7] used the cost-benefit analysis method to determine the optimal capacity allocation for P2G. Reference [8] established a multi-stage power supply, power grid, and natural gas network joint planning model, and proposed a natural gas power flow calculation method to verify the safety of the natural gas network. Reference [9] accurately modeled the natural gas system, and based on this, a dynamic model for joint planning of gas grids was established, which can not only obtain investment decisions for power grids and natural gas pipelines, but also decision results for pressurized stations and natural gas storage.

In view of the above problems, this paper proposes a comprehensive energy system scheduling method that considers the impact of P2G operating costs on the system's wind power acceptance capacity and operating economics. First, based on the establishment of the P2G operating cost and energy hub model, the impact of P2G operating cost on system economics and wind power acceptance capacity is analyzed, and a multi-objective day-to-day optimal scheduling model is established based on this. Then, an improved multi-objective particle swarm optimization algorithm is used to coordinate and optimize multiple targets. Finally, a 9-node energy hub system is used as an example for simulation experiments. The analysis of the results illustrates the impact of P2G operating costs on the system operation, and verifies the correctness of the proposed model.

2   P2G and Establishment of Energy Hub Model

2.1   The Principle of Power to Gas (P2G)

P2G refers to the process of synthesizing natural gas by electrolyzing water and absorbing CO_2. At present, this technology is relatively expensive, but P2G is used to absorb excess renewable energy output (especially when wind power is generated at night and the load is low). It is widely considered to be the most suitable scenario for P2G and has investment value.

The P2G process is divided into two phases: electrolyzed water and methanation. The water electrolysis stage uses alkaline electrolysis or proton exchange membrane technology to electrolyze water to form hydrogen and oxygen, namely:

$$2H_2O \rightarrow O_2 + 2H_2 \tag{1}$$

The hydrogen produced will be further utilized in the methanation stage. At this stage, the reaction speed of hydrogen production is fast, and it can better adapt to the fluctuation of renewable energy output. The amount of hydrogen obtained mainly depends on the capacity of the electrolyzer, the amount of power supply, and the efficiency of the electrolytic reaction. The methanation stage is the use of hydrogen produced by electrolyzed water to react with carbon dioxide to generate methane and water, namely:

$$CO_2 + 4H_2 \rightarrow CH_4 + 2H_2O \tag{2}$$

This stage is often achieved by chemical catalysis at high temperature and pressure. At present, the efficiency of the entire P2G process can reach 50% to 60%.

2.2 P2G Operating Costs

P2G operating costs include the fixed operating costs and the variable operating costs. Fixed operating costs include equipment maintenance costs, labor costs, and so on; variable operating costs refer to the costs required to generate a unit of natural gas, which directly affect the optimization and scheduling of the day before. Therefore, the P2G operating costs mentioned below refer to its variable operating costs, which mainly include electricity costs and raw material costs. Among them, the cost of electricity is directly proportional to the power consumption; the cost of raw materials is mainly the cost of carbon dioxide, which varies greatly depending on its source (such as carbon capture technology, biogas, etc.), and the value is between \$10 and \$1,000/t [10, 11].

In summary, P2G operating costs are expressed as:

$$C^{P2G} = C^E P^{P2G} \Delta t + \alpha C^M P^{NG} \Delta t \tag{3}$$

where C^{P2G} is the running cost of P2G technology; Δt is the running time of P2G equipment; C^E, α, and C^M are P2G electricity price, CO_2 coefficient and CO_2 price coefficient required to generate unit natural gas; P^{P2G} and P^{NG} are the electric power consumed by the P2G and the natural gas power generated respectively. The relationship between the two is shown in Eq. (4).

$$P^{NG} = \eta_{eg} P^{P2G} \tag{4}$$

where η_{eg} is the efficiency of P2G technology.

It can be seen that with different electricity prices and CO_2 sources, P2G operating costs are different. Therefore, when P2G operation cost is high, it must consider its impact on system scheduling operation.

2.3   Modeling of Energy Hub with P2G

P2G can couple the natural gas system and the power system to form a multi-energy system, so that multiple energy systems can be optimized for the elimination and abandonment of wind. The integrated energy system can provide P2G with wider operating flexibility.

Energy hub (EH) model is a modeling method often used in the research of integrated energy systems. EH is defined as an input-output port model used to describe various coupling relationships between different energy sources, networks, and loads in integrated energy. It is versatility and scalability. It is used in the planning and operation of integrated energy. This paper constructs a P2G-containing energy hub model as shown in Fig. 1: Input power and natural gas are converted and stored through P2G equipment, combined heat and power (CHP) units, gas boilers, and gas storage equipment to output electrical energy and thermal energy. Supply load demand. It should be noted that power networks and natural gas networks are generally interconnected in a large range, and thermal networks are generally limited to local small areas due to the nearby supply and demand characteristics and transmission delay characteristics. Therefore, in the model constructed in this paper, it is assumed that thermal energy is only transmitted inside the energy hub, that is, no thermal network is considered.

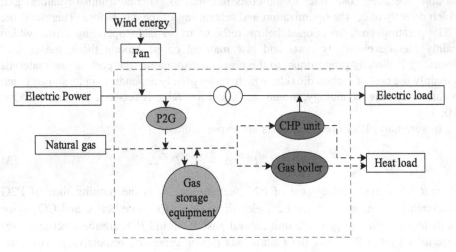

Fig. 1. Energy hub including a P2G device

3 Multi-objective Day-Ahead Scheduling Model for Integrated Energy Systems

Our expectation is to maximize wind power consumption. However, the actual situation is that when the system consumes more wind power, it will increase the P2G operating cost, which will affect the system's wind power acceptance capacity and operating economy to a certain extent. There is a certain contradiction between the two. At this time, if scheduling is performed with the objective of optimal system economics, it may result in reduced output of P2G equipment and reduced wind power acceptance capacity; while scheduling with the maximum wind power acceptance capacity as the goal, P2G operation costs increase, except for higher costs. In addition to the P2G operating costs, each device in the system may deviate from the economic operating point, resulting in poor system economics. Therefore, in this paper, a multi-objective optimized scheduling model is established on the basis of taking into account the P2G operation cost, in order to obtain an optimization result that takes into account the system operation economy and wind power acceptance capacity.

3.1 Objective Function

Goal 1 is the total operating cost of the integrated energy system F_g minimum, including thermal power costs, wind power costs, gas source output costs, gas storage equipment operating costs, and P2G raw material costs. It should be noted that for the unified dispatch agency, the cost of wind power is the cost that the system pays to the owner of the wind turbine, and the cost of P2G electricity consumption is already included in the cost of thermal power or wind power. Therefore, Objective 1 looks like this:

$$\min F_g = \sum_{t=1}^{T} \left[\sum_{i=1}^{N_{tu}} \left(a_i \left(P_{i,t}^{tu} \right)^2 + b_i P_{i,t}^{tu} + c_i \right) + \sum_{j=1}^{N_w} C_j^w P_{j,t}^w \right. $$
$$\left. + \sum_{k=1}^{N_{sp}} C_k^{sp} G_{k,t}^{sp} + \sum_{m=1}^{N_{gs}} C_m^{gs} Q_{m,t}^{out} + \sum_{m=1}^{N_{P2G}} \alpha C^M P_{m,t}^{NG} \right] \tag{5}$$

where N is the number of units and equipment in the system; T is the total number of dispatching periods; $P_{i,t}^{tu}$ is the output power of thermal power unit i during t; $P_{j,t}^w$ is the planned wind power received by wind turbine unit j during t; $G_{k,t}^{sp}$ is t The air flow rate output by air source k; a_i, b_i, and c_i are the parameters of the consumption characteristic curve of thermal power unit i; C_j^w, C_k^{sp}, and C_m^{gs} are the cost coefficients of wind power unit j, air source k, and gas storage equipment m.

Goal 2 is the maximum wind power capacity F_w:

$$\max F_w = \sum_{t=1}^{T} \sum_{j=1}^{N_w} P_{j,t}^w \tag{6}$$

3.2   Constraints

EH Internal Output Constraints

1) *CHP Unit Output Constraints*

$$0 \leqslant P_{m,t}^{\text{CHP}} \leqslant P_{m,\max}^{\text{CHP}} \tag{7}$$

$$0 \leqslant H_{m,t}^{\text{CHP}} \leqslant H_{m,\max}^{\text{CHP}} \tag{8}$$

where $P_{m,\max}^{\text{CHP}}$ and $H_{m,\max}^{\text{CHP}}$ are the upper limit of the power output and the upper limit of the heat output of the CHP unit in the m-th EH.

2) *Gas Boiler Output Constraints*

$$0 \leqslant H_{m,t}^{\text{gb}} \leqslant H_{m,\max}^{\text{gb}} \tag{9}$$

where $H_{m,\max}^{\text{gb}}$ is the upper limit of the heating output of the m-th EH gas boiler.

3) *P2G Output Constraints*

$$0 \leqslant G_{m,t}^{\text{P2G}} \leqslant G_{m,\max}^{\text{P2G}} \tag{10}$$

where $G_{m,\max}^{\text{P2G}}$ is the upper limit of P2G output in the m-th EH.

4) *Gas Storage Equipment Operating Constraints*

The gas storage equipment model includes the gas storage balance constraint shown in (9), the gas storage capacity constraint shown in (10), and the upper and lower limits of gas storage and out-gas power shown in (11) and (12).

$$S_{m,t} = S_{m,t-1} + Q_{m,t}^{\text{in}} \Delta t - Q_{m,t}^{\text{out}} \Delta t \tag{11}$$

$$S_{m,\min} \leqslant S_{m,t} \leqslant S_{m,\max} \tag{12}$$

$$0 \leqslant Q_m^{\text{out}} \leqslant Q_{m,t,\max}^{\text{out}} \tag{13}$$

$$0 \leqslant Q_m^{\text{in}} \leqslant Q_{m,t,\max}^{\text{in}} \tag{14}$$

where $S_{m,t-1}$ and $S_{m,t}$ are still the gas storage capacity of two adjacent periods in the m-th EH; $S_{m,\max}$ and $S_{m,\min}$ are the upper and lower limits of the gas storage equipment capacity; $Q_{m,t,\max}^{\text{out}}$ and $Q_{m,t,\max}^{\text{in}}$ are the gas release power and the gas storage power upper limit, respectively.

In order to reserve a certain adjustment margin for the next scheduling cycle, the gas storage volume after a period of operation is restored to the original gas storage volume, which means that the inflation volume in one cycle is equal to the deflation volume:

$$\sum_{t=1}^{T} Q_{m,t}^{in} = \sum_{t=1}^{T} Q_{m,t}^{out} \tag{15}$$

Power Network Constraints

1) *Node Power Balance Constraint*

$$P_{i,t}^{tu} - P_{i,t}^{e} - U_{i,t} \sum_{j\in i} U_{j,t}\left(G_{ij}\cos\theta_{ij,t} + B_{ij}\sin\theta_{ij,t}\right) = 0 \tag{16}$$

$$Q_{i,t}^{tu} - Q_{i,t}^{e} - U_{i,t} \sum_{j\in i} U_{j,t}\left(G_{ij}\sin\theta_{ij,t} - B_{ij}\cos\theta_{ij,t}\right) = 0 \tag{17}$$

where $Q_{i,t}^{tu}$ is the reactive output power of node i during period t; $P_{i,t}^{e}$ and $Q_{i,t}^{e}$ are the active and reactive power of the EH connected to node i during period t; $U_{i,t}$ and $\theta_{ij,t}$ are the voltage amplitude and node phase angle difference, respectively; G_{ij} and B_{ij} are the real and imaginary parts of the node admittance matrix, respectively; $j \in i$ represents the node connected to node i.

2) *Unit Output Constraints*

$$P_{i,min}^{tu} \leqslant P_{i,t}^{tu} \leqslant P_{i,max}^{tu} \tag{18}$$

$$Q_{i,min}^{tu} \leqslant Q_{i,t}^{tu} \leqslant Q_{i,max}^{tu} \tag{19}$$

$$0 \leqslant P_{i,t}^{w} \leqslant P_{i,t}^{wf} \tag{20}$$

where $P_{i,max}^{tu}$ and $P_{i,min}^{tu}$ are the upper and lower limits of active power output of unit i; $Q_{i,max}^{tu}$ and $Q_{i,min}^{tu}$ are the upper and lower limits of reactive power output of unit i; $P_{i,t}^{wf}$ is the predicted wind power value of wind turbine i during time t.

3) *Node Voltage Constraints*

$$U_{i,min} \leqslant U_{i,t} \leqslant U_{i,max} \tag{21}$$

where $U_{i,max}$ and $U_{i,min}$ are the upper and lower voltage limits of node i.

4) *Branch Flow Constraint*

$$\left|P_{kl,t}\right| \leqslant P_{kl,max} \tag{22}$$

where $P_{kl,max}$ is the upper current limit of branch kl.

Gas Network Constraints

1) *Node Traffic Balance Constraint*

$$G_{i,t}^{sp} - G_{i,t}^{gas} - \sum_{j \in i} f_{ij,t} = 0 \qquad (23)$$

where $G_{i,t}^{sp}$ is the output natural gas flow of the gas source connected to node i at time t; $G_{i,t}^{gas}$ is the natural gas flow of the input of the energy hub connected to node i; $\sum_{j \in i} f_{ij,t}$ is the sum of the flows of natural gas lines connected to node i at time t.

2) *Air Source Output Constraint*

$$G_{i,\min}^{sp} \leqslant C_{i,t}^{sp} \leqslant C_{i,\max}^{sp} \qquad (24)$$

where $C_{i,\max}^{sp}$ and $G_{i,\min}^{sp}$ are the upper and lower limits of the output of the air source point i, respectively.

3) *Gas Node Pressure Constraints*

$$\omega_{i,\min} \leqslant \omega_{i,t} \leqslant \omega_{i,\max} \qquad (25)$$

where $\omega_{i,\max}$ and $\omega_{i,\min}$ are the upper and lower pressure limits of natural gas node i.

4) *Constraint of Pressure Station*

Since natural gas will cause pressure loss due to pipe wall friction and the like during the transmission process, a pressure station is usually required for pressurization. The natural gas flow $f_{ij,t}$ between nodes i and j is the sum of the natural gas flow $f_{nj,t}$ from the pressure station exit n to node j and the natural gas flow $f_{in,t}^{c}$ consumed by the pressure station, that is:

$$f_{ij,t} = f_{nj,t} + f_{in,t}^{c} \qquad (26)$$

The flow $f_{in,t}^{c}$ consumed by the pressurization station is related to the natural gas flow at the outlet and the pressure between the two nodes, namely:

$$f_{in,t}^{c} = f_{nj,t} C_{ij}^{com} \left(\omega_{n,t} - \omega_{i,t} \right) \qquad (27)$$

where C_{ij}^{com} is the constant coefficient of the pressure station between nodes i and j.

5) *Pipeline Flow Constraints*

$$f_{nj,t} = \text{sgn}\left(\omega_{n,t}, \omega_{j,t} \right) C_{nj} \sqrt{\left| \omega_{n,t}^{2} - \omega_{j,t}^{2} \right|} \qquad (28)$$

$$\text{sgn}\left(\omega_{n,t}, \omega_{j,t}\right) = \begin{cases} 1 & \omega_{n,t} \geqslant \omega_{j,t} \\ -1 & \omega_{n,t} < \omega_{j,t} \end{cases} \tag{29}$$

$$f_{nj,\min} \leqslant f_{nj,t} \leqslant f_{nj,\max} \tag{30}$$

where: C_{nj} is the transmission coefficient of the transmission pipeline between node n and node j; $f_{nj,\max}$ and $f_{nj,\min}$ are the upper and lower limits of the flow of pipeline nj.

6) *Natural gas flow can be converted into power flow through its calorific value. The conversion relationship between the two is:*

$$P^{\text{gas}} = H_{\text{GV}} G^{\text{gass}} \tag{31}$$

where P^{gas} is the natural gas power flow; H_{GV} is the high heating value of natural gas, with a value of 39 MJ/m^3.

4   Multi-objective Coordinated Optimization Processing

4.1   Improved Multi-objective Particle Swarm Optimization Algorithm

By coordinating the multi-objectives in the above model, this paper uses an improved multi-objective particle swarm optimization algorithm. In solving multi-objective optimization problems, particle swarm has the advantages of simple algorithm, high search efficiency and fast convergence. For the basic multi-objective particle swarm optimization algorithm, there is no mutation and it is easy to fall into the problem of local optimization and convergence in advance. This article adds mutation operators to improve this situation. Latin hyper-cube sampling was used to generate the first generation particles, so that the particles could be searched in a larger feasible region.

4.2   Eclectic Solution

There is no optimal solution for multi-objective optimization, but many non-inferior solutions constitute the Pareto optimal solution set. Without knowing the decision makers' preferences and the actual situation, the fuzzy theory is used to select the solution with the greatest comprehensive satisfaction as a compromise solution, taking into account multiple goals. Define the fuzzy membership function as:

$$\mu_j = \begin{cases} 1, & f_j \leqslant f_j^{\min} \\ \frac{f_j^{\max} - f_i}{f_j^{\max}} - f_j^{\min}, & f_j^{\min} \leqslant f_j \leqslant f_j^{\max} \\ 0, & f_j \geqslant f_j^{\max} \end{cases} \tag{32}$$

where, f_j represents the value of the j-th objective function; $f_j^{\min}$ and $f_j^{\max}$ represent the minimum and maximum values of the j-th objective function. Then calculate the comprehensive membership U of each objective function according to (33), and choose the largest U as the compromise solution.

$$U = \sum_{j=1}^{J} \mu_j \tag{33}$$

5   Example Analysis

5.1   Study System Introduction

Aiming at the model and algorithm used in this paper, the 9-node EH integrated system is used as an example for programming calculation. As shown in Fig. 2.

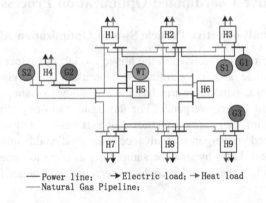

— Power line;   → Electric load; → Heat load
— Natural Gas Pipeline;

Fig. 2. Integrated energy system including 9 energy hubs

In Fig. 2, H1 to H9 are nine energy hubs. The internal structure of H5 is shown in Fig. 1. The remaining energy hubs have no P2G and gas storage equipment inside. H3, H4, and H9 are connected to the thermal power plants G1, G2, and G3, respectively, and H5 is connected to the wind farm WT. H3 and H4 are connected to the air source points S1 and S2.

Among them, the predicted wind power generation capacity is 7169 MW·h. Assume that the load is evenly distributed among the nine energy hubs. Electric power system, natural gas system, heat load, etc. are collectively classified as power unit measurement, and the power reference value is 100 MW, which is expressed by the standard value; the cost reference value is 4 USD/ (MW·h), which is expressed in financial units. Among the operating costs of P2G, $\alpha = 0.2$t/(MW·h), $C^M = 90$ USD/t, that is, P2G's raw material cost coefficient is 4.5 and it is expressed in financial units (mu).

In order to study the characteristic relationship between system operation economy and wind power acceptance capacity, the following four scenarios are set for comparative analysis.

Scenario 1:   There is no P2G in the system, and the goal is to minimize system operating costs.

Scenario 2:   There is P2G in the system, and the goal is to minimize system operating costs.

Scenario 3:   There is P2G in the system, and the goal is to maximize the wind power received by the system.

Scenario 4:   There is P2G in the system, and an improved multi-objective particle swarm optimization algorithm is used to coordinate and optimize multiple targets

5.2   Impact of P2G Raw Material Cost on System Scheduling Operation

In Scenario 1, F_w = 5495 MW·h, F_g = 3725.58 (expressed in financial units). In order to analyze the impact of P2G operating costs on the system's wind power acceptance capacity and operating economy, different P2G raw material cost coefficients (that is, different CO_2 price coefficients C^M) were optimized in Scenario 2 and compared with the optimization results in Scenario 1. As shown in Fig. 3, when the cost of P2G raw materials is not taken into account, wind power is accepted as much as possible because it is limited only by the physical capacity of the system. At this time, F_w = 7149 MW·h and F_g = 3593.46 (in financial units). Comparing with Scenario 1, it can be found that at this time P2G can significantly improve the system's wind power acceptance capacity and reduce system operating costs.

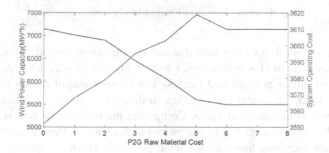

Fig. 3. Effect of P2G material cost on system dispatch

As the cost of P2G raw materials increases, the output of P2G gradually decreases, so the wind power acceptance of the system decreases, and the total operating cost of the system gradually increases. Comparing with Scenario 1, it can be found that at this time P2G can still increase the system's wind power acceptance capacity and reduce operating costs, but its effect is limited by its operating costs.

When the cost of P2G raw materials increases to 6 (expressed in financial units), starting P2G to accept wind power is not economical for the system, so P2G is not started, and the amount of wind power received is fixed at 5495 MW·h, which is the same as Scenario 1.

It can be seen that when P2G operating costs are high, it will affect the system's wind power acceptance capacity and operating economy to a certain extent. This conclusion also fully proves the necessity of taking P2G operation costs into account in the previous scheduling study.

5.3   Analysis of Optimization Result

This article compares and analyzes the optimization results of scenarios 2, 3, and 4. The wind power usage in different scenarios is shown in Fig. 4.

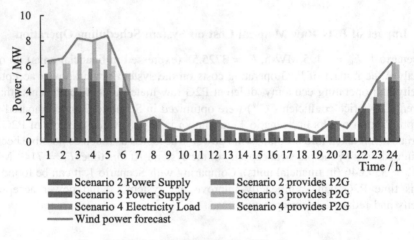

Fig. 4. Wind Power utilization in different scenarios

According to Fig. 4, it can be seen that the system can be roughly divided into two operating states at night (1 – 6 h, 23 – 24 h) and daytime (7 – 22 h). Due to the large surplus of wind power at night and the less wind power output during the day, P2G only starts at night. The wind power usage in different scenarios is shown in Fig. 4, where the power is the standard value. Comparing the different scenarios in Fig. 4, it can be seen that from Scenario 2 to Scenario 4 to Scenario 3, the P2G output at night increases, which increases the system's wind power acceptance capacity.

Compared with Scenario 2, the wind power capacity of Scenario 4 increased from 81.54% to 91.53%; compared with Scenario 3, the system operation cost also decreased. This shows that the improved multi-objective particle swarm optimization algorithm used in this paper deals with multi-objective optimization. While improving the wind power acceptance capacity of the system, it can also effectively ensure the economics of system operation, which fully validates the effectiveness of the model and method in this paper.

6   Conclusion

In the day-to-day scheduling of P2G-integrated energy systems, this paper considers the impact of P2G operating costs on the system's wind power acceptance capacity and operating economy, and proposes an improved multi-objective particle swarm optimization algorithm for multi-objective optimization. Compromises are made and multiple target needs are taken into account. The results of calculation examples show that higher P2G operating costs will affect the system's wind power acceptance capacity and operating economy to a certain extent, causing a certain contradiction between them; and the multi-objective model proposed in this paper can take into account the system operation Economical and wind power acceptance capacity, and can provide choices for scheduling decisions.

This paper does not consider the uncertainty of wind power generation in the modeling. In the future, we will take into account the randomness of wind power output, establish an uncertainty model, make full use of the energy storage capacity of natural gas and thermal systems, respond to random fluctuations in the operation of the system, and realize the adjustment of operation mode. And further consider the impact of P2G operating costs and network dynamic characteristics on the optimal operation of the system.

Acknowledgement. This work was supported by LiaoNing Revitalization Talents Program (XLYC1907138).

References

1. Chen, Z., Zhao, Z., Zhang, Y., Lin, X.: Optimized dispatching and high proportion wind power consumption of electricity-pneumatic interconnection system considering dynamic management. Autom. Electric Power Syst. **43**(09), 31–42 (2019)
2. Li, Y., et al.: Benefit analysis of electricity-gas-heating system with power-to-gas coordination and wind power benefit analysis. Grid Technol. **40**(12), 3680–3689 (2016)
3. Fan, M., Liu, J., Zhu, Z.: Low-carbon economic dispatching strategy for electricity-to-gas comprehensive energy system containing electricity to gas. J. Hydroelectric Energy **37**(10), 204–208 (2019)
4. Götz, M., et al.: Renewable power-to-gas: a technological and economic review. Renew. Energy **85**, 1371–1390 (2016)
5. Clegg, S., Mancarella, P.: Integrated modeling and assessment of the operational impact of power-to-gas (P2G) on electrical and gas transmission networks. IEEE Trans. Sustain. Energy **6**(4), 1234–1244 (2015)
6. Bucher, M.A., Haring, T.W., Bosshard, F.: Modeling and economic evaluation of Power ? gas technology using energy hub concept. In: 2015 IEEE Power & Energy Society General Meeting, pp. 1–5. IEEE (2015)

7. Jentsch, M., Trost, T., Sterner, M.: Optimal use of power-to-gas energy storage systems in an 85% renewable energy scenario. Energy Proc. **46**, 254–261 (2014)
8. Barati, F., Seifi, H., Sepasian, M.S.: Multi-period integrated framework of generation, transmission, and natural gas grid expansion planning for large-scale systems. IEEE Trans. Power Syst. **30**(5), 2527–2537 (2014)
9. Chaudry, M., Jenkins, N., Qadrdan, M.: Combined gas and electricity network expansion planning. Appl. Energy **113**, 1171–1187 (2014)
10. Parra, D., Zhang, X., Bauer, C., Patel, M.K.: An integrated techno-economic and life cycle environmental assessment of power-to-gas systems. Appl. Energy **193**, 440–454 (2017)
11. Vandewalle, J., Bruninx, K., D'haeseleer, W.: Effects of large-scale power to gas conversion on the power, gas and carbon sectors and their interactions. Energy Convers. Manage. **94**, 28–39 (2015)

Research on the Transformation Method of Power Grid Equipment Model Based on Improved BP Neural Network

Qiming Zhuang[1]([✉]), Changyong Yin[2], Dawei Li[3], and Yan Zhao[4]

[1] School of Electric Power, Shenyang Institute of Engineering,
Shenyang 110136, Liaoning, China
1179677489@qq.com
[2] College of International Education, Shenyang Institute of Engineering,
Shenyang 110136, Liaoning, China
[3] State Grid Chaoyang Electric Power Supply Company,
State Grid Liaoning Electric Power Supply Co., Ltd., Chaoyang 122000, China
[4] School of Renewable Energy, Shenyang Institute of Engineering,
Shenyang 110136, Liaoning, China

Abstract. With the rapid development of modern economy and the increase of urbanization rate, the scale of power grid is expanding, and the installed power grid equipment is increasing. Power grid equipment can produce different models based on different standards. Large scale power flow analysis of power grid needs to unify the model of power grid equipment. However, the traditional conversion method is low, which can not meet a large number of grid models generated by the development of power grid planning. Based on the functions of learning, storage and self-adaptive of artificial neural network, this paper proposes a large-scale transformation method of power grid equipment model based on Improved BP neural network. It can improve the efficiency of model transformation of traditional power grid equipment and meet the needs of rapid development of power grid.

Keywords: Power grid equipment models · Transformation method · Improved BP neural network

1 Introduction

With the rapid development of modern economy and the increase of urbanization rate, the demand for electricity continues to grow, and the scale of power grids in various countries is also expanding. Currently, the power system has undoubtedly developed into one of the most complex man-made industrial networks in the world. In the operation of modern power systems, various application systems, power companies and power companies exchange data more and more frequently, but the data standards they use are not necessarily the same. Therefore, when the power flow analysis of a large power system is performed on the dispatching side, the equipment model needs to be converted according to different standards according to the actual situation. Generally, the data needs to be converted into relatively general data.

© Springer Nature Singapore Pte Ltd. 2020
J. Qian et al. (Eds.): ICRRI 2020, CCIS 1335, pp. 397–408, 2020.
https://doi.org/10.1007/978-981-33-4929-2_27

Scholars have done a lot of research on the model transformation of power grid equipment. Some scholars in the Department of electrical engineering of Tsinghua University have analyzed the methods of disassembling and merging the power grid equipment model, and then put forward the conversion principles and specifications based on CIM power grid model. According to the specifications, they have carried out a number of disassembling and merging tests. The specifications have been successfully applied to the actual provincial power grid system dispatching in China [1, 2]. Scholars from put forward the combination method of power grid equipment model based on CIM model and the transformation scheme of regional power grid equipment model and provincial power grid equipment model respectively [3, 4]. Scholars from Jiangxi electric power dispatching center and Henan electric power test and Research Institute have given the conversion scheme including the equivalence of external network, and finally verified the feasibility of the method through practical cases [5]. The relevant companies and the school of electrical engineering of Southeast University have studied the key problems and solutions in model splicing respectively for the problems of insufficient calculation accuracy of the external network model established according to the equivalent model and the inability to share the model information of management centers at all levels in the scheduling field [6].

However, with the improvement of power grid construction, the scale of calculation tasks of power system calculation platform is gradually increasing, and the number of tasks is gradually increasing, which tends to be diversified and dynamic. The traditional manual transformation of power grid equipment model gradually exposes the problems of low efficiency and low accuracy. As a new subject, artificial neural network theory has been widely used in communication field (such as network self configuration, self optimization and self-management) and industrial field (such as defect identification and classification) in recent years. Its real-time, fault tolerance and learning characteristics provide effective theoretical and technical support for the solution of regional problems. At present, BP learning algorithm based on error inverse propagation is widely used. From the perspective of the development of neural network in pattern recognition, the optimization of large-scale network structure and network learning problems need to be further solved and improved [7, 8].

Based on the improved BP neural network, this paper studies a transformation method of different models of power grid equipment, and verifies the rationality and feasibility of the method through the actual data of power grid equipment model, striving to provide a more reliable equipment model for the stability calculation of power system and a more accurate theoretical basis for the evaluation of dispatching plan.

2   Background Technology of Model Transformation

2.1   Model Transformation of Power Grid Equipment

Based on different standards, power grid equipment models have different naming methods. For example, the standard GB/T3601-2017 grid secondary equipment model naming method is: the naming of relay protection equipment and automatic safety device is named by dispatching, and the description is based on the protected equipment, such as: Huazhong Geheyan plant/500 kV. Qingge line. Phase separation current differential protection. For example, the standard DL/T1171-2012 power grid secondary equipment model is named as: relay protection equipment and automatic safety device are named by natural name and described by the protected equipment, such as: Huazhong Geheyan plant/Qingge line. Phase separation current differential protection.

From the above naming method, it can be seen that based on different standards, although the naming method of the same power grid equipment model is different, some of the name items, such as power grid, voltage level, equipment and other parameters remain unchanged. Therefore, the transformation method of different models of power grid equipment is studied based on the learning, storage, adaptive and other functions of neural network [9].

2.2 Characteristics of BP Neural Network

According In power system analysis and calculation, the power grid equipment model is a physical or mathematical description of the characteristics of the power grid equipment, while the characteristics of the power grid equipment model reflect the law that the power grid equipment changes with the operating parameters of the system (mainly voltage and frequency), so there is an essential relationship between the power grid equipment model and the power grid equipment.

BP neural network can realize a mapping function from input to output, and mathematical theory has proved that it has the function of realizing any complex nonlinear mapping, especially suitable for solving internal mechanism complex problems [10]. The network can automatically extract "reasonable" solution rules by learning the case set with correct answers, that is, it has self-learning ability, and the network has certain promotion and generalization ability [11, 12].

However, in practical application, the traditional BP neural network gradually exposed the following shortcomings [13]:

(1) The convergence speed of learning algorithm is slow. When dealing with a more complex problem such as online detection of image defects, because of the high real-time requirements of the system, it requires the convergence speed of the learning algorithm to be faster.
(2) Easy to fall into local minimum. The gradient descent method is used in BP network. The training is that the slope from a certain starting point gradually reaches the minimum error value. For complex networks, the error function is a multi-dimensional surface, so in the process of training, it may fall into a local minimum. Although the traditional algorithm adds a new impulse item to avoid falling into local minimum, the effect is not obvious.
(3) There is no theoretical guidance for the selection of the number of layers and neurons in the hidden layer of the network, but the selection is based on experience. Therefore, the network often has a lot of redundancy, which also increases the time of network learning.
(4) The learning and memory of the network are unstable, and the newly added samples will affect the finished samples.

Based on the above characteristics of BP neural network, this paper uses the improved BP neural network to transform the power grid equipment model. In this paper, the transfer function of BP neural network is improved to improve the performance of the network.

3 Transformation Method of Power Grid Equipment Model Based on Improved BP Neural Network

3.1 Data Preprocessing

According to experience, there are many Chinese characters in power grid equipment model, but artificial neural network can't recognize Chinese characters [14]. So in this paper, the GB2312 code is suitable for the information exchange between Chinese character processing, Chinese character communication and other systems, and the Chinese characters included in the power grid equipment model are transformed into zone bit codes for processing [15, 16]. For example, a character "Huazhong" in a power grid equipment model can be converted into location code 27105448 as the data that can be processed by artificial neural network [17].

The zone bit codes obtained from the above-mentioned character conversion also needs data normalization processing. Data processing is based on the real and reliable sample data, so that the amplitude of input data changes in a certain range. Linear function transformation can be used for data normalization, and the formula is as follows:

$$Y_i = \frac{X_i - X_{\min}}{X_{\max} - X_{\min}} \tag{1}$$

Where X_i is the sample data i in the sample set, Y_i is the normalized sample, $X_{\max}$ is the sample with the largest value in the sample set, and $X_{\min}$ is the sample with the smallest value in the sample set. According to the literature, it is better to take 59005900 for $X_{\max}$ and 11001100 for $X_{\min}$. By normalizing the sample data, the mean value of the input signals of all samples is close to 0 or very small compared with its mean square deviation, which can avoid the phenomenon of neuron saturation, and accelerate the network learning speed.

3.2 Traditional BP Neural Network

Here, take 3-layer BP network as an example. Suppose that the number of neurons in input layer, hidden layer and output layer is N, M and L respectively, corresponding to any input sample vector is $\xi^k = (\xi_1^k, \xi_2^k, \ldots, \xi_N^k)$, $1 \leq k \leq P$, the actual output vector is $C^k = (C_1^k, C_2^k, \ldots, C_L^k)$, the expected output vector is $y^k = (y_1^k, y_2^k, \ldots, y_L^k)$, the connection weight from input layer to hidden layer is $w_{ij}(1 \leq i \leq N, 1 \leq j \leq M)$, the connection weight from hidden layer to output layer is $v_{ij}(1 \leq M, 1 \leq t \leq L)$, the output threshold of cells in hidden layer is $\theta_j(1 \leq j \leq M)$, the output threshold of cells in output layer is $\gamma_t(1 \leq t \leq L)$, $g(x)$ and $f(x)$ are hidden respectively Transfer function with layer and output layer. Let m be the number of iterations of the training network, and the link weight and the actual output are both functions of M [18].

Using the k-th input sample $\xi^k = (\xi_1^k, \xi_2^k, \ldots, \xi_N^k)$, connect the weight w_{ij} and threshold θ_j to calculate the input S_j^k of neurons in the hidden layer, and then use S_j^k to calculate the output O_j^k of neurons in the hidden layer through the transfer function $g(x)$:

$$
\begin{aligned}
S_j^k(m) &= \sum_{i=1}^{N} w_{ij}(m)\xi_i^k - \theta_j(m); \quad j = 1, 2, \ldots, M \\
O_j^k(m) &= g(S_j^k(m)); \qquad\qquad j = 1, 2, \ldots, M
\end{aligned}
\tag{2}
$$

Similarly, the output b_j^k of the hidden layer, the connection weight v_{ij} and the threshold γ_t are used to calculate the input L_t^k of each neuron in the output layer, and then the corresponding C_t^k of each neuron in the output layer is calculated by L_t^k through the transfer function $f(x)$:

$$
\begin{aligned}
L(m) &= \sum_{j=1}^{M} v_{jt}(m)O_j^k(m) - \gamma_t(m); \quad t = 1, 2, \ldots, L \\
C_t^k(m) &= f(L_t^k(m)); \qquad\qquad t = 1, 2, \ldots, L
\end{aligned}
\tag{3}
$$

For all training samples, the global mean square error function between the actual output value C_t^k and the ideal expected value y_t^k of the neural network is expressed as:

$$
MSE = E(m) = \frac{1}{2}\sum_{k=1}^{P} E_k(m) = \frac{1}{2}\sum_{k=1}^{P}\sum_{t=1}^{L} (e_t^k(m))^2 = \frac{1}{2}\sum_{k-1}^{P}\sum_{t-1}^{L} (y_t^k - C_t^k(m))^2
\tag{4}
$$

Where, $e_t^k(m) = y_t^k - C_t^k(m)$, represents the error of the number t neuron in the output layer at the number m iteration of the neural network. After obtaining the square sum error function, update the variable according to the gradient of each parameter of BP neural network and MSE [19, 20]. The formula is as follows:

$$
F(m+1) = F(m) - \eta\frac{\partial E(m)}{\partial F(m)}
\tag{5}
$$

Where η is the learning rate and F is the parameter to be updated, such as the output threshold θ of each unit in the hidden layer and the output threshold γ of each unit in the output layer [21].

For the traditional BP neural network, Sigmoid function is usually chosen as the transfer function of the output layer and the hidden layer:

$$
f(x) = g(x) = \frac{1}{1+e^{-x}} \quad -\infty < x < +\infty
\tag{6}
$$

In optimization theory, BP algorithm is a gradient descent method, and the learning rate of traditional BP algorithm is the step length of gradient descent method. In BP algorithm, the amplitude of each adjustment of network parameters is a learning rate η which is proportional to the network error function. In the traditional BP algorithm, whether in the hidden layer or in the output layer, the learning rate η is always fixed in the process of modifying the connection weight and threshold parameters.

3.3 Improved BP Neural Network

For the new improved algorithm, the hyperbolic tangent S-type function with α as the inclination parameter is chosen as the transfer function of the hidden layer, that is:

$$g(x) = \tanh(\alpha x) = \frac{e^{\alpha x} - e^{-\alpha x}}{e^{\alpha x} + e^{-\alpha x}} = \frac{1 - e^{-2\alpha x}}{1 + e^{-2\alpha x}} = \frac{2}{1 + e^{-2\alpha x}} - 1, \alpha > 0 \qquad (7)$$

The transfer function is a monotonically increasing bounded function of the inclination parameter α. its basic characteristics include:

$$\lim_{x \to -\infty} g(x) = -1 \quad \lim_{x \to \infty} g(x) = 1 \quad g(x) \underset{x \to \min}{\approx} \tfrac{1}{2}\alpha x \qquad (8)$$

$$g'(x) = \alpha \operatorname{sech}^2(\alpha x) = \alpha(1 - \tanh^2(\alpha x)) = \alpha[1 + g(x)][1 - g(x)] \qquad (9)$$

Figure 1 shows the transfer function curve when the tilt function α is 5, 1 and 0.5.

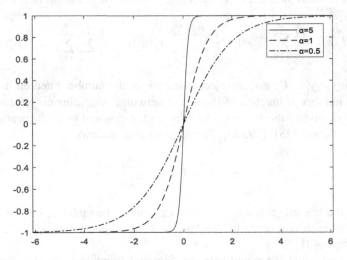

Fig. 1. Transfer function curve

Figure 2 shows the error rate when using $\tfrac{1}{2}\alpha x$ to approximate the transfer function.

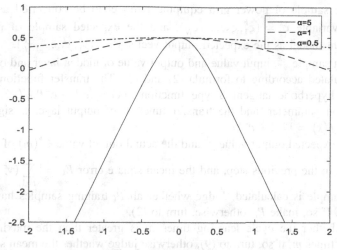

Fig. 2. Error rate curve

The calculation formula of error rate in Fig. 2 is:

$$\left(\frac{2}{1+e^{-2\alpha x}} - 1 - \frac{1}{2}\alpha x\right) \bigg/ \left(\frac{2}{1+e^{-2\alpha x}} - 1\right) \tag{10}$$

It can be seen from Fig. 1 that the smaller the α is, the closer the nonlinear transfer function $g(x)$ is to the linear function $\alpha x/2$, and the smaller the error rate of the linear approximation is. Based on the basic characteristics of the transfer function, the correction Δw_{ij} and threshold θ_j of the hidden layer are proportional to the nonlinear transfer function and the skew parameter α. In order to spend less training time and reduce the calculation amount of each training, it is expected to reduce the α of the transfer function, so that the nonlinear operation is replaced by linear operation; but reducing α makes the whole nonlinear network more linear network, which greatly reduces the advantages of multi-layer nonlinear network in defect detection and identification application to a certain extent, so it is necessary to select one transfer function with the best inclination Slope to balance the speed of network training and defect detection Effect.

3.4   Algorithm Description

The transformation method of power grid equipment model based on Improved BP neural network proposed in this paper is summarized as follows:

(1) Initialize connection weights (w_{ij} and v_{ij}), thresholds (θ_j and γ_t), learning rate η, and transfer function tilt parameter α. Set the expected error standard and the maximum number of iterations M. Let $C^k(m) = (C_1^k(m), C_2^k(m), \ldots, C_L^k(m)), 1 \le k \le P_i$ be the actual output of the network in the m-th iteration, $y^k = (y_1^k, y_2^k, \ldots, y_L^k), 1 \le k \le P_i$ be the expected output, and m be the number of iterations, so that $M = 0$.

(2) Take the sample of power grid equipment model to be converted as the input sample variable $\xi^k = (\xi_1^k, \xi_2^k, \ldots, \xi_N^k)$, and the expected sample of power grid equipment model as the expected output vector $y^k = (y_1^k, y_2^k, \ldots, y_L^k)$.

(3) For input sample ξ^k, input value and output value of hidden layer and output layer are calculated according to formula (2) and (3). The transfer function of hidden layer is hyperbolic tangent S-type function $g(x) = (e^{\alpha x} - e^{-\alpha x})/(e^{\alpha x} + e^{-\alpha x})$ of inclination parameter, and the transfer function of output layer is sigmoid type function $f(x) = 1/(1 + e^{-x})$.

(4) There is expected output value y^k and the actual output value $C^k(m)$ of the sample obtained in the previous step, and the mean square error $E_k = \frac{1}{2} \sum_{t=1}^{L} (y_t^k - C_t^k(m))^2$ of the sample is calculated. Judge whether all P_i training samples have finished learning. If so, make P_i; otherwise, turn to (2).

(5) Judge whether the cycle learning times m is greater than the maximum cycle iteration times m, if so, turn to (9), otherwise judge whether the mean square error of the m-th cycle iteration of the training sample $E(M)$ is less than the expected error standard ε; if so, turn to (7), otherwise turn to (6).

(6) Reset the learning rate η and tilt parameter α and return to (2).

(7) Under the condition of constant learning rate η, judge whether the number of times of change α exceeds 3, if so, turn to (8); otherwise, appropriately change the gradient α of transfer function and return (2).

(8) Comparing the mean square error $E(M)$ under the same learning rate η and different α parameter values, we get the α with the lowest mean square error.

(9) The algorithm ends.

4 Experiment

4.1 Neural Network Training

In this paper, the transformation between power grid equipment models is simulated, and the standard to be converted is set as standard 1, the standard to be obtained is standard 2, the standard 1 is GB/T3601-2017, and the standard DL/T1171-2012 is dl for simulation experiment. The neural network is constructed according to the model transformation method of power grid equipment described in Sect. 3. According to experience, the last character of the power grid equipment model must describe the equipment type, and other characters shall describe the application environment and parameters of the equipment, so there are:

$$N' = N - 1$$
$$L' = L - 1 \tag{11}$$

Where N' and L' are the input sample variables and expected output variables of BP neural network in this simulation respectively, N and L are the number of characters of power grid equipment model specified in standard 1 and standard 2 respectively.

Take the one grid equipment model sample of standard as the input sample variable $\xi^k = (\xi_1^k, \xi_2^k, \ldots, \xi_N^k)$, the other as the expected output vector $y^k = (y_1^k, y_2^k, \ldots, y_L^k)$, $1 \leq k \leq P_i$, and set the expected error standard $\varepsilon = 0.01$ within the group and the maximum cycle iteration number $M = 100$. According to the neural network training steps proposed in Sect. 2.4, the neural network is trained. The change curve of the mean square error function $E(m)$ with the training times is shown in the figure. It can be seen from the figure that at the 100th time, the mean square error function $E(100) < \varepsilon$ meets the expected error standard, and the neural network training is completed (Fig. 3).

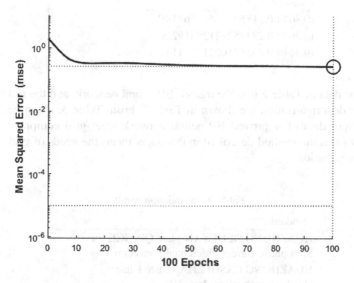

Fig. 3. $E(m)$ changes with training times

4.2 A Subsection Sample

In this paper, simulation analysis is carried out based on the data of a certain power grid equipment model. The data is shown in Table 1.

Table 1. Model data of power grid equipment in a certain place.

Standard 1
HUAZHONG.GEHEYAN plant/500 kV QINGGE line.Split phase current differential protection 1
HUAZHONG.ENSHI/1000 kV.ZHANGEN 1 line Quick disconnection device 1
HUAZHONG.JIUHE plant/10 kV GAOTIAN line.Transformer protection 1

The standard used is GB/T3601-2017, and now DL/T1171-2012 is used as the expected value for simulation. Firstly, the simulation data is preprocessed, the grid equipment model shown in Table 1 is transformed into location code according to Sect. 3.1, the data is normalized according to Formula 1, and each character is regarded as an input variable of BP neural network. The data after preprocessed is shown in Table 2.

Table 2. Model data of power grid equipment after data preprocessing.

Standard1
$(0.3640\ 0.3185\ 0.1064\ 0.6318)^T$
$(0.3640\ 0.27395\ 0.2128\ 0.9229)^T$
$(0.3640\ 0.4335\ 0.0021\ 0.3114)^T$

Input the data in Table 2 into the trained BP neural network, and the results and the results after denormalization are shown in Table 3. From Table 3, it can be concluded that the output data of improved BP neural network meet grid equipment standard, which shows that the method described in this paper meets the needs of grid equipment model transformation.

Table 3. Simulation result

Standard 2
HUAZHONG.GEHEYAN plant.QINGGE line Split phase current differential protection 1
HUAZHONG.ENSHI.ZHANGEN 1 line Quick disconnection device 1
HUAZHONG.JIUHE plant.GAOTIAN line. Transformer protection 1

5 Conclusion

With the rapid development of the modern economy, the demand for electricity continues to increase, and the scale of the power grid continues to expand. The current power system has undeniably developed into one of the most complex man-made industrial networks in the world. The stable, safe, and efficient operation of the power grid is inseparable from the support of power grid equipment model data and operating data. The information interaction between various application systems and the integration between applications have become problems that must be resolved as soon as possible. Therefore, how to perform the conversion between power grid equipment models based on different standards has important research significance for the analysis of power systems.

This paper based on artificial neural network, studies a conversion method of different models of power grid equipment. First, the data preprocessing method is determined according to the characteristics of the power grid equipment model, and then the structure of the BP neural network is determined according to the actual situation of the power grid equipment model that needs to be converted. Finally, the actual power grid equipment model data verifies that the method is reasonable and feasible.

BP neural network has some problems such as slow convergence speed and easy to fall into local minimum. On this basis, this paper improves the transfer function from input layer to hidden layer, proposes a method of power grid equipment model conversion based on Improved BP neural network. The effectiveness of the improved method is verified by the actual data of power grid equipment model.

At present, there are few researches on model conversion of power grid equipment in power systems. The power grid equipment model conversion method based on the improved BP neural network proposed in this paper fills up the gap in this part, provides a more reliable equipment model for the stability calculation of the power system, and provides a more accurate theoretical basis for the evaluation of dispatching plans.

Acknowledgement. This work was supported by Natural Science Foundation of Liaoning Province (2019-MS-239).

References

1. Liu, C., Sun, H., Yao, J.: Research on the application of public information model splitting and merging. Autom. Electric Power Syst. **28**(12), 51–59 (2004)
2. Zhang, H., Huang, H., Song, X.: Research on key technologies of grid model splicing in IEC 61970 standard. CICED **09**, 20–23 (2010)
3. Mi, W., Ling, X., Qian, J.: Application of CIM/XML based grid model merging method in Beijing Electric Power Company Dispatching System. Power Syst. Technol. **32**(10), 33–37 (2008)
4. Qin, C., Zhao, Z., Zhao, T.: Splicing of regional power grid model and provincial power grid model. J. Hehai Univ. **04**, 354–359 (2013)
5. Zou, G., Huang, W., Yao, Z.: Research on the realization method of Jiangxi power grid model splicing considering external network equivalence. Power Syst. Prot. Control **37**(13), 94–97 (2009)
6. Chen, G., Gu, Q.: The whole power grid model based on model splicing. Power Syst. Technol. **34**(12), 94–98 (2010)
7. Zhang, H., Lai, K., Dai, D.: Application of fuzzy neural network on parameter recognition of steel strip's nondestructive testing. Comput. Autom. Measur. Control **11**(1), 14–17 (2003)
8. Wang, Y., Cao, C.: Analysis of local minimization for BP algorithm and its avoidance methods. Comput. Eng. **28**(6), 35–37 (2002)
9. Wang, J., Ren, K., Hu, B.: PID control based on BP neural network. Ind. Control Comput. **24**(3), 72–73 (2011)
10. Zhang, G., Hu, Z.: Improved BP neural network model and its stability analysis. J. Cent. South Univ. (Sci. Technol.) **42**(1), 115–128 (2011)

11. Kamarthi, S.V., Pittner, S.: Accelerating neural network training using weight extrapolations. Neural Netw. **12**(9), 1285–1299 (1999)
12. Martin, F., Moller, S.: A scaled conjuegate gradient algorithm for fast supervised learning. Neural Netw. **6**(3), 525–533 (1993)
13. Chen, B., Wan, J., Wu, Y.: Pipeline leakage diagnosis method based on neural network and evidence theory. J. Beijing Univ. Posts Telecommun. **32**(1), 9–13 (2009)
14. Li, P., Li, X., Lin, S.: Critical review on synthesis load modeling. Proc. CSU-EPSA **20**(5), 56–64 (2008)
15. Xie, H., Ju, P.: Electric load modeling for wide area power system. Autom. Electric Power Syst. **32**(1), 1–5 (2008)
16. Ju, P., Liu, W., Xiang, L.: Automatic post-disturbance simulation based method for power system load modeling. Autom. Electric Power Syst. **37**(10), 60–64 (2013)
17. Yuan, R., Ai, Q., He, X.: Research on dynamic load modeling based on power quality monitoring system. IET Gener. Transm. Distrib. **7**(1), 46–51 (2013)
18. Yoshihiro, O., Talik, K.: A new type neural network PID control for nonlinear plants control. IEEE Trans. Neural Netw. **11**(4), 495–506 (2003)
19. Zhou, H., Zheng, P., Niu, B.: HGA-BP-based pattern classification method. J. Syst. Simul. **29**(8), 2243–2247 (2009)
20. Guan, X., Sun, Y., Cheng, L.: Correlation of loads to wide-area dynamic characters of power systems. Autom. Electric Power Syst. **32**(15), 7–11 (2008)
21. Gutierrez-Martinez, V.J., Cañizares, C.A.: Neural-network security-boundary constrained optimal power flow. IEEE Trans. Power Syst. **26**(1), 64–72 (2011)

Study on Operation Strategy of Thermal Storage in Thermal Power Plant Based on Continuous Discrete Hybrid Control Method

Xiaodong Chen[1], Xiangluan Dong[2(✉)], and Shunjiang Wang[1]

[1] State Grid Liaoning Electric Power Supply Co., Ltd., Shenyang, China
[2] School of Electric Power, Shenyang Institute of Engineering,
Shenyang 110136, Liaoning, China
820370889@qq.com

Abstract. With the continuous growth of power peak load and the rapid development of distributed new energy, the difficulty of power grid dispatching operation is increased, which poses a new major challenge to the power system regulation ability. The operation control target of high temperature solid electric heat storage unit is the temperature and flow control of the output heat energy. The operation control strategy of thermal power unit mainly focuses on how to conduct efficient heating, heat release control and follow the wind abandoning. In this paper, a single analog instruction continuous discrete hybrid control method based on the characteristics of electric storage load is presented. Through the Automatic Generation Control (AGC) regulation and control, the grid coordinated control of load control and regulating units in thermal power plant is carried out. Traditional AGC is limited to power supply control and adopts electric heating and heat storage devices with timing and framing control. In the combined regulation area of electric heating and heat release, the hybrid control method of single instruction continuous regulation and continuous discrete dynamic switching is proposed. Under the condition of small fluctuation of power grid frequency under the existing equipment and channel resources, the heat release and heat production can fully follow the absorption of wind abandoning.

Keywords: Electric heat storage · The control strategy · AGC

1 Introduction

The rapid development of wind power and other new energy sources at home and abroad plays an important role in the power grid. However, new energy sources such as wind power have the characteristics of large capacity and wide distribution. The result is that the problem of new energy consumption can't be solved effectively. The uncertainty and randomness of the new energy sources bring new challenges to the power network planning [1]. The traditional dispatching operation control mode can't meet this demand, which makes the load peak and valley difference increase further,

© Springer Nature Singapore Pte Ltd. 2020
J. Qian et al. (Eds.): ICRRI 2020, CCIS 1335, pp. 409–419, 2020.
https://doi.org/10.1007/978-981-33-4929-2_28

thus affecting the load characteristics of the system. Therefore, it is necessary to establish a new energy consumption oriented power network planning method [2]. Make the system more rotary standby to ensure the absorption of new energy such as wind power [3].

At present, with its mature market mechanism, the United States has become the largest wind power producer in the world [4]. In 2015, the wind power of 1×108 MWh was produced by 74.5 GW, and European countries realized efficient wind power consumption with the help of flexible power supply structure and power interconnection mode [5]. Among them, 42% of Danish power generation comes from wind power, setting a record of the highest proportion of wind power in the world, and the proportion of wind power in other European countries is generally more than 10% [6]. At the end of 2015, the total installed capacity of wind power worldwide was 432.9 GW, with China accounting for 145.4 GW, making it the largest market. However, the problem is that the utilization and consumption of wind power can't be effectively solved. In 2015, the amount of wind power abandoned in China reached 3.39×1010 kW h, and the direct economic loss exceeded 18 billion yuan [7]; The abandoned wind rate in Sanbei area has reached 40%. In the face of such energy problem, how to solve the problem of energy absorption has become an important issue [8].

In recent years, China's new energy grid-connected capacity and consumption have increased rapidly By the end of 2015, the average annual growth rate during the 12th five-year plan period was 34%. The installed capacity of solar power generation is 43.18 GW, with an average annual growth rate of 119% during the 12th five-year plan period. Among them, the cumulative installed capacity of wind power in the dispatching range of state grid company reached 116.64 GW, accounting for 91% of the country. The installed capacity of solar power generation reaches 39.73 GW, accounting for 92% of the country's total. State grid has become the world's largest grid with access to new energy capacity. In 2015, China's wind power generation capacity reached 185.1 TWh, with an average annual growth rate of 30% in the 12th five-year plan. Solar power generation grew at an average annual rate of 219% in the 12th five-year plan [9]. The growth rate of wind power and solar power generation was 28.7 percentage points higher than that of national power generation in the same period. The proportion of wind power generation in total power generation increased from 0.7% in 2010 to 3.23% in 2015, and the proportion of solar power generation increased from 0.003% to 0.688% [10].

The ratio of new energy installed capacity to maximum load (i.e. new energy source penetration rate) is 22% in China, higher than that of the United States (10%), lower than that of Denmark (93%), Spain (78%) and Portugal (63%), which is in the medium level. On the whole, China has made remarkable achievements in the development of new energy. The total installed capacity is the largest in the world, and the consumption volume has achieved rapid growth. However, under the condition that the overall permeability of new energy is not prominent, the wind and light power are increasing continuously, which attracts extensive attention from the society and has become the focus of academic research. How to reduce air abandoning and light abandoning needs to be further discussed from the mechanism in combination with China's reality, and the root of the problem is analyzed to find out scientific solutions.

Under the traditional mode, the dispatching realizes peak and frequency modulation by controlling the power supply to adapt to the load change, so as to ensure the balance between supply and demand of the power grid. Power supply in the grid in China, including thermal power, nuclear power, wind power, solar, hydropower, batteries, in recent years the rapid growth of the nuclear power, wind power and photovoltaic relatively slow growth, but load oversupply phenomenon is relatively serious, wind power and photovoltaic existence of intermittent characteristics at the same time, the operation of power grids bring bigger potential safety hazard, clean energy acceptance problem is outstanding [11]. Even in some areas in winter to abandon the wind reached 50%, even for greater use of clean energy, reduce thermal power, protect the ecological environment, and combining region winter heating, development construction electric heat storage system, electric heat storage system with flexible load characteristic, through coordinate grid scheduling, control of electric heat storage cast back, promote clean energy acceptance ability of power grid. The electric heat storage system in this study includes large capacity electric heat storage and distributed electric heat storage in thermal power plants. Through improving the real-time control method of AGC dispatching for large capacity power storage and heat storage, and realizing the coordinated dispatching of distributed power storage and heat storage through power grid dispatching.

2 Principle of Correlation of Electric Heat Storage

2.1 Operation Control Principle of Single Electric Heat Storage Unit

The single electric heat storage equipment includes four steps: heat storage, heat control, heat release and heat transfer [12].

Heat storage: the heating elements are embedded in the solid heat storage body at multiple points, and the heat generated after the heating elements are started is stored in the solid heat storage body. This medium can store heat energy with a maximum energy storage temperature of 800°.

Heat control: in order to make the stored heat be used effectively, the heat storage tank outer layer adopts ceramic fiber insulation body, so that the high temperature heat storage tank and the outside environment to achieve thermal insulation, in order to prevent heat loss (see Fig. 1).

Heat release: the high temperature heat energy stored in the solid heat storage pool is released by the built-in circulating fan. According to the set temperature requirements, the fan speed is controlled by the inverter in an orderly and stable manner and released outward. The fan is driven by a stepless speed regulating motor (see Fig. 2).

Heat transfer: the load need heat supply, equipment can be according to the pre-defined procedures, according to the set temperature and heating load, provided by the cycle of high temperature air, automatic frequency conversion fan through an air water heat exchanger to heat exchange, load circulating water pump by load will provide hot water to the terminal device (such as fan coil, radiator, or living hot water), the purpose of heating.

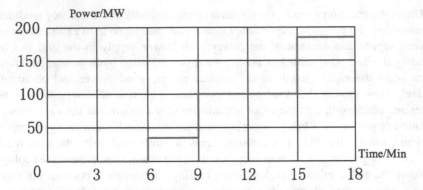

Fig. 1. Characteristics of heating control.

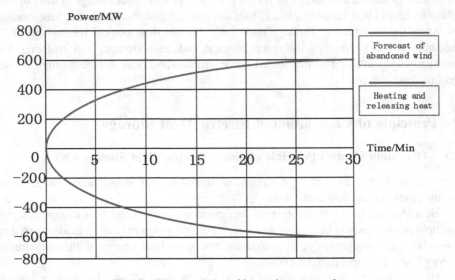

Fig. 2. Characteristics of heat release control.

2.2   Operation Control Principle of Single Electric Heat Storage Unit

When the working state of the automatic detecting device in power grid is in low work status or abandon the wind power system exist, another is the case in the preset time period or wind power grid trough abandon the wind, in both cases, the automatic control system will automatically turn on the high voltage switch, by the high voltage network heating unit power supply, heating units linked together by the level of resistance heating rods, converts electrical energy into heat energy at the same time be heat storage unit (regenerator) constantly absorbing, when the temperature of the heat storage unit to set the upper limit of temperature or grid off-peak end or wind power to refuse the period at the end of the wind, The heat control system cuts off the system switch, the high-voltage power grid stops the power supply, and the heating unit stops working.

When need external heating or hot water, the heat stored in the regenerator will play a role, through the heat exchange device and heat fan, first fan driven by heat regenerative high temperature gas flow in the body, and then let the high temperature gas through the heat exchanger to heat the water in the heat exchanger pipe for water heating in the heat exchanger to a certain temperature. In order to keep the water outlet temperature and the temperature of the return water stable, it is necessary to continuously check the temperature of the regenerator and control the water temperature by controlling the speed of the variable frequency fan through the control system (see Fig. 3).

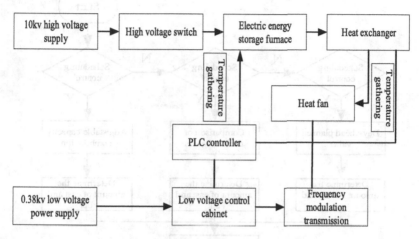

Fig. 3. Flow chart of heat storage body.

2.3 Electric Heat Storage Control Function

There are three control methods of electric heat storage: dispatching control, unconventional control and planning control. The dispatching control is to sum up any combination according to the capacity of each heat storage unit of the station, arrange all and values in order, and then manually determine the input/cut amount by the dispatcher. The algorithm is as follows: according to the measurement information of the electric heat storage boiler, determine the sequence of adjustable capacity of electric heat storage:

$$Preg = \{P1, P2, P3, P4, [P1+P2], [P1+P3], \cdots, [P1+P2+P3+P4]\} \quad (1)$$

Where: P1, P2, P3 and P4, representing the capacity of each heat storage unit of a certain field; [P1 + P2] and [P1 + P3] represent the adjustable capacity of each heat storage unit after free combination.

When under unconventional control, firstly compare the different gear positions of real-time unconventional and system preset unconventional, automatically put in/cut off an appropriate amount of electric heat storage devices to quickly balance the fluctuation of the power grid on the premise of fully considering the security of the power grid and leaving certain reserve. The plan control is to receive the plan value through the day-ahead plan, determine the input cut amount under the premise of fully considering the safety of the power grid and leaving certain reserve, and issue the control instruction after a series of instructions are verified. When the load is low, the electric heat storage boiler can be used as the load heat storage, which can balance the power generation and store the electric energy in the form of heat energy for standby, which can fill the valley of the power grid (see Fig. 4).

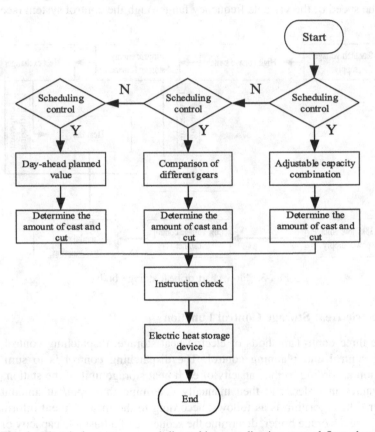

Fig. 4. Electric heat storage and dispatching coordination control flow chart.

3   Continuous Discrete Hybrid Control Method for Electric Heat Storage

3.1   Control Principle of AGC Thermal Power Unit

AGC system is an automatic power generation control system for three kinds of power sources and loads: thermal power, electric heat storage and battery. The system structure is shown in the following figure. Secondly, different regulation modes, such as automatic, manual and planned modes, are selected, and the power regulation demand is calculated according to the above grid operation data. In combination with the peak and valley conditions of the grid load, the starting conditions and regulation strategies of each type of energy participating in the regional regulation are determined. Finally, a regulation mechanism with thermal power regulation as the main part and thermal power storage as the auxiliary part is established to regulate various types of energy units in the power plant. In this way, the peak-valley difference of power grid load can be reduced to a certain extent, and unnecessary starting and stopping can be reduced, so as to realize peak-filling (or load transfer) and maximize the acceptance of clean energy (see Fig. 5).

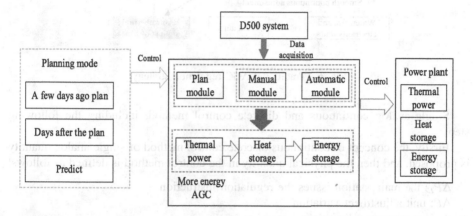

Fig. 5. AGC system structure diagram.

3.2 Traditional AGC Adjustment Strategy

Traditional AGC confined to power control, the timing tentering control electrical heating device for storing heat, there exist deviations in tie line, the system frequency deviation, can only achieve the wind generator for thermal power unit and smooth continuous adjustment, unable to realize the automatic control of electric heating, if use manual control, there will be followed, the retention time and the adjustment problems, cause peak shaving and abandon the wind given effect is poorer (see Fig. 6).

3.3 A Continuous Discrete Hybrid Control Method for Single Analog Instruction

By issuing a certain control quantity to AGC, the unit can automatically judge the continuous, discrete and mixed control modes of the control quantity according to the numerical value, and automatically adjust the heat input/cut storage. In this paper, the coordinated control system of wind abandon, heat production and heat release needs to solve the difficult problems of heat production and frequency, heat production and heat release, single instruction and multi-object.

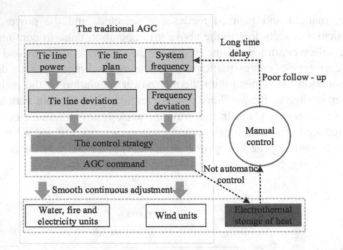

Fig. 6. Traditional AGC regulation strategy

Single analog continuous and discrete control method, including the following steps:

Firstly, the concept of continuous discrete control method of single analog quantity is proposed, and then the physical quantity in the control method is defined as follows:

ΔP_L: the main station issues the regulation instruction
ΔP: unit adjustment variation
$\Delta P'$: continuous adjustment
$\Delta P''$: discrete regulating quantity
ΔP_{MAX}: continuously adjust the threshold

Main station issues regulation instruction B; Each power plant receives instructions and issues a certain control quantity (analog quantity) to AGC. AGC selects a reasonable control method by analyzing the value of the control quantity.

When instruction B is issued, AGC carries out continuous control (see Fig. 7).

When instruction B is issued, AGC carries out continuous and discrete hybrid control. At this time:

$$\Delta P_L = \Delta P' + \Delta P'' \qquad (2)$$

$$\Delta P' = \min(\Delta P, \Delta P\max) \qquad (3)$$

As shown in the figure $t_2 - t_3$, through the control of the analog quantity issued by AGC, the single-instruction continuous and continuous discrete dynamic switching hybrid control is realized.

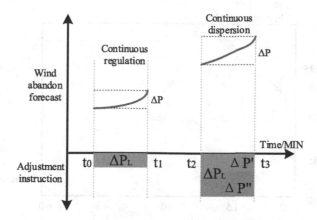

Fig. 7. Hybrid control method

4 Simulation

In this paper, a power plant is taken as an example. A power plant has a total of 280 MVA, which is composed of four electric heat storage groups with capacities of 60 MVA, 60 MVA, 80 MVA and 80 MVA respectively. Let X = 60 MVA and Y = 80 MVA respectively. Theoretically regular adjustment strategy of single reservoir group investment alone, first put into small capacity electric heat storage group, put in large capacity storage heat group, input process of 60 MVA, 60 MVA, 60 MVA, 60 MVA, 60 MVA, 80 MVA, 80 MVA, 80 MVA, exit process is also out of small capacity electricity thermal storage group first, then exit the large capacity storage heat group, exit process for 60 MVA, 60 MVA, 60 MVA, 60 MVA, 60 MVA, 80 MVA, 80 MVA, 80 MVA, but in the short-term prediction strategy and network loss under the optimal control strategy, The input mode may change greatly. The super-short term prediction is based on the actual situation of the load power supply, and the optimal strategy of network loss is based on the network loss after input (see Fig. 8).

The figure above illustrates that the frequency of the system fluctuates up and down due to wind abandoning. The frequency fluctuation diagram is obtained according to the actual measured frequency.

The following diagram shows the wind power following process. The blue line represents the power of electric heat storage and heating, the yellow line represents the wind abandoning power, and the black line represents the number of units invested in electric heat storage (see Fig. 9).

The diagram below shows the addition of electrical heat storage followed by the absorption of wind abandon. The two lines are very close to each other, indicating that these abandoned wind can be followed by absorption according to the amount of investment (see Fig. 10).

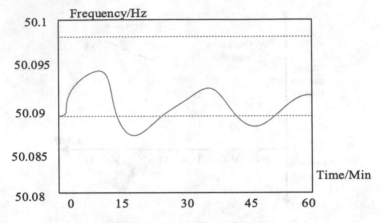

Fig. 8. Frequency fluctuation curve

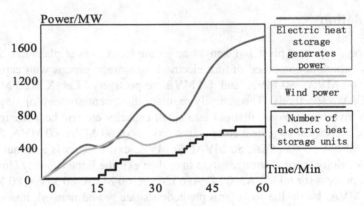

Fig. 9. Wind power follows the process

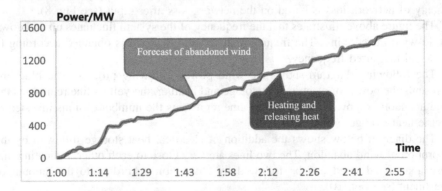

Fig. 10. Electric heat storage and heating power following process

5 Conclusion

In this paper, AGC hybrid control method is used to coordinate the control of thermal power and heat storage, which plays a positive role in promoting the new energy generation forms into the grid. For the coordinated control of distributed electric heat storage, its input is conducive to reducing the peak-valley difference of the load curve, adjusting the load peak-frequency modulation, and improving the consumption of new energy. The reasonable input of electric heat storage can effectively ensure the stable waveform of the load curve.

Acknowledgement. This paper was supported by Science and Technology Project of State Grid Corporation of China (SGLNXT00SJYYXX1900760).

References

1. Sun, R., Zhang, T., Liang, J.: Evaluation and application of wind power integration capacity in power grid. Autom. Electric Power Syst. **35**(4), 70–75 (2011)
2. Wang, Z.M., Su, A.L., Lu, S.: Analysis on capacity of wind power integrated into Liaoning power grid based on power balance. Autom. Electric Power Syst. **34**(3), 86–90 (2010)
3. Wei, L., Jiang, N., Yu, G.: Research on Ningxia power grid's ability of admitting new energy resources. Power System Technol. **34**(11), 176–181 (2010)
4. Zalba, B., Marin, J.M., Cabeza, L.E.: Review on thermal energy storage with phase change: materials, heat transfer analysis and applications. Appl. Therm. Eng. **23**(3), 251–283 (2003)
5. Jin, Y., Lee, W.P., Ding, Y.: A one-step method for producing microencapsulated phase change materials. Particuology **8**(6), 588–590 (2010)
6. Zhang, H.F., Ge, X.S., Ye, H.: Characteristics of the heat charge and discharge of the encapsulated phase charge materials. Acta Energiae Solaris Sin. **26**(6), 825–830 (2005)
7. Kurnia, J.C., Sasmito, A.P., Jangam, S.V.: Improved design for heat transfer performance of a novel phase change material (PCM) thermal energy storage (TES). Appl. Therm. Eng. **50**(1), 896–907 (2013)
8. Lv, Q., Li, L., Zhu, Q.S.: Comparison of coal-saving effect and national economic indices of three feasible curtailed wind power accommodating strategies. Autom. Electric Power Syst. **39**(7), 75–83 (2015)
9. Fragaki, A., Andersenb, A.N., Tokec, D.: Exploration of economical sizing of gas engine and thermal store for combined heat and power plants in the UK. Energy **33**(11), 1659–1670 (2008)
10. Lund, H., Mathiesen, B.V.: energy system analysis of 100% renewable energy systems—the case of Denmark in years 2030 and 2050. Energy **34**(5), 524–531 (2009)
11. Liu, M., Saman, W., Bruno, F.: Review on storage materials and thermal performance enhancement techniques for high temperature phase change thermal storage systems. Renew. Sustain. Energy Rev. **16**(4), 2118–2132 (2012)
12. Lv, Q., Jiang, H., Chen, T.Y.: Wind power accommodation by combined heat and power plant with electric boiler and its national economic evaluation. Autom. Electric Power Syst. **38**(1), 6–12 (2014)

Research on High Precision Adaptive Phasor Measurement Algorithm Based on Taylor Series and Discrete Fourier Transform

Guangfu Wang[1], Huanxin Guan[2], Peng Jin[3], and Yan Zhao[2(✉)]

[1] School of Electric Power, Shenyang Institute of Engineering,
Shenyang 110136, Liaoning, China
m18265537369@163.com
[2] School of Renewable Energy, Shenyang Institute of Engineering,
Shenyang 110136, Liaoning, China
[3] State Grid Liaoning Electric Power Co., Ltd.,
Shenyang 110136, Liaoning, China

Abstract. The application of HVDC power transmission, flexible ac power transmission and large-scale grid connection of new energy have introduced a large number of power electronic equipment to the power grid. This results in frequent subsynchronous oscillations, and the harmonic disturbance of power grid presents a broadband trend, which seriously affects the accuracy of phasor measurement. To solve the problem that the traditional phasor algorithm can not satisfy the demand of precision and speed in the dynamic process, this paper proposes a comprehensive adaptive phasor algorithm based on Taylor series and discrete Fourier transform (DFT). For steady state and dynamic state, the time domain algorithm and the frequency domain algorithm are designed respectively. Specifically, the time-domain algorithm USES two adjacent data Windows for DFT analysis, and is simplified according to specific accuracy requirements to calculate the frequency and phasor. The Taylor series expansion of the power signal model is carried out by the frequency domain algorithm, and the phasor, frequency and frequency change rate are calculated by the fundamental wave and each harmonic content of a data window. Finally, simulation analysis and experimental test results show that the measurement accuracy and response performance of the proposed algorithm are better than that of the traditional algorithm and the corresponding commercial synchronous phasor measuring device in both steady state and dynamic state, and meet the practical application requirements.

Keywords: New energy grid · Phasor measurement · Discrete Fourier transform · Adaptive phasor algorithm

1 Introduction

With the large-scale development and utilization of renewable energy and the development of smart power grid, China has now built a super-large and complex interconnected power system [1]. It has the highest level of transmission voltage and

J. Qian et al. (Eds.): ICRRI 2020, CCIS 1335, pp. 420–433, 2020.
https://doi.org/10.1007/978-981-33-4929-2_29

accommodates the largest amount of renewable energy in the world, and poses a great risk for safe operation [2]. Large-scale renewable energy that used to grid introduced a number of inverter, inverter, unified power flow controller of new power electronic equipment [3]. In the power system of power supply, power grid and load each link has obvious power electronic characteristics, cause subsynchronous oscillation disturbance phenomenon such as more frequent, presents the trend of the broadband domain grid harmonic interference, therefore improve the monitoring ability of power grid distur-bance of different frequency components, reduce the risk of power grid operation [4]. In recent years, the wide area measurement system (WAMS) based on the synchronous phasor measurement device (PMU) has been widely used in the fields of power system dynamic process monitoring, online identification, security and stability analysis and wide area control [5]. With the deepening of WAMS application research, PMU requires more and more synchronous phasor measurement, and the accuracy and rapidity of its phasor algorithm will directly affect the reliability of related application functions [6].

At present, scholars at home and abroad have put forward a variety of synchronous phasor measurement algorithms, including discrete Fourier transform (DFT) method, dynamic phasor method, wavelet transform method, digital filter method and so on [7]. Among them, DFT algorithm is widely used because of its fast computation speed and strong harmonic suppression ability [8]. Literature [9] proposed a DFT calculation results are modified phasor measurement algorithm, compared with traditional DFT algorithm, the algorithm significantly improves the calculation precision, such as dynamic process response speed is limited, but for mutations and simplify the process done approximation is bigger, when great moment, amplitude changes in the frequency deviation meets the requirement of accuracy [10]. Recently, literature [11] proposed an algorithm based on the frequency domain dynamic model, which used the response of different frequency point filters of the same data window to modify the DFT estimation results, and improved the response speed of dynamic processes such as mutation. To solve the problem of weak noise suppression ability and low accuracy in fault process, literature [12] further proposes phasor algorithm using time-frequency information, which gives consideration to accuracy and speed requirements to some extent, but they do not give the method of frequency and harmonic suppression. As for the actual PMU, the PMU from the major manufacturers in China and the relevant test results abroad also show that the phasor algorithm used in various kinds of devices is still insufficient in the application range and practicability [13].

In view of the above problems, this paper also considers the speed and accuracy requirements of PMU algorithm, and proposes a comprehensive adaptive phasor algorithm: (1) improve the time domain algorithm, propose the use of Taylor series expansion to solve the original approximate processing problem, so as to significantly improve the accuracy of phasor measurement, still meet the national standard and IEEE standard requirements when the frequency deviation is large; (2) an improved frequency-domain algorithm is proposed to calculate the phasor, frequency and fre-quency change rate by calculating the fundamental wave and harmonic content on a data window. A kind of adaptive switching logic is proposed, which takes into account the calculation accuracy of steady state and the dynamic performance of transient state. By referring to the test methods given by national standards and IEEE standards, the advantages of the proposed algorithm over the traditional DFT and the algorithm proposed in literature [9, 11] are verified through simulation. Furthermore, the physical experiment platform was used to test the measurement performance of the actual PMU developed based on the algorithm in this paper, and the measurement accuracy and response performance of the PMU under dynamic conditions were comprehensively compared with that of a mainstream commercial PMU.

2   Signal Modeling

Considering that the amplitude and phase Angle of the dynamic power signal may change with time, this paper adopts the complex signal to represent the dynamic phasor of the power signal. The power signal can be expressed as:

$$P(t) = a(t)e^{j\theta(t)} \tag{1}$$

$$x(t) = P(t)e^{j2\pi f_0 t} + P^*(t)e^{-j2\pi f_0 t} \tag{2}$$

Where, $a(t)$ and $\theta(t)$ represent the polynomials of the amplitude and phase Angle of the power signal respectively, f_0 is the rated frequency, $*$ and represents the conjugate.

The steady-state calculation mode uses the first-order model, assuming that the amplitude and frequency deviation are constant in the calculation period, that is: $a(t) = a$, $\theta(t) = \theta_0 + \theta_1 t$, Where a, θ_0, θ_1 are respectively polynomial coefficients of amplitude and phase Angle. According to the specific accuracy requirements, the Taylor series expansion is simplified, and the DFT results of two adjacent data Windows are modified, with less computation and strong harmonic suppression ability, which meets the steady-state accuracy requirements.

In the dynamic calculation mode, it is assumed that the amplitude and phase Angle in the phasor model are k-order models (K is a natural number, in special case K = 0, corresponding to the traditional DFT algorithm), that is $a(t) = \sum_{i=0}^{K} a_i t^i$, $\theta(t) = \sum_{i=0}^{K} \theta_i t^i$. This can better reflect the dynamic characteristics of the signal. The phasor, frequency and frequency change rate are calculated by the fundamental wave and harmonic content of a data window, so as to meet the requirement of fast response of dynamic process.

3   Algorithm Synthesis Implementation

3.1   Time Domain Algorithm

In the steady-state computing mode, the original approximate processing problem is solved by the time-domain algorithm and the Taylor series expansion. Specifically, the discrete signal model is obtained by sampling formula (2) at N points every period, and the DFT transformation of the signal model is carried out. In this paper, the rectangular window is selected. Thus, the phasor of the power signal can be obtained as:

$$X = a_c e^{j\theta_c} = a e^{j\theta_0} e^{j\frac{\theta_1 (N-1)}{2Nf_0}} A(1 + e^{-jC}B)$$ (3)

$$\begin{cases} A = \frac{1}{N}\sin(\frac{\theta_1}{2f_0}) / \sin(\frac{\theta_1}{2Nf_0}) \\ B = \sin(\frac{\theta_1}{2Nf_0}) / \sin(\frac{\theta_1 + 4\pi f_0}{2Nf_0}) \\ C = 2\theta_0 + (\theta_1 + 2\pi f_0)(N-1)/(Nf_0) \end{cases}$$ (4)

Where, a_c and θ_c respectively represent the calculated amplitude and phase Angle of DFT before correction.

To solve the problem that the approximate error of traditional DFT correction algorithm is large, the time domain algorithm proposed in this paper controls the error by expanding the Taylor series. To be specific, the error requirements under strict conditions of national standard and IEEE standard are considered comprehensively. In the process of approximate derivation, the approximate error should be within the order of 10^{-5} and below. Taking the approximation of B as an example, the Taylor series in the numerator and denominator are expanded to order 1 and order 3 respectively and simplified. As shown in Eq. (5), in the adverse case of sampling at 32 points per cycle and frequency deviation within 5 Hz, the maximum error after simplification is 1.5×10^{-6}, which is acceptable.

$$B \approx \frac{3N^2\theta_1}{4\pi(3N^2 - 2\pi^2)f_0 + (3N^2 - 6\pi^2)\theta_1}$$ (5)

Simplify A and $1 + e^{-jC}B$ by following a similar principle. On this basis, the relationship between the phasor after correction and the direct calculation of the phasor by DFT before correction can be obtained as follows:

$$a \approx \frac{a_c}{\frac{24f_0^2 - \theta_1^2}{24f_0^2}\sqrt{1 + B^2 + 2B\cos C}}$$ (6)

$$\theta_0 \approx \theta_c + D - \frac{\theta_1(N-1)}{2Nf_0}$$ (7)

Where: $D = B\sin C/(1 + B\cos C)$

The Taylor series expansion of formula (7) can be obtained as follows:

$$\frac{D}{B} \approx F + GD + \frac{2}{3}GD^3 - \frac{4}{3}FD^4$$ (8)

Where: $F = \sin(2\theta_c - 2\pi/N)$, $G = \cos(2\theta_c - 2\pi/N)$.

Under the same adverse circumstances, the error after the elimination of the higher order term is acceptable, and the approximate D is substituted into Eq. (7). The data window with k points is calculated like DFT, and the quadratic equation with one variable about B is obtained by combining with the assumed signal model, and then θ_1 is obtained. Substitute θ_1 into Eq. (7) to obtain the corrected phase Angle; The amplitude of a after correction is obtained by formula (6). The corrected frequency is $f = f_0 + \theta_1/(2\pi)$. Finally, the phasor and frequency are smoothed according to the corrected frequency.

3.2 Frequency Domain Algorithm

Aiming at algorithms in the literature [11] gave no spectrum and frequency variation, and the harmonic interference when unable to obtain the phasor problem correctly, this article proposed frequency-domain algorithm and Taylor series expansion of the power signal model, and through a data window base wave frequency and frequency variation and phasor calculation, harmonic content, can be in a data window to calculate the frequency and frequency variation, and solve the problem of containing harmonic cases phasor calculation. Specifically, formula (1) is expanded into the form of real part and imaginary part of order K through Taylor series:

$$P(t) = \sum_{k=0}^{K} R_k t^k + j \sum_{k=0}^{K} I_k t^k = R(t) + jI(t) \tag{9}$$

Available:

$$\begin{cases} a_0 &= \sqrt{R_0^2 + I_0^2} \\ \tan\theta_0 &= \frac{I_0}{R_0} \\ \theta_1 &= \frac{R_0 I_1 - R_1 I_0}{R_0^2 + I_0^2} \\ \theta_2 &= \frac{R_0 I_2 - R_2 I_0}{R_0^2 + I_0^2} + \frac{(R_0 I_1 + R_1 I_0)(R_1 I_0 - R_0 I_1)}{(R_0^2 + I_0^2)^2} \end{cases} \tag{10}$$

By sampling at N points every period of formula (2), the signal model is discretized, and then DFT with coefficient of $e^{-jg_k\frac{2\pi}{N}n}$, where g_k is the set variable, is carried out after the signal model is windowed. In this paper, a rectangular window is selected, and the complex domain equation is:

$$x_k = \frac{1}{N} \sum_{n=0}^{N-1} (P(n)e^{j\frac{2\pi n}{N}} + P^*(n)e^{-j\frac{2\pi n}{N}})e^{-jg_k\frac{2\pi}{N}n}$$

$$\frac{2}{N} \sum_{n=0}^{N-1} (\bar{R}(n)\cos\frac{2\pi n}{N} + \bar{I}(n)\sin\frac{2\pi n}{N})e^{-jg_k\frac{2\pi}{N}n} \tag{11}$$

Where: X_k is the calculation result of the K times Fourier transform; $\bar{R}(n) = \sum_{k=0}^{K} \bar{R}_k n^k$, $\bar{I}(n) = \sum_{k=0}^{K} \bar{I}_k n^k$, where, $\bar{R}_k = R_k / (Nf_0)^k$, $\bar{I}_k = I_k / (Nf_0)^k$.

Expand the complex domain equation into the form of real part and imaginary part:

$$X_k = \frac{2}{N} M_k P \tag{12}$$

Where, $X_k = [X_{kR}, X_{kI}]^T$ is the vector formed by the real part X_{kR} and the imaginary part X_{kI} of the K times Fourier transform calculation result; $M_k = [M_{k0}, M_{k1}, \cdots, M_{kK}]$ is the coefficient of the system, and its element expression is shown in Eq. (13). $P = [\bar{R}_0, \bar{I}_0, \bar{R}_1, \bar{I}_1, \cdots, \bar{R}_K, \bar{I}_K]^T$ is the signal model parameter.

$$M_{ki} = \begin{bmatrix} \sum_{n=0}^{N-1} n^i \cos\frac{2\pi n}{N} \cos\frac{2\pi ng_k}{N} & -\sum_{n=0}^{N-1} n^i \sin\frac{2\pi n}{N} \cos\frac{2\pi ng_k}{N} \\ -\sum_{n=0}^{N-1} n^i \cos\frac{2\pi n}{N} \cos\frac{2\pi ng_k}{N} & \sum_{n=0}^{N-1} n^i \sin\frac{2\pi n}{N} \sin\frac{2\pi ng_k}{N} \end{bmatrix} \tag{13}$$

When $k = 0, 1, \cdots, K$ simultaneous equations, get:

$$X = \frac{2}{N} MP \tag{14}$$

Where: $X = [X_0^T, X_1^T, \cdots, X_K^T]^T$, $M = [M_0^T, M_1^T, \cdots, M_K^T]^T$.

It should be noted that the selection of g_k in DFT coefficient should make the matrix M^{-1} condition number as small as possible to avoid ill-condition. At the same time, the power system harmonic interference should be suppressed. In order to suppress the common odd harmonic interference, 1,2,4..., that is, find the fundamental wave and the 2nd and 4th harmonic content of power signal in a data window. Since g_k can be determined in advance, the matrix M and its inverse M^{-1} can be calculated offline. In combination with Eq. (10) and (14), the amplitude at the calculation point is a_0 and the phase Angle is θ_0. When the order K is ≥ 1, the frequency deviation is $\theta_1/2\pi$. When $K \geq 2$, the rate of frequency change can be obtained as θ_2/π.

Theoretically, the higher the order K is, the higher the accuracy will be. However, in practical application, the higher the order K is, the higher the requirement for PMU software and hardware resources will be. Therefore, in engineering application, it is necessary to select according to the actual situation.

3.3 Verification Algorithm and Comprehensive Algorithm Implementation

In view of the problem that the existing algorithm can not take into account both the calculation accuracy of steady state and the dynamic performance of transient state, the algorithm proposed in this paper can realize the adaptive switching of calculation mode through verification under different conditions, so as to take into account the requirements of fast response and high precision.

Specifically, by calculating the phasor, the theoretical calculated values of each sampling point are deduced, and compared with the actual measured points. If the total deviation is small, the test passes. For example, for the time-domain algorithm, the theoretical calculation value of each extraction point is deduced from θ_0, θ_1 and a after correction, and the density of the extraction point is determined according to the actual situation:

$$x_c(n_i) = a \cos(\theta_0 + \frac{n_i\theta_1}{Nf_0} + \frac{2\pi n_i}{N}) \tag{15}$$

Where: $n_i(i = 1, 2. \cdots, L)$ is the selected point, and $1 \leq n_i \leq N$, where L is the total number of samples; $x_c(n_i)$ is the theoretical value of the extracted point.

The total deviation between the theoretical calculated value and the measured value is:

$$\varepsilon = \sum_{i=1}^{L} |x_c(n_i) - x_m(n_i)| \tag{16}$$

Where: $x_m(n_i)$ is the measured value of the corresponding sampling point.

The overall flow of the comprehensive algorithm is shown in Fig. 1, mainly including the following steps: (1) assume that the power signal is a first-order model, solve the phasor through the time domain algorithm; (2) through the comparison of the back value and the measured value, check the calculation result is correct, if the check through, the end, if not, it means that the current may be in a sudden change and other dynamic process, into the next step; Assuming that the power signal is a k-order model, the phasor, frequency and frequency change rate are solved by frequency domain algorithm. (4) check whether the calculation results of the frequency-domain algorithm are correct. If the calibration fails, the failure may occur in the case that the equal-frequency algorithm in this data window is also unable to deal with. In order to avoid large fluctuations, the results of the time-domain algorithm are still used.

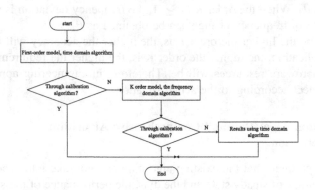

Fig. 1. Comprehensive phasor measurement algorithm to achieve the flow

4 Simulation Analysis

The national standard and IEEE standard specify the measurement method and precision requirement of PMU under steady state and dynamic condition in detail. According to the reference standards in literature [14], devices of four major PMU manufacturers in China were tested, and the test results showed that all four failed to fully meet the standard requirements, and some indicators even had a large deviation.

In this paper, the performance of the algorithm was comprehensively tested and analyzed by reference to the standard in MATLAB, and compared with the traditional DFT algorithm (algorithm 1), the DFT correction algorithm (algorithm 2) proposed in literature [9], and the DFT correction algorithm (algorithm 3) proposed in literature [11] using the response of different frequency point filters. Due to the limitation of space, the steady-state performance and the dynamic performance of the step response under the three important indexes of frequency deviation, harmonic influence and frequency linear change are analyzed emphatically in this paper. In the simulation process, the sampling frequency of each algorithm is equal to 6400 Hz, and the interval between two adjacent data Windows in the time-domain algorithm in this paper is 64 points. The phasor model of the frequency-domain algorithm adopts the third-order model.

4.1   Frequency Deviation Analysis

When the frequency deviation is 5 Hz, the corresponding maximum error, average absolute error and root mean square error of each algorithm are shown in Table 1. It can be seen that when the frequency deviates from the rated frequency, the error of algorithm 1 and 2 is large. The measurement accuracy of this algorithm is high and meets the requirements of national standards. The reason is that the unsynchronized sampling of algorithm 1 leads to the spectrum leakage, and algorithm 2 makes a large approximation in the process of simplification. However, under the premise of ensuring the accuracy, the algorithm in this paper deduces the approximation by controlling the number of Taylor series expansion terms, so the accuracy is high.

Table 1. Error of each algorithm when frequency deciation is 5 Hz

Algorithm	Angle error (°)			Amplitude error (%)		
	Maximum	Mean absolute	Root mean square	Maximum	Mean absolute	Root mean square
1	20.5899	17.9382	18.0240	−5.2690	2.6560	3.0882
2	−5.7974	3.5874	3.9897	−1.6936	1.1305	1.2126
3	−3.009	0.1444	0.1801	−0.4.89	0.1927	0.2416
This paper	−0.0022	0.0009	0.0012	0.0072	0.0071	0.0071

4.2   Harmonic Influence Analysis

According to the national standard, the algorithms are analyzed under the frequency deviation of 0.5 Hz and the superposition of 2–13 harmonics of rated amplitude of 20%. The simulation results are shown in Table 2. It can be seen from Table 2 that when the frequency is shifted by 0.5 Hz with serious harmonic interference, the error of algorithm 3 is large. The accuracy of the algorithm in this paper is better than that of algorithm 1 and 2, and it meets the requirements of the latest national standard. Due to the limitation of the signal model and algorithm principle of algorithm 3, the fundamental phasor of the power signal cannot be acquired correctly. In this paper, the harmonic suppression capability of DFT algorithm is integrated, and the frequency deviation is high precision.

Table 2. Error of each algorithm under harmonic influence

Algorithm	Angle error (°)			Amplitude error (%)		
	Maximum	Mean absolute	Root mean square	Maximum	Mean absolute	Root mean square
1	2.4059	1.7921	1.8334	3.831	0.444	0.777
2	−2.0174	0.4138	0.5425	3.777	0.395	0.747
3	178.9468	40.0514	55.2037	210.990	49.748	62.768
This paper	0.0040	0.0015	0.0016	−0.005	0.002	0.000

4.3 Frequency Slope Response Analysis

According to IEEE standard, the slope test signal with frequency deviation of 5 Hz and rate of change of 1 Hz/s is applied to test the accuracy of each algorithm. The mathematical expression of the input signal is as follows:

$$x(t) = a \cos(2\pi f_0 t + 2\pi \Delta f t + \pi t^2 + \pi/2) \tag{17}$$

When a is equal to 1, it's equal to 50 Hz. The error results of each algorithm are shown in Table 3. It can be seen that when the signal frequency range is wide and the frequency linear change is fast, the algorithm in this paper still maintains high accuracy and is better than the other three algorithms. Among them, the maximum frequency error is 0.001 Hz and the maximum frequency change rate error is -0.0088 Hz/s, which all meet the requirements of IEEE standard.

Table 3. Error of each algorithm during ramp change of frequency

Algorithm	Angle error (°)			Amplitude error (%)		
	Maximum	Mean absolute	Root mean square	Maximum	Mean absolute	Root mean square
1	21.7653	18.6971	18.7951	−5.63	2.71	3.18
2	−6.1125	3.7240	4.1378	−1.89	1.23	1.32
3	−0.3546	0.1637	0.2042	−0.48	0.22	0.27
This paper	0.0537	0.0446	0.0449	−0.01	0.01	0.01

4.4 Step Response Analysis

In order to test the response performance of the algorithm to dynamic processes such as mutation, 90° phase Angle step signal was applied according to the national standard:

$$x(t) = \begin{cases} a\,\cos(2\pi f_0 t + \pi/6)\, t < 40 \text{ ms} \\ a\,\cos(2\pi f_0 t + \pi/6 + \pi/2)\, t \geq 40 \text{ ms} \end{cases} \tag{18}$$

The step response curves of each algorithm are shown in Fig. 2.

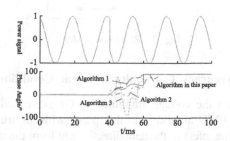

Fig. 2. Step response curves of each algorithm

According to the definition of response time by the national standard, the response time of the algorithm in this paper is about 20 ms, which is faster than algorithm 2 (30 ms) and has higher accuracy than algorithm 3. This is because the algorithm in this paper ADAPTS to the frequency domain algorithm after the step response, and only needs one period of data. Algorithm 2 needs two data Windows, so the algorithm in this paper has a faster response speed than algorithm 2, while algorithm 3 fluctuates greatly in the data window where the fault is due to the limitation of principle. At the same time, it also shows the correctness of the proposed algorithm.

5 Experimental Test

Based on the embedded software and hardware platform, PMU is developed. The developed PMU converts large voltage and current signals into small voltage signals through voltage and current transformers. Then, according to the global positioning system's second pulse, the voltage signals are sampled at regular intervals in time synchronization through 16-bit analog digital converter, and the sampled signals are processed in the digital signal processor chip. The experimental test platform used for this test is shown in Fig. 3.

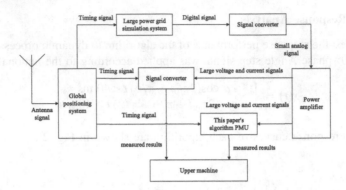

Fig. 3. PMU test system

5.1   Measurement Accuracy Test Under Dynamic Condition

The curve of comparing the voltage amplitude of the slave node measured by commercial PMU and the PMU algorithm in this paper with the truth value is shown in Fig. 4(a). The truth value refers to the data directly sent from the simulation system. By calculation, the average absolute error of the amplitude of commercial PMU is 0.17%, while the algorithm in this paper is 0.14%. Figure 4(b) shows the curve comparing the relative phase Angle measured by the commercial PMU and the PMU algorithm in this paper with the true value. By calculation, the average absolute error of the phase Angle of the commercial PMU is 0.156°, while the algorithm in this paper is 0.062°. According to Fig. 4 and the calculation error, in the dynamic process, both PMUs meet the requirements of the national standard. The algorithm in this paper is closer to the truth value than the commercial PMUs, but due to the limitation of hardware conditions, the accuracy improvement level is not significant.

5.2   Fast Response Performance Test Under Dynamic Conditions

In order to objectively compare the response performance of the commercial PMU and the algorithm in this paper under dynamic conditions, 34 short-circuit simulation experiments were carried out in succession. The test results show that the phasor cannot be measured correctly 16 times in the short circuit process of the commercial PMU, and the algorithms in this paper are all normal. Figure 5(a) shows the curve of amplitude and truth value comparison of slave stations measured by commercial PMU and the algorithm in this paper. Figure 5(b) shows the curve of relative phase Angle and true value of the horizontal leach-derived station measured by commercial PMU and the algorithm in this paper. It can be seen that in the short-circuit process, the commercial PMU has a point with amplitude of 0 and a large fluctuation of relative phase Angle, which is not consistent with the true value, indicating that the algorithm in this paper has a faster response performance than the commercial PMU in some transient processes.

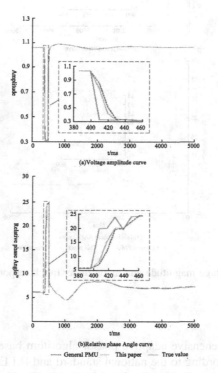

Fig. 4. Curves of voltage magnitude and angle

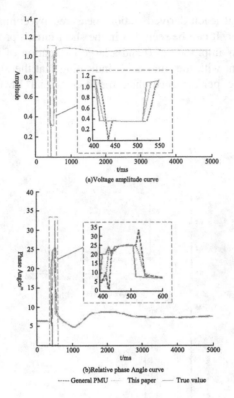

Fig. 5. Curves of voltage magnitude and angle under quick response performance testing

6 Conclusion

In this paper, a comprehensive adaptive phasor algorithm based on Taylor series and DFT is proposed. According to the national standard and IEEE standard, the proposed algorithm is comprehensively simulated in MATLAB. Based on the algorithm in this paper, PMU is implemented and compared with a mainstream commercial PMU on a digital - physical hybrid simulation platform. Theoretical simulation and experimental test results show that the proposed algorithm has superior performance, mainly in the following aspects.

1) in the steady-state calculation mode, time-domain algorithm is adopted to simplify the calculation by Taylor series expansion under the premise of analysis to ensure the accuracy, which has high accuracy and strong harmonic suppression ability.
2) in the dynamic computing mode, the frequency domain algorithm is adopted to calculate the amplitude, phase Angle, frequency and frequency change rate only through the fundamental wave and each harmonic content of a data window, with fast response speed and harmonic suppression ability.

3) through the calculation of the phasor, the strategy of reversely deducing the theoretical calculation value of each sampling point and comparing it with the measured value can be adopted to switch the calculation mode adaptively, taking into account the accuracy and rapidity requirements.

4) the implementation, accuracy and fast response performance of the actual PMU are better than the commercial PMU compared, indicating that the proposed algorithm can meet the requirements of practical application.

Acknowledgement. This work was supported by Natural Science Foundation of Liaoning Province (2019-MS-239).

References

1. Imdadullah, S., Amrr, M.: A comprehensive review of power flow controllers in interconnected power system networks. IEEE Access **26**(8), 18036–18063 (2020)
2. Liu, B.: An AC–DC hybrid multi-port energy router with coordinated control and energy management strategies. IEEE Access **24**(7), 109069–109082 (2019)
3. Pegoraro, P., Brady, K., Castello, P.: Compensation of systematic measurement errors in a pmu-based monitoring system for electric distribution grids. IEEE Trans. Instrum. Meas. **68** (10), 3871–3882 (2019)
4. Narduzzi, C., Bertocco, M., Frigo, G.: Fast-TFM—multifrequency phasor measurement for distribution networks. IEEE Trans. Instrum. Meas. **67**(8), 1825–1835 (2018)
5. Zhao, J., Zhang, G., Jabr, R.: Robust detection of cyber attacks on state estimators using phasor measurements. IEEE Trans. Power Syst. **32**(3), 2468–2470 (2017)
6. Moghimi, M., Xu, W.: Online determination of external network models using synchronized phasor data. IEEE Trans. Smart Grid **9**(2), 635–643 (2018)
7. Kabiri, M., Amjady, N.: A new hybrid state estimation considering different accuracy levels of pmu and scada measurements. IEEE Trans. Instrum. Meas. **68**(9), 3078–3089 (2019)
8. Sun, L.: Optimum placement of phasor measurement units in power systems. IEEE Trans. Instrum. Meas. **68**(2), 421–429 (2019)
9. Wang, M., Sun, Y.: A DFT-based method for phasor and power measurement in power systems. Autom. Electric Power Syst. **29**(2), 20–24 (2005)
10. Koteswara, A., Soni, K., Sinha, S.: Accurate phasor and frequency estimation during power system oscillations using least squares. IET Sci. Meas. Technol. **13**(7), 989–994 (2019)
11. Fu, L., Han, W.: Dynamic phasor estimator based on frequency-domain model. Proc. CSEE **35**(6), 1371–1378 (2015)
12. Jain, S., Singh, N.: A fast harmonic phasor measurement method for smart grid applications. IEEE Trans. Smart Grid **8**(1), 493–502 (2017)
13. Fernandes, E.: Application of a phasor-only state estimator to a large power system using real PMU Data. IEEE Trans. Power Syst. **32**(1), 411–420 (2017)

Research on Multi-energy Trading Mechanism of Energy Internet

Jiaxin Liu[1]([⊠]), Dong Zhang[2], Peng Yin[3], and Yan Zhao[2]

[1] School of Electric Power, Shenyang Institute of Engineering,
Shenyang 110136, Liaoning, China
176886394@qq.com
[2] School of Renewable Energy, Shenyang Institute of Engineering,
Shenyang 110136, Liaoning, China
[3] Construction Branch of Liaoning Electric Power Co., Ltd.,
Shenyang 110136, Liaoning, China

Abstract. There are multi-energy trading and multi-market players in the energy Internet, so it is one of the urgent problems to establish a reasonable multi-market player cost-benefit model and trading mechanism. Mainly for the energy Internet, its trading scenarios are three market trading entities: multi-energy users, low-energy users and public energy networks. The integrated energy system, as a user, can flexibly conduct transactions of electricity, gas and heat types and consumption according to the needs of buyers. Therefore, the multi-energy users with the energy hub first meet their own load needs and then quote the remaining energy; public energy network according to the network service fee form quotation. According to the quotation of the former two energy users, choose to trade electricity, gas and heat with the multi-energy users or the public energy network. Under this trading mechanism, a three-party non-cooperative game revenue model with electric thermal coupling is established, which proves the existence of Nash equilibrium in the game model. By means of example analysis, the range of Nash equilibrium in the multi-energy trading system of electricity, gas and heat is obtained. This paper explains the physical significance of Nash equilibrium and analyzes the effects of the return parameters on the distribution of Nash equilibrium in game theory.

Keywords: Energy Internet · Integrated energy system · Multi-market entity · Game theory · Nash equilibrium

1 Introduction

In the traditional electricity market, only electricity trading exists, and energy trading is relatively single. It is a new trend for the development of energy market in the future to formulate a new market trading model and encourage electricity, gas and thermal energy to participate in market trading. The integrated energy system can realize the conversion of electricity, gas and heat. The development and application of energy hub is conducive to the comprehensive utilization of diversified energy sources, which has attracted the attention of all countries in the world. In the two-way energy hub, the three networks of power system, natural gas system and thermal system are interwoven in the

© Springer Nature Singapore Pte Ltd. 2020
J. Qian et al. (Eds.): ICRRI 2020, CCIS 1335, pp. 434–449, 2020.
https://doi.org/10.1007/978-981-33-4929-2_30

physical structure, and highly integrated through the developed information technology at the information level. As the number of energy sources and market participants increases, the optimal operation of integrated energy systems and the trading strategies of participants will become more and more complex. In the future energy market [1], the production optimization method and pricing mechanism of integrated energy system are the focus of manufacturers. It should be emphasized that the users of this paper mainly refer to the large industrial and commercial users with electricity, gas and heat demands. In the interaction, the behaviors of the integrated energy system, multi-energy user and low-energy user will affect the interests and interaction strategies of other participants. As a rational participant, when formulating the trading mechanism contract, the integrated energy system will fully analyze the behavior of other participants in the market to formulate the optimal mechanism to maximize its own benefits. It can be seen that in the interaction of multiple multi-energy systems containing energy hub, they act as different interest subjects and pursue the maximization of their respective interests. This is a game problem involving multiple stakeholders. Therefore, the equilibrium interaction strategy of the integrated energy system in the market environment is worth studying [2].

In terms of games related to the comprehensive energy system, literature [3–5] established a static non-cooperative game model for the interactions between distributed energy stations, natural gas companies and power grid companies, in which the peak-regulating function of distributed energy stations in the comprehensive energy system was emphatically analyzed. In [6] studied the coupling demand response of multiple distributed energy stations to electricity and natural gas, and established the potential game model for distributed energy stations, natural gas companies and power grid companies to study their equilibrium interaction strategy, which is mainly about the amount of natural gas and electricity exchanged. However, the game models in the above studies are static game models which only take the energy trading volume as the interactive variable and do not consider the price as the interactive variable. In [7] summarized key technologies such as multi-energy system planning, control and trading on the basis of multi-energy flow hybrid modeling. There are still few studies on the transaction between multi-types of energy and the competition between multi-market players in the existing literature, especially on the benefit model of multi-market players such as energy suppliers, power grids or thermal networks and users.

Therefore, this paper proposes a multi-market revenue model of energy Internet based on game theory, and proves the existence of Nash equilibrium, aiming at the three-party game players in the energy Internet with electricity, gas and heat. By containing more energy users, less energy users and public energy grid case analysis, a kind of miniature energy system for the gas and heat energy after joining the Nash equilibrium of the model, calculated the electricity, gas, heat energy market the benefits of each market subjects, and quantitative analysis of the users and the public energy grid pricing, cost of energy market impact of Nash equilibrium.

2 Energy Hub Model

As the interface of energy, energy hub plays an important role in energy integration and energy conversion. Energy hub is an important part of the energy Internet. Through the interconnection of multiple energy hubs, comprehensive energy management can be realized. Energy hub can be regarded as a black box, energy can flow from the input port to the output port, the ports of the input energy hub are connected to each other through the energy link, and the output ports of the energy hub are connected to a certain energy area. Traditional models include energy transfer, energy center energy conversion and storage, they can only from the flow of energy from the input port to the output port. On the energy Internet, however, not only is the energy region connected to the output port of the energy hub composed of electrical and thermal loads, but also has distributed power supply. The new energy structure hub with smart housing (see Fig. 1). The house is equipped with electrical loads, gas loads, thermal loads and solar cells that are self sufficient to generate electricity from photovoltaic power and interact with other energy fields through the energy hub. energy center solid state transformer (SST) battery, electrothermal boiler (EB), micro-turbine (MT), heat exchanger, heat accumulator (HS) and so on to achieve the generation of electricity and heat energy, the conversion of electricity and heat energy, the storage of electricity and the use of heat energy.

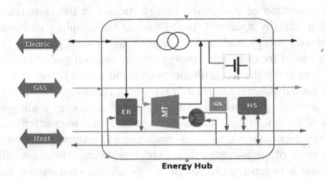

Fig. 1. Energy hub structure

Different from previous studies on energy hubs, this paper proposes a novel energy hub that can realize two-way transmission of electric energy and thermal energy. When the intelligent user has surplus electric energy, the electric energy can be transferred to the hub input port and converted into heat energy through the electric boiler. We assume that if the inlet temperature of the heating pipe network is higher than the outlet temperature, the heat energy flows from the input port to the output port, and if the inlet temperature of the heating pipe network is lower than the outlet temperature, the heat energy flows from the output port to the input port.

2.1 Energy Conversion

Firstly, the model of conversion equipment in energy hub is created, which is respectively electric to heat, gas to electricity, gas to heat and electric to gas (electric hydrogen production).

Electric Energy to Heat Energy

As the heating equipment in the thermal pipe network, the electric boiler can convert the electric energy into heat energy to provide energy for the thermal pipe network. When the electric boiler with the electric power of P_{EB}^e is input, Its thermal power $P_{h,EB}$ is expressed as follows:

$$P_{h,EB} = \eta_{EB} P_{e,EB} \tag{1}$$

Where $P_{e,EB}$ is the input power of the electric boiler; η_{EB} is the thermal efficiency of electric boiler.

Electric Energy to Gas

Electricity to gas is achieved by electric hydrogen production technology, which can be converted into gas by electrolysis of water. The relationship between gas and power supply is as follows:

$$P_{E2G}^g = \eta_{E2G} P_{E2G}^e \tag{2}$$

Where P_{E2G}^g is the output power of gas; η_{E2G} is the electric to gas efficiency; P_{E2G}^e is input power for electricity.

Gas to Electric Energy, Gas to Heat Energy

In the micro power grid, the micro gas turbine provides power supply and heat supply functions in the electric power system and the power system respectively. The relationship between the power supply and heat supply and the gas is as follows:

$$\begin{cases} P_{MT}^e = \sigma_{g2e} \cdot \rho \cdot P_{MT} \\ P_{MT}^h = \eta_{g2h} \cdot H_u \cdot \rho \cdot P_{MT} \end{cases} \tag{3}$$

Where P_{MT}^e is the output electric power of the micro gas turbine; σ is the electrical conversion efficiency of micro gas turbine; P_{MT}^h is the output thermal power of the micro gas turbine; η_{g2h} is the gas thermal efficiency of micro gas turbine; H_u is the low calorific value of natural gas (heat released by burning a gas of a given volume or mass).

2.2 Energy Storage

As an important part of energy hub, energy storage device can improve energy efficiency and reduce energy cost. Energy storage mainly includes electricity storage and heat storage. Surplus energy or cheap energy can be stored in energy storage devices for a certain period of time and released when energy is insufficient. Unlike other devices, energy storage devices need to consider many parameters, including charging and discharging states, charging and discharging energy, power limit, and charged states. The energy storage of the model can be expressed as:

$$
\begin{cases}
P_{e,ES} = \kappa_{e,ES}^{charge} P_{e,ES}^{charge} - \kappa_{e,ES}^{discharge} P_{e,ES}^{discharge} \\
P_{h,HS} = \kappa_{h,HS}^{charge} P_{h,HS}^{charge} - \kappa_{h,HS}^{discharge} P_{h,HS}^{discharge}
\end{cases}
\tag{4}
$$

Where $P_{e,ES}$ and $P_{h,HS}$ represents the power of electricity and heat storage, $\kappa_x^{discharge}$ and κ_x^{charge} represents charging and discharging states, is a variable of 0–1 (1 represents the working state, 0 represents the non-working state), P_x^{charge} and $P_x^{discharge}$ represents the charging and discharging power, Where x represents the storage of type energy. It can be electrical storage or it can be thermal storage. According to the conditions of the energy storage equipment, the following restrictions shall be imposed on the storage of energy saving transformation:

$$
\begin{cases}
\kappa_x^{charge} + \kappa_x^{discharge} \leq 1 \\
\kappa_x^{charge} \cdot \kappa_x^{discharge} = 0
\end{cases}
\tag{5}
$$

$$
\begin{cases}
\kappa_x^{charge} P_{x,min}^{charge} \leq P_{x,min}^{charge} \leq \kappa_{x,max}^{charge} P_{x,max}^{charge} \\
\kappa_x^{discharge} P_{x,min}^{discharge} \leq P_{x,min}^{discharge} \leq \kappa_{x,max}^{discharge} P_{x,max}^{discharge}
\end{cases}
\tag{6}
$$

$$
E_x^{t+1} = E_x^t(1 - \delta_x) + (\eta_x^{charge} P_x^{charge} - \frac{P_x^{discharge}}{\eta_x^{discharge}})\Delta t
\tag{7}
$$

Where $P_{x,min}^{discharge}$ and $P_{x,max}^{discharge}$ respectively represent the upper limit of charging and the lower limit of discharging energy, δ_x represent energy loss energy storage, η_x^{charge} and $\eta_x^{discharge}$ respectively represent the energy efficiency of charging and discharging, E_x represents energy storage setting energy, energy storage devices are subject to the following restrictions:

$$
E_x^{min} \leq E_x \leq E_x^{max}
\tag{8}
$$

Where E_x^{min} and E_x^{max} represents the minimum energy storage and maximum energy storage.

3   The Main Body of Energy Internet Game with Energy Hub

In the energy Internet, the operation efficiency and economic performance of the multi-energy system can be improved by taking the multi-energy system containing the energy hub as the user and formulating the transaction mechanism [8].

For N users, some more energy systems due to internal is intermittent energy sources (e.g., wind power, photovoltaic power generation), or its control ability is limited, in unit time, and the amount of energy is not necessarily produced by its internal load balancing, foreign present certain energy surplus (referred to as "more energy users") or the deficiency (known as "little energy users"). In order to ensure the normal operation of their own load, users with less energy need to purchase from multi-energy users or public platforms. In order to maximize their own economic interests, multi-energy users hope to sell their surplus energy to less energy users or public platforms. Based on these requirements, it is considered to introduce a market mechanism in the energy Internet, so that users of the integrated energy system can conduct electricity, gas and heat consumption transactions [9].

Each user in the energy Internet may belong to different interest subjects, with different operation objectives, user needs, control means, etc., but they are closely related to each other and have certain interests. Game theory is an effective tool to solve the conflict between different stakeholders [10–12]. Therefore, the general non-cooperative game model can be established:

$$G = \langle \Gamma : (B_i); (u_i) \rangle \qquad (9)$$

Game Party
There are multiple energy users with energy surplus in time period T, as well as public energy networks.

Game Strategy
The game strategy of multi-energy user p is its selling price p^e, selling price p^h, selling price p^g;

The game strategy of public energy network is the unit price of service charge for energy transaction between users s^d;

The policy set of multi-energy user p is $[0,p^e_{max}]$, $[0,p^h_{max}]$, $[0,p^g_{max}]$;

The strategy set for the public energy grid is $[0,s_{d\ max}]$, $p^e_{max} > 0$, $p^h_{max} > 0$, $p^g_{max} > 0$, $s_{d\ max} > 0$;

Less energy users will choose different ways of trading based on the prices of more energy users and public energy networks and the types and amounts of surplus energy.

Game Benefits
Each participant hopes to achieve the maximum return under the Nash equilibrium condition through the market trading behavior. See the following for the specific game income model.

3.1 Energy Internet Trading Mechanism

In this paper, the main players of energy market transaction of energy Internet mainly include multi-energy users, low-energy users and public energy network.

At the initial moment of time period T, N multi-energy users and the public energy network offer prices to the low-energy users according to the load demand of the low-energy users.

In order to meet the demand of their own electrical heat load, the less energy users choose the party with the lowest cost to conduct the transaction.

When the less energy user trades with the selected multi-energy user, the multi-energy user USES the mode of heat fixed power operation, first trades the heat, then trades the electric quantity, finally trades the gas quantity; If the multi-energy users cannot meet the energy load demand of the small-energy users, the small-energy users will continue to purchase the remaining energy from the public energy network; If a multi-energy user has too much energy, the energy produced by the multi-energy user is sold to the public energy network, which then charges both parties for the appropriate over-grid service.

When the energy types and consumption required by the less energy users are insufficient, the more energy users will convert their surplus energy according to the demand of the less energy users, and then sell to the less energy users.

Supplementary hypothesis: all participants are completely rational, in a short period of time, the participants' load or power generation, gas, heat, will not change, according to the production capacity of the order, trading potential users will be the priority to participate in the transaction. In the transaction process, only the influence of electricity price and electricity quantity is considered, and other factors are not considered to affect the transaction behavior.

3.2   The Payoff Function for Game Players

Multi-energy Users
Trading income $u_{M,tra}$
Multi-energy users gain from trading electricity, gas and heat as follows:

$$
\begin{aligned}
u_{M,tra} = &\sum_{j=1}^{N} P_{m,j} p_{me,j} + \sum_{j=1}^{N} G_{m,j} p_{mg,j} + \sum_{j=1}^{N} Q_{m,j} p_{mh,j} \\
&+ \sum_{j=1}^{N} P_{m,R} p_{me,R} + \sum_{j=1}^{N} G_{m,R} p_{mg,R} + \sum_{j=1}^{N} Q_{m,R} p_{mh,R}
\end{aligned}
\tag{10}
$$

Where, N number of multi-energy users; $P_{m,j}$ and $p_{me,j}$ are the electricity quantity and unit electricity price of the number of j multi-energy user and the less energy user; $G_{m,j}$ and $p_{mg,j}$ are the volumes and prices per unit of gas traded between the number of j multi-energy user and small-energy user; $Q_{m,j}$ and $p_{mh,j}$ are the heat and heat price per unit traded between the number of j multi-energy user and small-energy user, respectively. $P_{m,R}$ and $p_{me,R}$ are the residual power of the number of j multi-energy user and the transaction price of the same public energy network; $G_{m,R}$ and $p_{mg,R}$ are the residual gas quantity of the number of j multi-energy user and the gas price traded with the public energy network; $Q_{m,R}$ and $p_{mh,R}$ are the surplus heat of the number of j multi-energy user and the heat price of trading with the public energy network, respectively.

Service charge $u_{M,ser}$

$$
u_{M,ser} = -\sum_{j=1}^{N} P_{m,j} p_{e,ser} - \sum_{j=1}^{N} G_{m,j} p_{g,ser} - \sum_{j=1}^{N} Q_{m,j} p_{h,ser}
\tag{11}
$$

Where $p_{e,ser}$ is the service fee charged by the public energy network for the electricity transaction between the multi-energy users and the less-energy users; $p_{g,ser}$ is the service fee charged by the public energy network for gas volume transactions between multi-energy users and small-energy users; $p_{h,ser}$ is the service fee charged by the public energy network for heat transaction between multi-energy users and small-energy users.

Other income $u_{M,oth}$
Other benefits of multi-energy users are expressed only in terms of costs, not other influencing factors.

$$u_{M,oth} = -\sum_{j=1}^{N} P_m q_{m,e} - \sum_{j=1}^{N} Q_m q_{m,h} - \sum_{j=1}^{N} G_m q_{m,g} \qquad (12)$$

Where P_m and $q_{m,e}$ are respectively the generation capacity and unit generation cost of multi-energy user j; Q_m and $q_{m,h}$ are respectively the production gas quantity and unit production gas cost of multi-energy user j; G_m and $q_{m,g}$ are respectively calorific value and unit heating cost of multi-energy user j.

In summary, the revenue function of multi-energy user j is:

$$\begin{aligned}
u_M = u_{M,tra} + u_{M,ser} + u_{M,oth} = \\
(\sum_{j=1}^{N} P_{m,j} p_{me,j} + \sum_{j=1}^{N} G_{m,j} p_{mg,j} + \sum_{j=1}^{N} Q_{m,j} p_{mh,j} \\
+ \sum_{j=1}^{N} P_{m,R} p_{me,R} + \sum_{j=1}^{N} G_{m,R} p_{mg,R} + \sum_{j=1}^{N} Q_{m,R} p_{mh,R} \\
- \sum_{j=1}^{N} P_{m,j} p_{e,ser} - \sum_{j=1}^{N} G_{m,j} p_{g,ser} - \sum_{j=1}^{N} Q_{m,j} p_{h,ser} \\
- \sum_{j=1}^{N} P_m q_{m,e} - \sum_{j=1}^{N} Q_m q_{m,h} - \sum_{j=1}^{N} G_m q_{m,g})
\end{aligned} \qquad (13)$$

Less Energy Users
Trading income $u_{U,tra}$

$$\begin{aligned}
u_{U,tra} = -(\sum_{j=1}^{N} P_{m,j} p_{me,j} + \sum_{j=1}^{N} G_{m,j} p_{mg,j} + \sum_{j=1}^{N} Q_{m,j} p_{mh,j} \\
+ \sum_{j=1}^{N} P_{m,R} p_{me,R} + \sum_{j=1}^{N} G_{m,R} p_{mg,R} + \sum_{j=1}^{N} Q_{m,R} p_{mh,R}) \\
- P_{ue} p_{ue} - G_{ug} p_{ug} - Q_{uh} p_{uh}
\end{aligned} \qquad (14)$$

Where P_{ue} and p_{ue} respectively represent the electricity quantity and electricity price of the less energy users' transaction with the public energy network; G_{ug} and p_{ug} are the gas volume and price of the transaction between the less energy users and the public energy network respectively; Q_{uh} and p_{uh} are the heat and heat prices of the transactions between the energy users and the public energy network respectively.

Service charge $u_{U,ser}$

$$u_{U,ser} = -\sum_{j=1}^{N} P_{m,j} p_{e,ser} - \sum_{j=1}^{N} G_{m,j} p_{g,ser} - \sum_{j=1}^{N} Q_{m,j} p_{h,ser} \tag{15}$$

Other income $u_{U,oth}$

Less energy users have no other benefits, namely $u_{U,oth} = 0$.

To sum up, the revenue function of less energy users is:

$$u_U = u_{U,tra} + u_{U,ser} + u_{U,oth} =$$

$$-(\sum_{j=1}^{N} P_{m,j} p_{me,j} + \sum_{j=1}^{N} G_{m,j} p_{mg,j} + \sum_{j=1}^{N} Q_{m,j} p_{mh,j}$$

$$+ \sum_{j=1}^{N} P_{m,R} p_{me,R} + \sum_{j=1}^{N} G_{m,R} p_{mg,R} + \sum_{j=1}^{N} Q_{m,R} p_{mh,R}) \tag{16}$$

$$-P_{ue} p_{ue} - G_{ug} p_{ug} - Q_{uh} p_{uh}$$

$$- \sum_{j=1}^{N} P_{m,j} p_{e,ser} - \sum_{j=1}^{N} G_{m,j} p_{g,ser} - \sum_{j=1}^{N} Q_{m,j} p_{h,ser}$$

Public Energy Networks

Trading income $u_{G,tra}$

$$u_{G,tra} = -(\sum_{j=1}^{N} P_{m,R} p_{me,R} + \sum_{j=1}^{N} G_{m,R} p_{mg,R} + \sum_{j=1}^{N} Q_{m,R} p_{mh,R})$$

$$+ P_{ue} p_{ue} + G_{ug} p_{ug} + Q_{uh} p_{uh} \tag{17}$$

Service charge $u_{G,ser}$

$$u_{G,ser} = 2(-\sum_{j=1}^{N} P_{m,j} p_{e,ser} - \sum_{j=1}^{N} G_{m,j} p_{g,ser} - \sum_{j=1}^{N} Q_{m,j} p_{h,ser}) \tag{18}$$

Other income $u_{U,oth}$

Other benefits of a public energy network are recorded as its own costs only:

$$u_{G,oth} = \sum_{j=1}^{N} P_{m,R}c_e - P_{ue}c_e$$

$$+ \sum_{j=1}^{N} G_{m,R}c_g - G_{ug}c_g \tag{19}$$

$$+ \sum_{j=1}^{N} Q_{m,R}c_h - Q_{uh}c_h$$

Where c_e is the cost unit price of electricity sold by the public energy network; c_g is the unit cost of the amount of gas sold by the public energy network; c_h is the unit cost of heat sold by the public energy network.

In summary, the revenue function of the public energy network is:

$$u_G = u_{G,tra} + u_{G,ser} + u_{G,oth} =$$

$$-\left(\sum_{j=1}^{N} P_{m,R}p_{me,R} + \sum_{j=1}^{N} G_{m,R}p_{mg,R} + \sum_{j=1}^{N} Q_{m,R}p_{mh,R}\right)$$

$$+ P_{ue}p_{ue} + G_{ug}p_{ug} + Q_{uh}p_{uh}$$

$$+ 2\left(-\sum_{j=1}^{N} P_{m,j}p_{e,ser} - \sum_{j=1}^{N} G_{m,j}p_{g,ser} - \sum_{j=1}^{N} Q_{m,j}p_{h,ser}\right) \tag{20}$$

$$+ \sum_{j=1}^{N} P_{m,R}c_e - P_{ue}c_e + \sum_{j=1}^{N} G_{m,R}c_g - G_{ug}c_g + \sum_{j=1}^{N} Q_{m,R}c_h - Q_{uh}c_h$$

3.3   Nash Equilibrium Solution for Multi-player Transactions

The game problem in this paper is solved by searching the whole strategy space according to the definition of Nash equilibrium. For the game model, firstly, it can be theoretically proved that the game has a Nash equilibrium. When the public grid collects service fees $p_{ser} \in [0, +\infty]$ and quotes from multi-energy users $p_{M,j} \in [0, +\infty]$, This shows that the set of strategy variables of the players is continuously unbounded, and Eqs. (21) and (22) are the coordinate expressions of the two endpoints of the line segment region of the Nash equilibrium of the game.

$$\begin{cases} p_{ser1} = \dfrac{p_{ug} - p_{fg} + \frac{Q_{f,j}}{P_{f,j}\eta}(p_{ug} - c_g)}{2} \\[4mm] p_{M,j1} = \dfrac{p_{ug} + p_{fg} - \frac{Q_{f,j}}{P_{f,j}\eta}(p_{ug} + c_g) - \frac{2Q_{f,j}p_{fg,j}}{P_{f,j}}}{2} \end{cases} \tag{21}$$

$$
\begin{cases}
p_{ser2} = \dfrac{p_{ug} - p_{fg} + \frac{Q_{f,j}}{P_{f,j}\eta}}{2} \\[4mm]
p_{M,j2} = \dfrac{p_{ug} + p_{fg} + \frac{Q_{f,j}}{P_{f,j}\eta}\left(\frac{p_{ug}}{\eta} - 2p_{fq,j}\right)}{2}
\end{cases}
\tag{22}
$$

Where $\left(p_{ser1}, p_{M,j1}\right)$ is the leftmost vertex of the line segment of the Nash equilibrium region; $\left(p_{ser2}, p_{M,j2}\right)$ is the rightmost vertex of the line segment in the Nash equilibrium region; $\left(p_{ser2}, p_{M,j1}\right)$ is the vertex of the rectangular region of the Nash equilibrium point of the game problem; η is the thermoelectric conversion efficiency.

In this game problem, the strategy players can choose is a continuous variable, and the Nash equilibrium of continuous strategy game can be solved by analyzing the optimal reaction function of players to the strategy combinations of other players. The payoff function of the game participants is a piecewise function, so the optimal reaction function of each participant is considered piecewise, and the Nash equilibrium of the game can be obtained by simultaneous solution.

According to the above analysis, MATLAB is used for programming to solve the Nash equilibrium of the system (see Fig. 2).

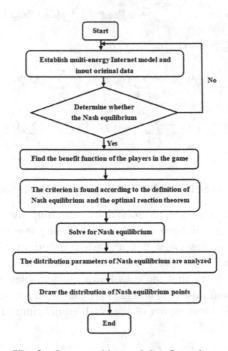

Fig. 2. Game problem solving flow chart

4 Example Analysis

In a small multi-energy system, large users, thermal power plants and power grid companies trade electric energy and thermal energy with each other, forming a game system. At some point, suppose the system is too short at this point, and the price and load requirements of each participant remain unchanged. Only one energy hub participates in the transformation, that is, N = 1. According to historical experience, set the following data, the units of heat have been converted from kJ to MW, Less load demand from energy users $P_{m,1}$ = 5 MW, Thermal load demand $Q_{m,1}$ = 3 MW, Gas load demand $G_{m,1}$ = 2 MW, Combined costs for multiple energy users c_m = 0.5 yuan/(kW·h). The average cost of purchasing electricity from public energy networks cg = 0.6 yuan/(kW·h). User's thermoelectric conversion efficiency η = 0.9, Public energy networks sell consumer prices for less energy $p_{GL,e}$ = 1.2 yuan/(kW·h), The public grid buys the multi-energy user tariff $p_{GM,e}$ = 0.2 yuan/(kW·h), Common energy network sells multi-energy users hot price $p_{GM,q}$ = 1 yuan/(kW·h), Sell to multi-energy users strategic tariff p_{MGe} ∈ [0,2] yuan/(kW·h), Public energy networks charge transaction fees p_{ser} ∈ [0,1.5] yuan/(kW·h).

4.1 Distribution of Nash Equilibrium Points

Set the policy step size to 0.05 yuan/(kW·h) and 0.01 yuan/(kW·h). Figure 3 shows the Nash equilibrium of the game.

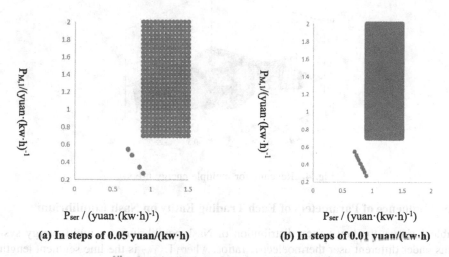

(a) In steps of 0.05 yuan/(kw·h)   (b) In steps of 0.01 yuan/(kw·h)

Fig. 3. Distribution of Nash equilibrium points

The composition of the wired region and the rectangular region of Nash equilibrium (see 4.1). Where, the upper $(p_{ser} = 0.7, p_{M,1} = 0.7)$ to $(p_{ser} = 0.9, p_{M,1} = 0.5)$ part of the line segment $p_{M,1} + p_{ser} = 1.4$, and the rectangle area are $0.7 \leq p_{M,1} \leq 2$ and $0.9 \leq p_{ser} \leq 1.5$. Consumers with less energy buy 5 MW·h of electricity and 3 MW·h of heat from consumers with more energy. Public energy networks charge a certain amount of transaction fees. In the rectangular range, low-energy users buy 25/3 MW·h of electricity from the public grid, while the public grid buys extra 5 MW·h from the multi-energy users.

The leftmost vertex A_1 of the segment is (0.7, 0.7). The right vertex A_2 is (0.9, 0.5).

In summary, the vertex A3 in the lower left corner of the rectangle region of Nash equilibrium is (0.9, 0.7). When multi-energy user quote $p_{M,1}$ and public energy grid service quote p_{ser} meet $p_{M,1} + p_{ser} = 1.4$, $0.5 \leq p_{M,1} \leq 0.7$ and $0.7 \leq p_{ser} \leq 0.9$, High energy users and low energy users trade directly; When $0.7 \leq p_{M,1} \leq 2$ is satisfied and $0.9 \leq p_{ser} \leq 1.5$ is satisfied, Multi-energy users and low-energy users trade with public energy networks respectively; In the rest of the cases, less energy users cannot make deals with more energy users or public energy networks.

4.2 Comparison of Profit Results of Market Entities

The following figure is the benefit function diagram of thermal power plants, power grid companies and large users under different strategies (see Fig. 4, Fig. 5, Fig. 6). In the whole strategic space, when $p_{M,1} + p_{ser} < 1.4$ is satisfied, less energy users are more inclined to directly deal with more energy users, and the public energy network charges certain service fees. When $p_{M,1} + p_{ser} > 1.4$ is satisfied, less energy users are more likely to trade with the public energy grid.

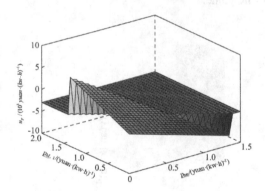

Fig. 4. Revenue for multiple energy users

4.3 Influence of Parameters of Each Trading Entity on Nash Equilibrium

Table 1 shows the key vertex distribution of Nash equilibrium for multi-energy systems under different user thermoelectric ratios, where L_{A1A2} is the line segment length between point A_1 and point A_2. As can be seen from Table 1, when the thermoelectric ratio of multi-energy user units is less than 0.6, the thermoelectric ratio does not affect the market transaction behavior. When the thermoelectric ratio is greater than 0.6 and less than 1.5, as the thermoelectric ratio increases, the Nash equilibrium rectangle area gets smaller and smaller, while the line segment area gets larger and larger, indicating that less energy users are more inclined to trade with more energy users. When the

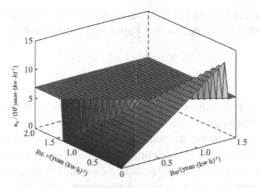

Fig. 5. Revenue from the public energy grid

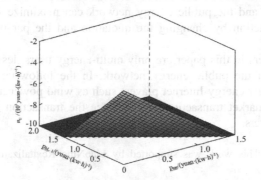

Fig. 6. Less revenue for energy users

thermoelectric ratio of the unit is greater than 1.5 and less than 3, the Nash equilibrium rectangle region disappears, indicating that at this time, the less energy users will not trade with the public energy network, and can only choose to trade with the more hot energy users; When the heat/power ratio of the unit is greater than 3 (the margin in the table indicates that the data has crossed the boundary), the Nash equilibrium line segment region disappears, and thereafter there will be no Nash equilibrium in the multi-energy system market [13].

Table 1. Influence of thermoelectric ratio on the range of Nash equilibrium points

Thermoelectric than	A_1	A_2	A_3	L_{A1A2}
0 0.6	(0.9,0.5)	(0.9,0.7)	(0.9,0.7)	0.2828
0.8	(1.03,0.43)	(1.03,0.43)	(1.03,0.7)	0.3771
1	(1.17,0.37)	(1.17,0.37)	(1.17,0.7)	0.4714
1.5	(1.50,2)	(1.50,2)	(1.50,0.7)	0.7071
3				0

According to the analysis in Table 1, 0.6 is the critical value of thermoelectric ratio. In the analysis of other examples in this paper, the thermoelectric ratio is 0.6.

5 Conclusion

Based on the revenue function of supply and demand among the multi-market entities of energy Internet, this paper studies the multi-energy transaction mode with energy hub. By using the game theory, the multi-market revenue model of energy Internet is established, the existence of the Nash equilibrium point is proved, and the distribution of the Nash equilibrium point is analyzed. The distribution range of Nash equilibrium points in a multi - energy system is studied. Combined with the example analysis, the quantitative effects of five parameters of the cost-benefit function of energy Internet on Nash equilibrium distribution and market transaction results are studied. The results show that the user and the public energy network can maximize the revenue in the multi-energy transaction by changing the quotation and the parameters of the cost-benefit function.

The game players in this paper are only multi-energy users, less-energy users and the three parties of the public energy network. In the follow-up research, we can consider adding other energy Internet players such as wind power and photovoltaic to participate in the market transaction, and analyze the transaction mode under more energy market players.

Acknowledgement. This work was supported by LiaoNing Revitalization Talents Program (XLYC1907138).

References

1. Dong, Z.Y.F., Zhao, J.H.S., Wen, F.S.T.: From smart grids to the energy Internet: Basic concepts and research framework. Power Syst. Autom. **38**(15), 1–11 (2014)
2. Wei, F.F., Jing, Z.X.S., Wu, P.Z.T.: A Stackelberg game approach for multiple energies trading in integrated energy systems. Appl. Energy **200**, 315–329 (2017)
3. Sheikhi, A.F., Rayati, M.S., Bahrami, S.T.: integrated demand side management game in smart energy hubs. IEEE Trans. Smart Grid **6**(2), 675–683 (2015)
4. Sheikhi, A.F., Bahrami, S.S., Ranjbar, A.M.T.: An autonomous demand response program for electricity and natural gas networks in smart energy hubs. Energy **89**, 490–499 (2015)
5. Sheikhi, A.F., Rayati, M.S., Bahrami, S.T.: A cloud computing framework on demand side management game in smart energy hubs. Int. J. Electrical Power Energy Syst. **64**, 1007–1016 (2015)
6. Bahrami, S.F., Sheikhi, A.S.: From demand response in smart grid toward integrated demand response in smart energy hub. IEEE Trans. Smart Grid **7**(2), 650–658 (2016)
7. Qian, A.I.F., Ran, A.O.S.: Key technologues and challenges for multi-energy complementarity and optimization of integrated energy system. Autom. Electric Power Syst. **42**(4), 2–10 (2018)
8. Zhang, X.H.F., Chen, Z.Q.S.: energy consumption performance of combined heat cooling and power system. Proc. CSEE **27**(5), 93–98 (2007)

9. Kang, Y.B.F., Zhang, J.G.S., Zhang, Y.T.: Study on current status, barriers and recommendations of China's CHP/DHC market development. Energy of China **30**(10), 8–13 (2008)
10. Bozchalui, M.C.F., Hashmi, S.A.S., Hassen, H.T.: Optimal operation of residential energy hubs in smart grids.IEEE Trans. Smart Grid **3**(4), 1755–1766 (2012)
11. Le Blond, S.F., Li, R.S., Li, F.T.: Cost and emission savings from the deployment of variable electricity tariffs and advanced domestic energy hub storage management. In: 2014 IEEE PES General Meeting. pp. 1–5. IEEE (2014)
12. Chicco, G.F., Mancarella, P.S.: Matrix modelling of small-scale trigeneration systems and application to operational optimization.Energy **34**(3), 261–273 (2009)
13. Mengelkamp, E., Gärttner, J., Rock, K., Kessler, S., Orsini, L., Weinhardt, C.: Designing microgrid energy markets: a case study: The Brooklyn microgrid. Appl. Energy **210**, 870–880 (2018)

Author Index

Printed in the United States
By Bookmasters

Printed in the United States
By Bookmasters